Principles of Antenna Theory

Principles of Antenna Theory

Kai Fong Lee

The Chinese University of Hong Kong

John Wiley & Sons

Chichester · New York · Brisbane · Toronto · Singapore

Library of Congress Cataloging in Publication Data:

Lee Kai Fong.
Principles of antenna theory.

Includes bibliographical references.
1. Antennas (Electronics) I. Title.
TK7871.6.L43 1984 621.38′028′3 83-7042
ISBN 0 471 90167 9

British Library Cataloguing in Publication Data:

Lee, Kai Fong
Principles of antenna theory.

1. Antennas (Electronics)
I. Title
621.38′ 028′3 TK7871.6

ISBN 0 471 90167 9

Typeset by Thomson Press (India) Limited, New Delhi
Printed by Page Bros (Norwich) Ltd.

To My Wife

Contents

Preface

Among the previously available books on antennas, there did not appear to be a text with a resonably comprehensive coverage of antenna principles that could form a bridge between the introductory material provided in undergraduate texts on electromagnetic theory and the more advanced texts intended primarily for graduate students. This book attempts to fill this gap in the literature. As such, the emphasis of the book is on simple exposition rather than exhaustive coverage. However, each of the topics covered is given detailed treatment, although the mathematics used seldom goes beyond the standard syllabus of an applied mathematics course for engineers at the junior undergraduate level.

The book is the outcome of notes prepared for a term course that the author has been teaching for the past several years to senior students majoring in electronics at the Chinese University of Hong Kong. In addition to being suitable for a senior-level elective course on antennas, the book can also be used for self-study by practising engineers in industry and government who belatedly find that they need to know something about antennas but, for some reason, have not previously acquired the knowledge. The book is also suitable for postgraduate students of several disciplines who do not intend to specialize in antenna engineering but would like to have a simple, concise introduction to the theory with adequate detail.

Chapter 1 provides an overall view of the book via a brief qualitative description of the types of antennas that will be discussed. Chapter 2 is concerned with the wave equation and its solution suitable for wire antennas. Chapters 3–6 are devoted to a comprehensive treatment of the dipole antenna. In these chapters, the reader is exposed not only to the various theoretical and practical aspects of the dipole antenna but also to detailed discussions of the basic ideas of reciprocity, mutual coupling, and the image principle. The theory of arrays is discussed in Chapter 7. The role of parasitic elements, the Yagi antenna, and the corner reflector are treated in Chapter 8. Chapter 9 deals with the cross-dipole, the loop antenna, the helical antenna, and frequency-independent antennas. Chapter 10 is concerned with aperture antennas, principally the horn, the slot, and the parabolic reflector. Worked examples, problems and selected bibliographies are provided at the end of each chapter. References (works specifically referred to in the text) are listed at the end of the book.

The reader is assumed to have had an introductory course on electromagnetic theory, which includes material on transmission lines, Maxwell's equations, and plane-wave propagation. He is assumed to have a working knowledge of vector calculus and Fourier analysis. Knowledge of complex integration and Bessel functions is helpful in understanding several of the more advanced sections of the book. The bulk of the book, however, can be read without prior exposure to these mathematical topics.

In the course of writing this book, the author has had the benefit of discussion and encouragement from his colleague, Dr J. S. Dahele, of the Department of Electronics, the Chinese University of Hong Kong.

Kai Fong Lee

August 1982
Hong Kong

CHAPTER 1
Introduction

An antenna (also known as an aerial) is defined as a means for radiating or receiving radio waves. The radio waves that an antenna radiates or receives propagate in free space rather than in a guiding structure. Antennas are used in three principal areas: communication, radar, and radio astronomy. In radar, which is an electronic device for the detection and location of objects, a directional transmitting antenna and a directional receiving antenna are indispensable parts of the system. In radio astronomy, antennas of large physical dimensions providing the required narrow beamwidth are necessary in order to pick up and locate the very weak electromagnetic radiation from extraterrestrial sources. In communication, an antenna is required in situations in which it is impossible, impracticable, or uneconomical to provide guiding structures between the transmitter and the receiver. For example, antennas must be used in communication with a satellite and with a moving vehicle such as a rocket, an aeroplane or a car. In broadcasting, where the aim is to send energy out in literally all directions, it is economical to use antennas since one transmitting terminal can serve an unlimited number of receivers. In situations where antennas and guiding structures are both feasible, it is usually the amount of attenuation suffered by the signal that determines the choice. In general, transmission of high-frequency waves over long distances favours the use of antennas, while small distances and low frequencies favour the use of transmission lines.

This book is concerned with the introduction to the theory and operation of some of the common antennas used in practice. While attempts will be made to emphasize the physical principles, an adequate understanding of the subject, even at the introductory level, must unavoidably involve a fair amount of mathematical deduction from Maxwell's equations. Before doing this, it is instructive in these initial pages to provide an overall view of the book via a brief qualitative description of the types of antennas that will be discussed.

The simplest and most widely used antenna is the dipole, which consists of two linear conductors separated by a small gap, usually located at the centre. The two sides of the gap (the terminals) are connected to a signal generator for the transmitting case and to a load in the receiving case, as illustrated in Figure 1.1. It turns out that the characteristics of an antenna are the same regardless of

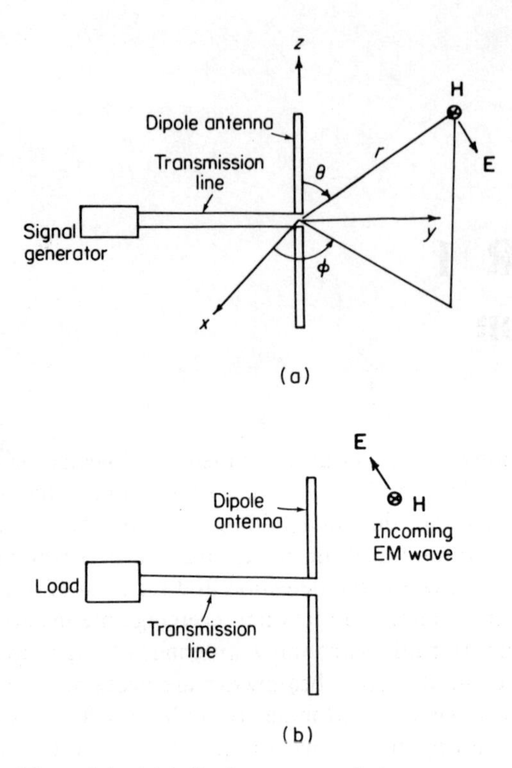

Figure 1.1 (a) A dipole as a transmitting antenna; (b) a dipole as a receiving antenna

whether it is used as a transmitter or as a receiver. In the rest of this chapter, the antenna will be regarded as a transmitting device for ease of discussion.

Of all the characteristics of an antenna, perhaps the single most important one is its radiation pattern. This is the variation of the magnitude of the electric or magnetic field as a function of direction at a distance far from the antenna. If the dipole is very short compared to the wavelength (λ), the patterns in the x–z and y–z plans vary as $\sin\theta$, while the pattern in the x–y plane is a circle, i.e. independent of ϕ. The electric field is along the θ-direction and the magnetic field is along the ϕ-direction, with the magnitude of the formal equal to that of the latter multiplied by the intrinsic impedance of the surrounding medium. The strongest radiation occurs perpendicular to the axis of the dipole. The strength of the electric or magnetic field decreases to $1/\sqrt{2}$ of its maximum value at the angles $\theta = 45^\circ$ and 135°. The half-power beamwidth of a very short dipole is therefore 90°. The power radiated is proportional to the square of the length-to-wavelength ratio, so that a very short dipole is not an efficient radiator.

In order to increase the power radiated, the length of the dipole has to be comparable to wavelength. The most widely used dipole is approximately half a wavelength long. The pattern of a half-wave dipole is similar to that of a very short dipole—it is a figure of eight in the y–z and x–z planes and a circle in the

x–y plane. The half-power beamwidth in the figure of eight is about 78°, which is narrower than that of the very short dipole.

If the length of the dipole is increased to one wavelength, the pattern is still similar, the only difference being that the half-power beamwidth in the figure of eight is now 47°, so that a further narrowing of the beam is achieved by increasing the length from 0.5λ to 1λ. However, this process of increasing the length to obtain a narrower beam works only up to about 1.2λ. Beyond 1.2λ, the pattern, instead of remaining a figure of eight, becomes multilobed. Figure 1.2 illustrates the patterns for several dipole lengths. Note that for the 1.5λ dipole, there are several main lobes, the maxima of which occur at oblique angles with respect to the dipole axis.

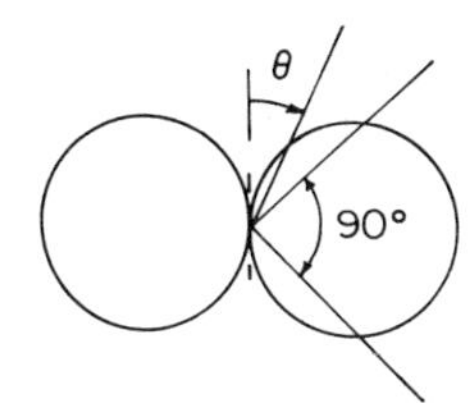

(a) Very short dipole

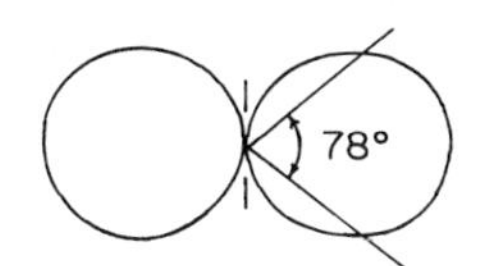

(b) Half-wave dipole

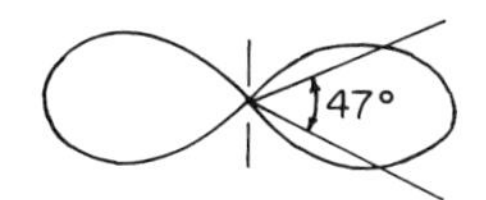

(c) One wavelength dipole

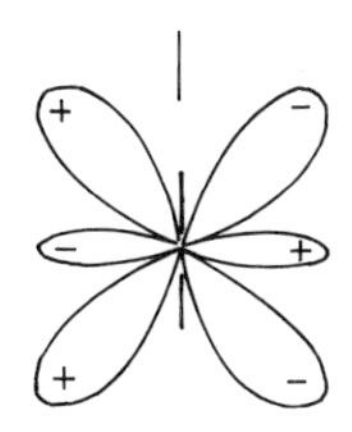

(d) One-and-a-half wavelength dipole

Figure 1.2 The patterns of several dipoles as functions of the angle θ measured from the axis. They are independent of the azimuthal angle ϕ

In many applications, e.g. radar, it is necessary for the antenna to concentrate its radiation into a very narrow beam in one direction only. While this cannot be achieved with a single dipole, it can be obtained by using an array of similar dipoles, all of which are fed by a source. The pattern of an array normally consists of a main beam and several side lobes, with the levels of the side lobes considerably smaller than that of the main beam. If the elements are equispaced, with their centres lying along the x-axis, as shown in Figure 1.3(a), the width of the main beam in the x–y and x–z planes can be made arbitrarily small by simply adding more elements to the array. The width of the beam in the y–z plane, however, is the same as that of the individual elements and is quite broad. A linear array of this type is said to produce a 'fan beam', which is narrow in two principal planes but wide in the third. To obtain a 'pencil beam', which is narrow in all three principal planes, a two-dimensional or planar array is required. Such an arrangement is shown in Figure 1.3(b).

An array of dipoles can also be used to approximate a given pattern. By

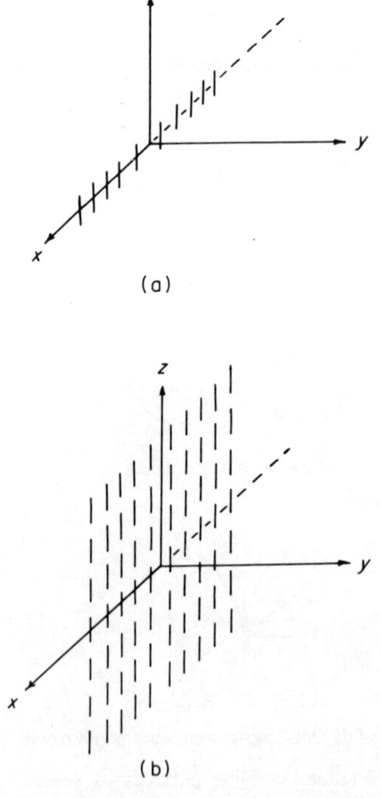

Figure 1.3 (a) A linear array of dipoles; (b) a planar array of dipoles. All elements are fed

feeding the elements with currents of appropriate amplitudes and phases and by suitably choosing the inter-element spacing, the mean square error between the pattern of the array and the desired pattern can be minimized.

The pattern of a dipole can be modified by placing a passive or parasitic conductor near it, as shown in Figure 1.4(a). Although the parasitic element is not connected to a source, a current is induced in it due to the radiation from the dipole. The total radiation is the sum of the driven and the parasitic elements. Depending on the length and spacing of the latter, it can act either as a reflector enhancing radiation in the direction of the dipole (negative x) or as a director enhancing radiation in its own direction (positive x). An antenna consisting of a driver, a reflecting element, and one or more directing elements is called a Uda–Yagi array, or a Yagi for short (Figure 1.4b). Being an array with only one driven element, it does not require the complicated feeding networks that are associated with an array of fed elements. However, there is a limit to the number of elements that can be added to a Yagi since the induced current on a parasitic element becomes progressively smaller as its distance from the driven element increases. It ceases to play an effective role if it is placed too far from the driven element.

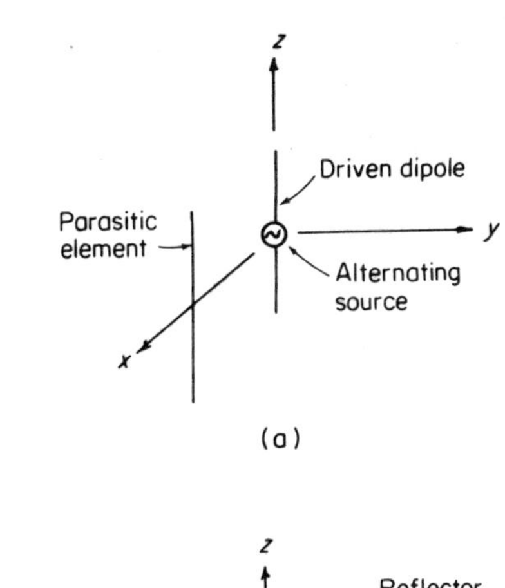

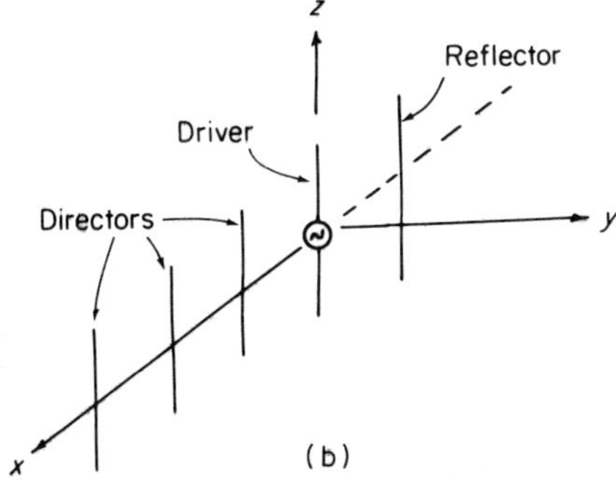

Figure 1.4 (a) A driven dipole and a parasitic element; (b) a Uda–Yagi antenna

While a thin conductor can act as a reflector it is highly sensitive to frequency. The frequency dependence is reduced if a plane conducting sheet is used instead. The effectiveness of the reflecting sheet can be further enhanced if it is bent into two sheets intersecting at an angle, as shown in Figure 1.5(a). The resulting structure is known as a corner reflector antenna. The corner reflector has the limitation that, even if the sides are infinite in extent, there is an upper limit to the directiveness of the resultant radiation. This limitation is removed if the shape of the reflector is parabolic instead of planar. For a parabolic reflector antenna, also known as a dish, the beamwidth is inversely proportional to the diameter of the dish, provided that the feed antenna is located at the focus of the paraboloid, as illustrated in Figure 1.5(b). In theory, the beamwidth can be made as narrow as one wishes by increasing the dish diameter d.

The ideas of shaping the pattern of a radiator by means of arrays, parasitic elements, and reflecting surfaces are general and not limited to the dipole, which is used in the above description for convenience only. Other basic radiating elements that we shall consider in this book are the loop antenna, the cross-dipole, the helical antenna, the slot antenna, and the microwave horn. We shall also consider a class of structures known as frequency-independent antennas.

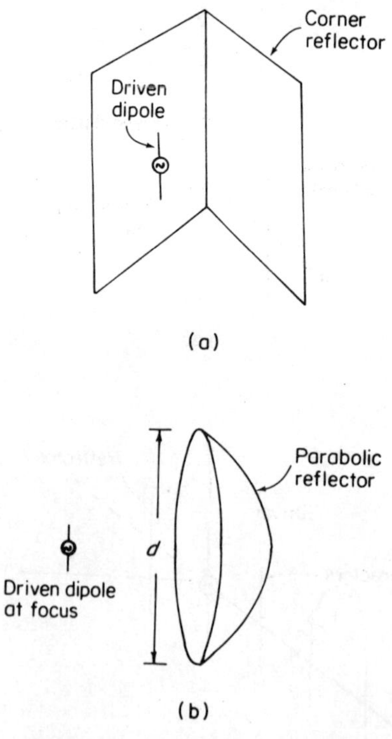

Figure 1.5 (a) A corner reflector antenna; (b) a parabolic reflector antenna

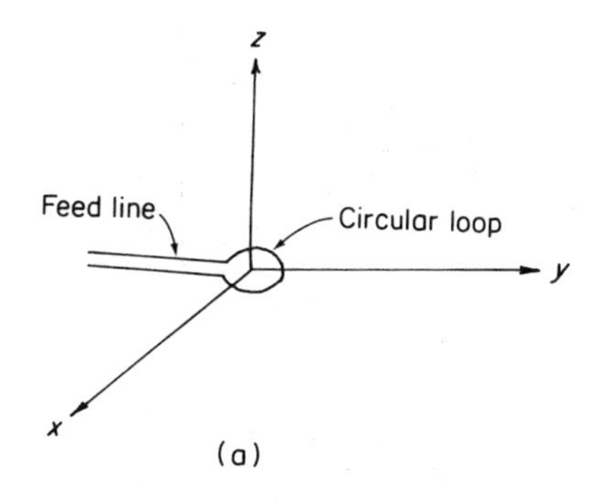

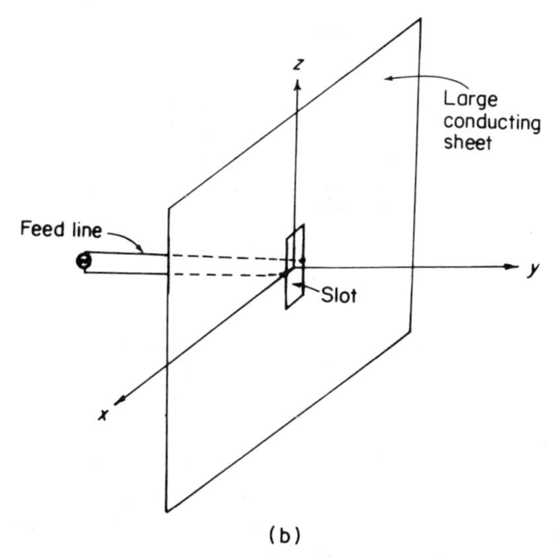

Figure 1.6 (a) A circular loop antenna in the x–y plane; (b) a slot antenna

Figure 1.6 shows a circular loop antenna lying in the x–y plane. If the radius is small compared to the wavelength, its radiation pattern is similar to a very short z-directed dipole, except that the electric and magnetic fields are interchanged. Partly for this reason, the small loop in the x–y plane and the z-directed very short dipole are said to be dual structures. Another radiator with properties closely related to a dipole is the slot antenna. When a narrow slot is cut in a large conducting sheet of metal and fed by a source connected to two opposite points at the centre of the slot, as shown in Figure 1.6(b), the structure radiates with a pattern that is similar to that of a dipole of the size and orientation of the slot radiating in free space, except that the electric and magnetic fields are interchanged. The slot and the dipole constitute an example of a pair of complementary antennas. In general, for a metal antenna, a complementary antenna is one that results when the metal is replaced by air and air replaced by metal.

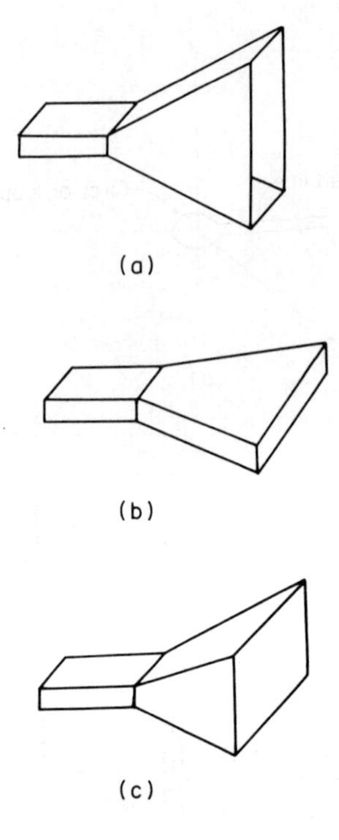

Figure 1.7 (a) An E-plane sectoral horn; (b) an H-plane sectoral horn; (c) a pyramidal horn

At microwave frequencies (10^9–10^{12} Hz), an efficient radiator can be obtained by simply flaring one or both of the walls of a rectangular waveguide operating in the dominant mode. The resultant structures are examples of a class of antennas known as horn antennas (Figure 1.7).

The basic radiators mentioned above all produce linearly polarized waves. This means that the locus of the electric field vector always lies along a straight line as time varies. In communications with satellites and space vehicles, linear polarization has the disadvantage that its direction is affected by a phenomenon known as Faraday rotation, which occurs in the Earth's ionosphere. Since the electron density of the ionosphere is variable, it is difficult to predict the direction of the electric field of a linearly polarized wave as it arrives at the location of the satellite. If the rotation is such that its direction at the satellite cannot induce a voltage at the terminals of the receiving antenna (e.g. perpendicular to the axis of a dipole), no signal will be received. This problem can be overcome if the radiation is circularly polarized instead of linearly polarized.

One of the methods of producing circularly polarized radiation is by means of two mutually perpendicular half-wave dipoles fed by currents that are equal in amplitude but 90° out of phase. Such an antenna is called a cross-dipole or turnstile antenna (Figure 1.8a). Another method is to use a corner reflector with a dipole whose axis is tilted at an angle α to the apex line (Figure 1.8b). A third method is to use a helical antenna (Figure 1.8c). When the diameter of the helix is small compared to the wavelength, it radiates in the 'normal' mode, the pattern of which is maximum in directions normal to the axis of the helix. When the diameter of the helix is comparable to the wavelength, it radiates in the 'axial' mode, with maximum radiation occurring along the axis of the helix.

Of all the basic radiators mentioned above, the helix operating in the axial mode has the largest bandwidth. The ratio of the upper to lower frequencies at

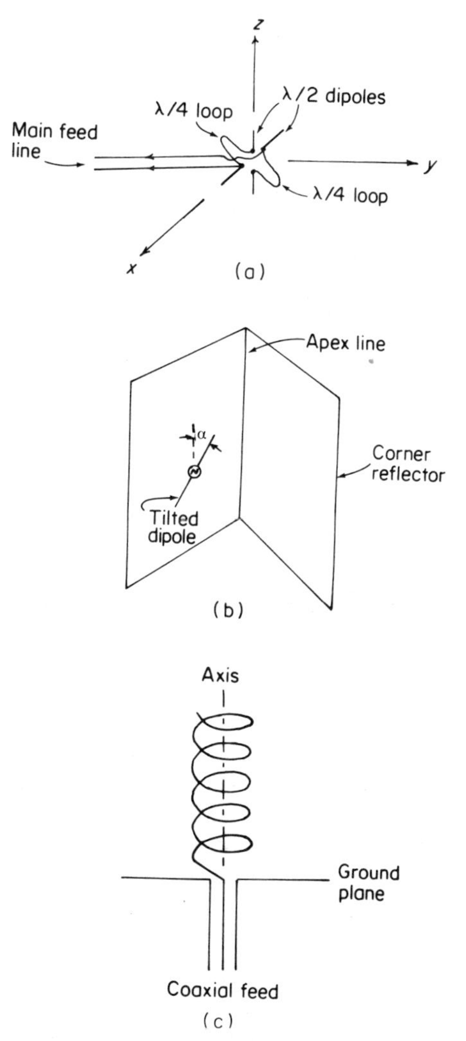

Figure 1.8 Three types of antennas capable of radiating circularly polarized waves: (a) the cross-dipole; (b) the corner reflector with tilted dipole; (c) the helical antenna

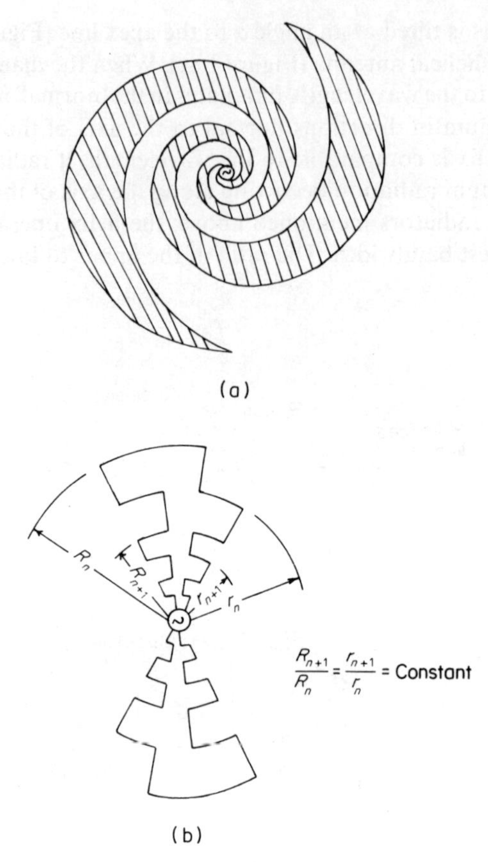

Figure 1.9 Two examples of frequency-independent antennas: (a) equiangular spiral antenna; (b) log-periodic toothed antenna

the 3 dB points is of the order of 1.8:1. The features that render a radiator frequency-dependent are their characteristic lengths. There is a class of structures defined primarily by angles rather than lengths. These structures can obtain bandwidths of 10:1 with ease and are known as frequency-independent antennas, two of which are shown in Figure 1.9.

The antennas described above are listed in Table 1.1, under the headings of basic radiators and composite antennas. By a composite antenna we mean a radiating structure that incorporates one or more methods of shaping the pattern of a basic radiator.

From an analytical viewpoint, antennas can be broadly divided into two categories, based on the equivalence theorem of electromagnetism, which permits the radiation from a source to be calculated either from a knowledge of its conduction current density or from a knowledge of the tangential electromagnetic field on a closed surface surrounding the source. Those structures for which the distributions of current are relatively simple and can be obtained quite

Table 1.1 List of antennas discussed in the book

Basic radiators	Composite antennas
Dipole	Array of active elements:
Loop	Linear array—fan beam
Slot	Planar array—pencil beam
Horn	Array with parasitic elements
Cross-dipole	(Uda–Yagi antenna)
Helix	Flat sheet reflector
Frequency-independent	Corner reflector
antennas	Parabolic reflector

accurately are known as wire antennas. Those structures for which the distribution of current is complicated and/or difficult to obtain, but for which the electromagnetic field configuration over a specific closed surface surrounding the antenna can be postulated with reasonable accuracy, are known as aperture antennas. In the radiators mentioned above, the dipole, the loop, the cross-dipole, the helix, and the frequency-independent antennas belong to the category of wire antennas. The horn, the slot, and the parabolic reflector belong to the category of aperture antennas.

The plan of the book is as follows. Chapter 2 discusses the wave equation and its solution suitable for wire antennas. Chapters 3–6 are devoted to a comprehensive treatment of the dipole antenna. In these chapters, the reader is exposed not only to the various theoretical and practical aspects of the dipole antenna but also to detailed discussions of the basic ideas of reciprocity, mutual coupling, and the image principle. The theory of arrays is discussed in Chapter 7. The role of parasitic elements, the Yagi antenna, and the corner reflector are treated in Chapter 8. Chapter 9 deals with the cross-dipole, the loop antenna, the helical antenna, and frequency-independent antennas. In Chapter 10, the horn antenna, the slot antenna, and the parabolic reflector antennas are discussed.

CHAPTER 2
The Wave Equation and its Solution

2.1 THE WAVE EQUATION

The behaviour of antennas is primarily governed by Maxwell's equations. In MKS units, they read

$$\nabla \times \mathbf{E}(\mathbf{r}, t) = -\frac{\partial \mathbf{B}(\mathbf{r}, t)}{\partial t} \tag{2.1}$$

$$\nabla \times \mathbf{H}(\mathbf{r}, t) = \frac{\partial \mathbf{D}(\mathbf{r}, t)}{\partial t} + \mathbf{J}(\mathbf{r}, t) \tag{2.2}$$

$$\nabla \times \mathbf{D}(\mathbf{r}, t) = \rho(\mathbf{r}, t) \tag{2.3}$$

$$\nabla \times \mathbf{B}(\mathbf{r}, t) = 0 \tag{2.4}$$

where

$$\mathbf{B}(\mathbf{r}, t) = \mu \mathbf{H}(\mathbf{r}, t) \tag{2.5}$$

$$\mathbf{D}(\mathbf{r}, t) = \varepsilon \mathbf{E}(\mathbf{r}, t) \tag{2.6}$$

In the above equations, $\mathbf{r}$ stands for position and t stands for time. $\mathbf{E}(\mathbf{r}, t)$ is the electric field intensity, $\mathbf{H}(\mathbf{r}, t)$ is the magnetic field intensity, $\mathbf{D}(\mathbf{r}, t)$ is the electric flux density, $\mathbf{B}(\mathbf{r}, t)$ is the magnetic flux density, $\mathbf{J}(\mathbf{r}, t)$ is the conduction current density, $\rho(\mathbf{r}, t)$ is the volume charge density, μ is the permeability, and ε is the permittivity. Unless otherwise stated, we shall assume the medium to be a loseless dielectric whose properties are time-independent and homogeneous, so that μ and ε are constants. In this chapter, we address ourselves to the problem of finding $\mathbf{E}(\mathbf{r}, t)$ and $\mathbf{H}(\mathbf{r}, t)$ when $\mathbf{J}(\mathbf{r}, t)$ and $\rho(\mathbf{r}, t)$ are given. The latter two quantities, of course, are not independent but are related by the continuity equation

$$\frac{\partial \rho(\mathbf{r}, t)}{\partial t} = -\nabla \cdot \mathbf{J}(\mathbf{r}, t) \tag{2.7}$$

As they stand, Maxwell's equations are a coupled set of partial differential equations for the fields $\mathbf{E}$ and $\mathbf{H}$. It is possible to obtain two uncoupled differential equations for $\mathbf{E}$ and $\mathbf{H}$ by direct manipulation of (2.1)–(2.4). The resultant equations (problems 2 and 3 of section 2.7) involve the spatial and time derivatives of **J** and ρ as source terms. Because of their complexities, these equations are seldom used in practice.

In what follows, we show that, by introducing two auxiliary functions **A** and Φ, from which **E** and **H** can be deduced, Maxwell's equations can be reduced to two uncoupled equations involving **A** and Φ, the source terms of which are **J** and ρ respectively. These equations are simpler to solve than those governing **H** and **E**.

We begin by observing that, since the divergence of the curl of any vector is identically equal to zero, we can, according to (2.4), set **B** to be the curl of a vector **A**:

$$\mathbf{B}(\mathbf{r}, t) = \nabla \times \mathbf{A}(\mathbf{r}, t) \tag{2.8}$$

Equation (2.8) defines the magnetic vector potential $\mathbf{A}(\mathbf{r}, t)$. Substitution of (2.8) into (2.1) yields

$$\nabla \times \left(\mathbf{E}(\mathbf{r}, t) + \frac{\partial \mathbf{A}(\mathbf{r}, t)}{\partial t} \right) = 0 \tag{2.9}$$

Since the curl of the gradient of any scalar is identically equal to zero, (2.9) allows us to define a scalar function Φ such that

$$\mathbf{E}(\mathbf{r}, t) + \frac{\partial \mathbf{A}(\mathbf{r}, t)}{\partial t} = -\nabla \Phi(\mathbf{r}, t)$$

or

$$\mathbf{E}(\mathbf{r}, t) = -\nabla \Phi(\mathbf{r}, t) - \frac{\partial \mathbf{A}(\mathbf{r}, t)}{\partial t} \tag{2.10}$$

The function Φ is known as the electric scalar potential. Note that for static problems, $\partial/\partial t = 0$ and $\mathbf{E} = -\nabla \Phi$, which is the familiar relation between the electrostatic field and the electrostatic potential.

In terms of $\mathbf{A}(\mathbf{r}, t)$ and $\Phi(\mathbf{r}, t)$, (2.2) and (2.3) become

$$\nabla^2 \mathbf{A}(\mathbf{r}, t) - \mu\varepsilon \frac{\partial^2 \mathbf{A}(\mathbf{r}, t)}{\partial t^2} - \nabla\left(\nabla \cdot \mathbf{A}(\mathbf{r}, t) + \mu\varepsilon \frac{\partial \Phi(\mathbf{r}, t)}{\partial t} \right) = -\mu \mathbf{J}(\mathbf{r}, t) \tag{2.11}$$

$$\nabla^2 \Phi(\mathbf{r}, t) + \frac{\partial}{\partial t}(\nabla \cdot \mathbf{A}(\mathbf{r}, t)) = -\frac{\rho(\mathbf{r}, t)}{\varepsilon} \tag{2.12}$$

In deriving (2.11), we have made use of the vector identity $\nabla \times (\nabla \times \mathbf{A}) = \nabla(\nabla \cdot \mathbf{A}) - \nabla^2 \mathbf{A}$, where ∇^2 is the Laplacian operator.

By introducing the functions **A** and Φ, which are related to **B** and **E** according to (2.8) and (2.10), we have reduced the set of four Maxwell's equations for **B** and **E** to two equations for **A** and Φ. But they are still coupled equations. We can now uncouple (2.11) and (2.12) by exploiting the fact that, so far, only the curl of **A** is specified. There are an infinite number of vector functions whose curl is the same and give the same **B** field. To make **A** unique, one has to specify its divergence, and this may be done according to convenience. For our problem, if we choose the divergence of **A** such that

$$\nabla \cdot \mathbf{A}(\mathbf{r}, t) = -\mu\varepsilon \frac{\partial \Phi(\mathbf{r}, t)}{\partial t} \tag{2.13}$$

then (2.11) and (2.12) will be decoupled and become

$$\nabla^2 \mathbf{A}(\mathbf{r}, t) - \frac{1}{c^2}\frac{\partial^2 \mathbf{A}(\mathbf{r}, t)}{\partial t^2} = -\mu \mathbf{J}(\mathbf{r}, t) \tag{2.14}$$

$$\nabla^2 \Phi(\mathbf{r}, t) - \frac{1}{c^2}\frac{\partial^2 \Phi(\mathbf{r}, t)}{\partial t^2} = -\frac{\rho(\mathbf{r}, t)}{\varepsilon} \tag{2.15}$$

where

$$c^2 = 1/\mu\varepsilon \tag{2.16}$$

Thus the components of the vector potential **A** and the scalar potential Φ satisfy an equation of the form

$$\left(\nabla^2 - \frac{1}{c^2}\frac{\partial^2}{\partial t^2}\right)\psi(\mathbf{r}, t) = -F(\mathbf{r}, t) \tag{2.17}$$

which is known as the time-dependent inhomogeneous wave equation. The relation (2.13) between **A** and Φ is known as the Lorentz condition.

To summarize the development so far, we have, by introducing the two potential functions according to (2.8), (2.13), and (2.10), reduced the problem to that of solving a time-dependent inhomogeneous wave equation for **A**. Once **A** is found, we can obtain the electromagnetic fields at any point outside the source region (where $\mathbf{J} = 0$) by performing the operations

$$\mathbf{H}(\mathbf{r}, t) = \mu^{-1}[\nabla \times \mathbf{A}(\mathbf{r}, t)] \tag{2.18}$$

$$\mathbf{E}(\mathbf{r}, t) = \varepsilon^{-1}\int \nabla \times \mathbf{H}(\mathbf{r}, t)\,\mathrm{d}t \tag{2.19}$$

The discussion up to this point is for the general case of arbitrary time variation. Of particular importance is the special case of sinusoidal time variations, also known as time harmonic variation. In this case, we can write

$$\begin{aligned} \mathbf{J}(\mathbf{r}, t) &= \mathrm{Re}[\mathbf{J}(\mathbf{r})e^{j\omega t}] \\ \mathbf{E}(\mathbf{r}, t) &= \mathrm{Re}[\mathbf{E}(\mathbf{r})e^{j\omega t}] \end{aligned} \tag{2.20}$$

and similarly for the other quantities. In (2.20), ω is the angular frequency and Re stands for "taking the real part of". The quantities $\mathbf{J}(\mathbf{r})$, $\mathbf{E}(\mathbf{r})$, etc., are known as phasors. From (2.14), (2.15), and (2.18)–(2.20), the phasors of **A**, Φ, **H**, and **E** satisfy the following equations:

$$\nabla^2 \mathbf{A}(\mathbf{r}) + k^2 \mathbf{A}(\mathbf{r}) = -\mu \mathbf{J}(\mathbf{r}) \tag{2.21}$$

$$\nabla^2 \Phi(\mathbf{r}) + k^2 \Phi(\mathbf{r}) = -\rho(\mathbf{r})/\varepsilon \tag{2.22}$$

$$\mathbf{H}(\mathbf{r}) = \nabla \times \mathbf{A}(\mathbf{r})/\mu \tag{2.23}$$

$$\mathbf{E}(\mathbf{r}) = \nabla \times \mathbf{H}(\mathbf{r})/j\omega\varepsilon \tag{2.24}$$

In (2.21) and (2.22),

$$k = \omega/c = 2\pi f/c = 2\pi/\lambda \tag{2.25}$$

where λ is the wavelength.

The components of $\mathbf{A}(\mathbf{r})$ and $\Phi(\mathbf{r})$ thus satisfy an equation of the form

$$\nabla^2\psi(\mathbf{r}) + k^2\psi(\mathbf{r}) = -F(\mathbf{r}) \tag{2.26}$$

which is the time-independent inhomogeneous wave equation, also known as the Helmholtz equation.

2.2 SOLUTION OF THE WAVE EQUATION BY INTUITIVE DEDUCTION

The rigorous solution to the time-dependent inhomogeneous equation, (2.17), can be found by applying standard methods of applied mathematics. The procedure is rather lengthy and will be postponed until sections 2.3–2.5. It is instructive first to demonstrate how the solution can be reached by intuitive deduction. To do this, let us consider equation (2.15) for the scalar electric potential, since it is useful to associate the dependent variable with a physical quantity. We begin by considering (2.15) for the case when the source term on the right-hand side is a time-varying point charge of strength $Q(t)$ localized at the origin. Mathematically, this is represented by $Q(t)\delta(\mathbf{r}')$, where $\delta(\mathbf{r}')$ is the Dirac delta function, so that (2.15) becomes

$$\nabla^2\Phi(\mathbf{r}, t) - \frac{1}{c^2}\frac{\partial^2\Phi(\mathbf{r}, t)}{\partial t^2} = -\frac{Q(t)\delta(\mathbf{r}')}{\varepsilon} \tag{2.27}$$

In (2.27), we have used the symbol $\mathbf{r}'$ on the right hand side to denote the source point. The unprimed $\mathbf{r}$ on the left hand side denotes the field point.

At any point outside the origin, Φ satisfies the homogeneous wave equation

$$\nabla^2\Phi(\mathbf{r}, t) - \frac{1}{c^2}\frac{\partial^2\Phi(\mathbf{r}, t)}{\partial t^2} = 0 \tag{2.28}$$

For a point charge at the origin, the function $\Phi(\mathbf{r}, t)$ will be a function only of the distance r. Using spherical coordinates (r, θ, ϕ) with $\Phi(\mathbf{r}, t)$ independent of θ and ϕ, (2.28) becomes

$$\frac{1}{r^2}\frac{\partial}{\partial r}\left(r^2\frac{\partial\Phi}{\partial r}\right) - \frac{1}{c^2}\frac{\partial^2\Phi}{\partial t^2} = 0$$

which can be rewritten

$$\frac{1}{r}\frac{\partial^2}{\partial r^2}(r\Phi) - \frac{1}{c^2}\frac{\partial^2\Phi}{\partial t^2} = 0$$

or

$$\frac{\partial^2 R}{\partial r^2} - \frac{1}{c^2}\frac{\partial^2 R}{\partial t^2} = 0 \tag{2.29}$$

where

$$R = r\Phi \tag{2.30}$$

Equation (2.29) is the one-dimensional wave equation in the variable r. Its solution is

$$R(r, t) = f(t \pm r/c) \tag{2.31}$$

where f is any differentiable function. That (2.31) does satisfy (2.29) can be readily verified by direct substitution. The positive sign corresponds to an inward-travelling wave with velocity c and the negative sign to an outward-travelling wave. Selecting the latter, we have a solution of (2.28) in the form of an outgoing spherical wave whose amplitude varies as $1/r$:

$$\Phi(r, t) = (1/r) f(t - r/c) \tag{2.32}$$

To determine what function f is required to make (2.32) satisfy the condition that the source is a point charge of strength $Q(t)$ at the origin, we examine the situation for very small r. Then $-r/c$ can be neglected in (2.32) and the solution for Φ becomes

$$\Phi = f(t)/r \qquad (r \to 0) \tag{2.33}$$

As $r \to 0$, the differential equation (2.27) simplifies in the following manner. Near the origin, the $1/r$ dependence of Φ causes the space derivatives to become very large. Since the time derivative does not depend on r, it is reasonable to assume that the term $\partial^2\Phi/\partial t^2$ in (2.27) can be neglected in comparison with $\nabla^2\Phi$. Thus (2.27) is equivalent to

$$\nabla^2\Phi = -Q(t)\delta(r')/\varepsilon \tag{2.34}$$

as $r \to 0$.

Equation (2.34) will be recognized as Poisson's equation for the electrostatic potential Φ due to a point charge Q at the origin, except that Q is now a function of time. The solution to (2.34) is simply the Coulomb potential, namely,

$$\Phi(\mathbf{r}, t) = \frac{Q(t)}{4\pi\varepsilon r} \tag{2.35}$$

By comparing (2.35) with (2.33), we see that we require

$$f(t) = \frac{Q(t)}{4\pi\varepsilon} \tag{2.36}$$

so that (2.32) becomes

$$\Phi(\mathbf{r}, t) = \frac{1}{4\pi\varepsilon r} Q(t - r/c) \tag{2.37}$$

Equation (2.37) is the solution to (2.27) for a point source. It shows that the effect of the $\partial^2\Phi/\partial t^2$ term, which arises from the time variation of the charge, is that the potential at a point $\mathbf{r}$ at time t is due to the value of the charge not at t but at an earlier time $t - r/c$. This time $t - r/c$ is called the retarded time, and expresses the fact that the effect of any time variation or disturbance at the source will take a time r/c to reach the point r. The potential given by (2.37) is called the retarded electric scalar potential.

For an extended source, characterized by the charge density $\rho(\mathbf{r}', t)$, we can think of it as made up of the sum of many point sources, one for each volume element $d\tau'$ located at r', and each with the source strength $\rho(\mathbf{r}', t)d\tau'$. This element will give rise to a potential of the amount

$$d\Phi(\mathbf{r}, t) = \frac{1}{4\pi\varepsilon}\frac{\rho(\mathbf{r}', t - |\mathbf{r} - \mathbf{r}'|/c)d\tau'}{|\mathbf{r} - \mathbf{r}'|} \tag{2.38}$$

at the field point $\mathbf{r}$. Since (2.15) is linear, the resultant field is the superposition of the fields from all such source elements. Hence the solution to (2.15) for an extended source is

$$\Phi(\mathbf{r}, t) = \frac{1}{4\pi\varepsilon}\iiint\frac{\rho(\mathbf{r}', t - |\mathbf{r} - \mathbf{r}'|/c)d\tau'}{|\mathbf{r} - \mathbf{r}'|} \tag{2.39}$$

where integration is over all regions where $\rho \neq 0$.

For sinusoidal time variation, the phasor of the electric scalar potential is governed by the Helmholtz equation (2.22). It can be solved by intuitive deduction in a manner similar to the above. The details are left as an exercise (problem 6 in section 2.7). The result is

$$\Phi(\mathbf{r}) = \frac{1}{4\pi\varepsilon}\iiint\frac{\rho(\mathbf{r}')\exp(-jk|\mathbf{r} - \mathbf{r}'|)}{|\mathbf{r} - \mathbf{r}'|}d\tau' \tag{2.40}$$

Comparing (2.40) with (2.39), we see that, for sinusoidal time variation, the retarded time $t - |\mathbf{r} - \mathbf{r}'|/c$ is replaced by the retarded phase $-k|\mathbf{r} - \mathbf{r}'|$, as expected.

Since each component of the magnetic vector potential $\mathbf{A}$ satisfies the same equation as Φ, except for the constants on the right-hand side, it follows that the solutions to (2.14) and (2.21) are respectively

$$\mathbf{A}(\mathbf{r}, t) = \frac{\mu}{4\pi}\iiint\frac{\mathbf{J}(\mathbf{r}', t - |\mathbf{r} - \mathbf{r}'|/c)}{|\mathbf{r} - \mathbf{r}'|}d\tau' \tag{2.41}$$

and

$$\mathbf{A}(\mathbf{r}) = \frac{\mu}{4\pi}\iiint\frac{\mathbf{J}(\mathbf{r}')\exp(-jk|\mathbf{r} - \mathbf{r}'|)}{|\mathbf{r} - \mathbf{r}'|}d\tau' \tag{2.42}$$

The potentials given by (2.41) and (2.42) are called the retarded magnetic vector potentials.

The following comments on the development of this section are in order:

(1) The solution (i.e. (2.31)) to the one-dimensional homogeneous wave equation (2.29) is handled in the 'backward' way, in which the solution is first written down, without derivation, and the student is asked to verify that it does satisfy the equation. This is the standard practice used in textbooks on electromagnetic theory, very few of which solve the equation by the 'forward' method of logical mathematical deductions.

(2) In (2.27), the term $\partial^2\Phi/\partial t^2$ is assumed to be negligible compared to the term $\nabla^2\Phi$ as r becomes small. This is not mathematically rigorous since there is no reason to impose an upper bound on the term $\partial^2\Phi/\partial t^2$.

(3) The solution (i.e. (2.35)) to Poisson's equation (2.34) is written down by appealing to Coulomb's law, rather than obtained by logical mathematical deduction.

It is clear from the above comments that the arguments and assumptions used in this section are quite acceptable to a physicist or an engineer but hardly rigorous from a mathematical viewpoint. The question naturally arises as to whether the wave equation can be solved strictly as a problem in mathematics. This will be the subject of the next three sections. While the procedure is somewhat lengthy, it is of some interest partly because it is seldom done in textbooks and partly because it serves as an excellent example of how several methods of applied mathematics that the student learns at the junior level are collectively used to solve a real physical problem, namely, radiation. For the less mathematically minded student, however, sections 2.3–2.5 can be omitted without loss of continuity.

2.3 GREEN'S FUNCTION FOR THE WAVE EQUATION

The time-dependent inhomogeneous wave equation (2.17) is a linear, inhomogeneous partial differential equation. The inhomogeneous term, $-F(\mathbf{r}, t)$, is the source or excitation and is assumed to be known. The function $\psi(\mathbf{r}, t)$ is the response to the excitation and is the unknown to be solved. To find the response to an arbitrary excitation, we can first solve for the response to a point source located at position $\mathbf{r}'$ and time t'. A distributed source in space and time can be regarded as a continuous sum (integration) of point sources of appropriate strengths. Since the equation is linear, the total response, i.e. the response to the sum, is equal to the sum of the responses to the individual point sources. This method of solution is familiar in linear system theory, and leads to the result that the response of a linear system is expressed as the convolution of the source excitation and the impulse response of the system. To review how this result comes about, consider a linear and time-invariant system designated by the block L in Figure 2.1. The system has the characteristic that, when an impulse $\delta(t)$ is applied to the input, a response $h(t)$ appears at the output. Suppose now that an arbitrary signal $x(t)$ is applied to the input, producing a response $y(t)$ at the output. We wish to express the output $y(t)$ in terms of the input $x(t)$ and the impulse response of the system $h(t)$. To solve this problem, consider first an

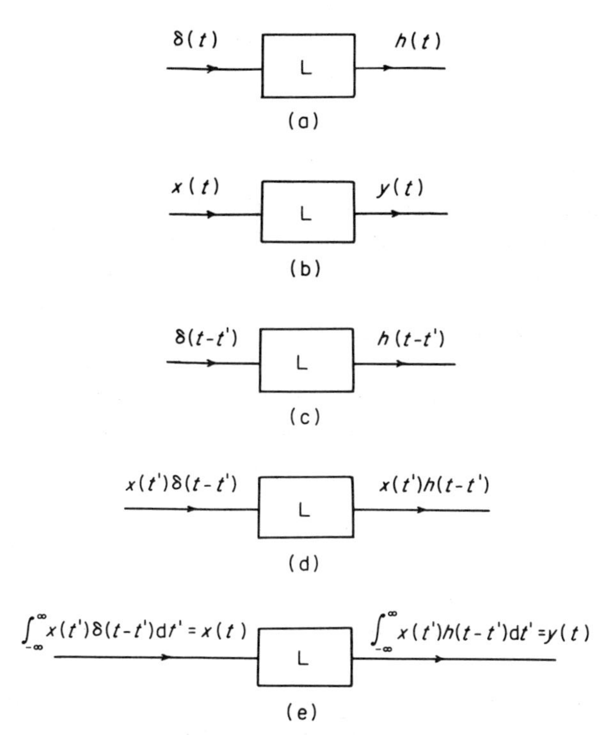

Figure 2.1 Relation between the response to an arbitrary signal and the impulse response of a linear system

impulse applied at a time $t = t'$, so that the input is $\delta(t - t')$, as shown in Figure 2.1(c). Since the system is time-invariant, the output will be $h(t - t')$. Also, since the system is linear, an input $x(t')\delta(t - t')$, where $x(t')$ is the value of the signal at $t = t'$, will produce an output $x(t')h(t - t')$, as shown in Figure 2.1(d). Finally, consider an input consisting of the continuous sum of impulses of strengths $x(t')\delta(t - t')$, where t' varies over all possible values. Such an input will simply be the signal $x(t)$. The response to $x(t)$ is by definition the output $y(t)$. It is the superposition of all the responses to the impulses that make up $x(t)$, as shown in Figure 2.1(e). Hence we have

$$y(t) = \int_{-\infty}^{\infty} x(t')h(t - t')\,\mathrm{d}t' \tag{2.43}$$

The integration in (2.43) is known as the convolution integral of the two functions $x(t)$ and $h(t)$.

Comparing the linear system with the differential equation (2.17), we see that the signal $x(t)$ corresponds to $-F(\mathbf{r}, t)$, and the output $y(t)$ corresponds to $\psi(\mathbf{r}, t)$. Since we now have a spatial coordinate in addition to a time coordinate, the impulse $\delta(t - t')$ would correspond to $\delta(\mathbf{r} - \mathbf{r}')\delta(t - t')$, where $\mathbf{r}'$ indicates the position and t' the instant at which the impulse is applied. The response to such an impulse (point source) will be denoted by the function $G(\mathbf{r} - \mathbf{r}'; t - t')$, which is analogous to $h(t-t')$ in linear system theory. The function $G(\mathbf{r}-\mathbf{r}';t-t')$ therefore satisfies the differential equation

$$\left(\nabla^2 - \frac{1}{c^2}\frac{\partial^2}{\partial t^2}\right)G(\mathbf{r}-\mathbf{r}';t-t') = \delta(\mathbf{r}-\mathbf{r}')\delta(t-t') \tag{2.44}$$

For an unbounded medium, we require G and its first-order spatial derivatives to vanish at infinity. From the physical principle of causality, which states that the effect (response) cannot precede the cause (source), we impose on G the initial conditions that G and $\partial G/\partial t$ vanish at $t = 0$, since we assume $t' > 0$.

The function $G(\mathbf{r}-\mathbf{r}';t-t')$ is called the Green's function of the time-dependent wave equation. It gives the response at the position $\mathbf{r}$ at time t due to a point source located at position $\mathbf{r}'$ at time t'. If we can obtain the solution for G, then the response $\psi(\mathbf{r}, t)$ due to an arbitrary source $-F(\mathbf{r}, t)$ will, in analogy to the result of linear system theory, be given by a convolution-like integral:

$$\psi(\mathbf{r}, t) = -\iiiint F(\mathbf{r}', t)G(\mathbf{r}-\mathbf{r}';t-t')\,\mathrm{d}t'\,\mathrm{d}\tau' \tag{2.45}$$

where $\mathrm{d}\tau'$ is an element of volume in space. Three of the integrations in (2.45) are concerned with spatial coordinates and one integration concerns time.

2.4 SOLUTION OF THE GREEN'S FUNCTION FOR THE WAVE EQUATION

We shall solve (2.44) by means of integral transforms. The idea is to reduce the partial differential equation for G to an algebraic equation for the transform of G. After solving for the transform of G, we can obtain the desired function G by applying the rules of inverse integral transforms.

Let us use a rectangular coordinate system and apply to (2.44) a three-dimensional Fourier transform in the spatial variables x, y, z, and a Laplace transform in the time variable t. Noting that

$$\delta(\mathbf{r}-\mathbf{r}') = \delta(x-x')\delta(y-y')\delta(z-z')$$

and applying the boundary and initial conditions for G, we obtain

$$(-k^2 - s^2/c^2)G_{ks} = \exp(\mathrm{j}\mathbf{k}\cdot\mathbf{r}')\exp(-st') \tag{2.46}$$

where

$$G_{ks} = \int_0^\infty G_k \exp(-st)\,\mathrm{d}t \tag{2.47}$$

$$G_k = \int_{-\infty}^{\infty}\int_{-\infty}^{\infty}\int_{-\infty}^{\infty} G\exp(\mathrm{j}k_1x + \mathrm{j}k_2y + \mathrm{j}k_3z)\,\mathrm{d}x\,\mathrm{d}y\,\mathrm{d}z \tag{2.48}$$

$$\mathbf{k} = (k_1, k_2, k_3) \tag{2.49}$$

$$k^2 = k_1^2 + k_2^2 + k_3^2 \tag{2.50}$$

The function G_k is the three-dimensional Fourier transform of G and G_{ks} is the Laplace transform of G_k.

Equation (2.46) is an algebraic equation for G_{ks}, which can be immediately solved. To obtain G_k from G_{ks}, we apply the rule for inverse Laplace transforms. This yields

$$G_k = -\frac{c^2}{2\pi \mathrm{j}} \exp(\mathrm{j}\mathbf{k}\cdot\mathbf{r}') \int_{\mathrm{L}} \frac{\exp[s(t-t')]}{s^2 + k^2c^2}\, \mathrm{d}s \tag{2.51}$$

In (2.51), L is the Bromwich contour in the complex s-plane. It is a straight line parallel to the imaginary s-axis to the right of the singularities of the integrand, extending from $-\infty$ to ∞.

Since the integrand has simple poles at $s = +\mathrm{j}kc$, L is any vertical line to the right of the imaginary axis, as shown in Figure 2.2. The contour integral can be evaluated by means of the residual theorem. For $t > t'$, we close the contour by a semi-finite circle on the left-hand side. By evaluating the residues at the two poles, we obtain

$$G_k = -ck^{-1} \sin[kc(t-t')] \exp(\mathrm{j}\mathbf{k}\cdot\mathbf{r}') \qquad t > t' \tag{2.52}$$

For $t < t'$, the contour must be closed on the right-hand side to ensure that there is no contribution on the part of the semi-infinite circle. Since in this case there is no singularity inside the closed contour, we have

$$G_k = 0 \qquad t < t' \tag{2.53}$$

To obtain the desired function G from its three-dimensional Fourier transform G_k, we apply the rule for inverting Fourier transforms. For $t > t'$, we have

$$G(\mathbf{r}-\mathbf{r}'; t-t') = -\frac{c}{(2\pi)^3} \int_{-\infty}^{\infty} \int_{-\infty}^{\infty} \int_{-\infty}^{\infty} k^{-1} \sin[kc(t-t')] \exp[-\mathrm{j}\mathbf{k}\cdot(\mathbf{r}-\mathbf{r}')]\, \mathrm{d}k_1\, \mathrm{d}k_2\, \mathrm{d}k_3 \tag{2.54}$$

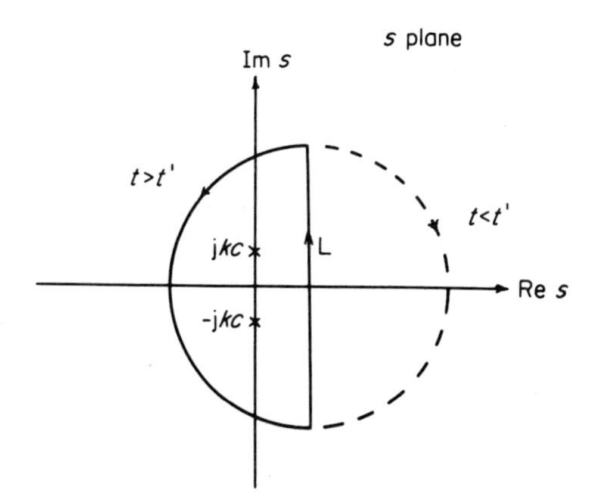

Figure 2.2 Contour integral for inverting the Laplace transform in equation (2.51)

Evaluation of the integral in (2.54) is facilitated by working in a spherical coordinate in the k_1, k_2, k_3 space. The volume element $\mathrm{d}k_1\,\mathrm{d}k_2\,\mathrm{d}k_3$ is then equal to $k^2 \sin\theta\,\mathrm{d}\theta\,\mathrm{d}\phi\,\mathrm{d}k$. If we orient the coordinate system so that θ is measured from the direction of the vector $(\mathbf{r}-\mathbf{r}')$, then

$$\mathbf{k}\cdot(\mathbf{r}-\mathbf{r}') = k|\mathbf{r}-\mathbf{r}'|\cos\theta$$

and (2.54) becomes

$$G(\mathbf{r}-\mathbf{r}';t-t') = -\frac{c}{(2\pi)^3}\int_0^\infty\int_0^\pi\int_0^{2\pi} \sin[kc(t-t')]k \exp(-\mathrm{j}k|\mathbf{r}-\mathbf{r}'|\cos\theta)\sin\theta\,\mathrm{d}k\,\mathrm{d}\theta\,\mathrm{d}\phi \tag{2.55}$$

The ϕ and θ integrations can be readily performed, giving

$$G(\mathbf{r}-\mathbf{r}';t-t') = -\frac{c}{(2\pi)^3}\int_0^\infty \frac{2\sin[kc(t-t')]\sin(k|\mathbf{r}-\mathbf{r}'|)}{|\mathbf{r}-\mathbf{r}'|}\mathrm{d}k \tag{2.56}$$

Expressing the sines as exponentials and noting that the integrand is even, we may write

$$G(\mathbf{r}-\mathbf{r}';t-t') = \frac{c}{16\pi^2|\mathbf{r}-\mathbf{r}'|}\int_{-\infty}^{\infty}\{\exp[\mathrm{j}kc(t-t')]-\exp[-\mathrm{j}kc(t-t')]\} \cdot[\exp(\mathrm{j}k|\mathbf{r}-\mathbf{r}'|)-\exp(-\mathrm{j}k|\mathbf{r}-\mathbf{r}'|)]\mathrm{d}k \tag{2.57}$$

Recalling the well known representation of the delta function

$$\delta(p) = \frac{1}{2\pi}\int_{-\infty}^{\infty}\exp(\mathrm{j}px)\,\mathrm{d}x \tag{2.58}$$

we have

$$G(\mathbf{r}-\mathbf{r}';t-t') = \frac{c}{8\pi|\mathbf{r}-\mathbf{r}'|}\{\delta[|\mathbf{r}-\mathbf{r}'|+c(t-t')]+\delta[-|\mathbf{r}-\mathbf{r}'|-c(t-t')] - \delta[|\mathbf{r}-\mathbf{r}'|-c(t-t')]-\delta[-|\mathbf{r}-\mathbf{r}'|+c(t-t')]\} \tag{2.59}$$

Since we have assumed $t > t' > 0$, the arguments of the first two delta functions can never be zero and therefore these delta functions vanish. The other delta functions are identical because the function is even. Making use of the identity

$$\delta(ax) = \frac{1}{|a|}\delta(x) \tag{2.60}$$

we finally have

$$G(\mathbf{r}-\mathbf{r}';t-t') = -\frac{1}{4\pi|\mathbf{r}-\mathbf{r}'|}\delta\left(t-t'-\frac{|\mathbf{r}-\mathbf{r}'|}{c}\right) \tag{2.61}$$

Equation (2.61) shows that the solution to the time-dependent wave equation with a delta-function excitation is also a delta function. If we think of the Green's function G as the response to a pulse at $(\mathbf{r}', t')$, then we see that this response is itself a (spherical) pulse expanding with a velocity c and damped by the reciprocal of its radius. At a point $\mathbf{r}$ in space, the response is felt at the time $t' + |\mathbf{r} - \mathbf{r}'|/c$, which is $|\mathbf{r} - \mathbf{r}'|/c$ seconds after the time at which the excitation is applied.

2.5 SOLUTION OF THE WAVE EQUATION

The solution to the inhomogeneous wave equation with a distributed source (2.17)

$$\left(\nabla^2 - \frac{1}{c^2}\frac{\partial^2}{\partial t^2}\right)\psi(\mathbf{r}, t) = -F(\mathbf{r}, t)$$

is (2.45)

$$\psi(\mathbf{r}, t) = -\iiiint F(\mathbf{r}', t')G(\mathbf{r} - \mathbf{r}'; t - t')\,\mathrm{d}t'\,\mathrm{d}\tau'$$

Substituting (2.61) into (2.45) and using the sifting property of the delta function to perform the t' integration, we obtain

$$\psi(\mathbf{r}, t) = \frac{1}{4\pi}\iiint \frac{F(\mathbf{r}', t - |\mathbf{r} - \mathbf{r}'|/c)}{|\mathbf{r} - \mathbf{r}'|}\,\mathrm{d}\tau' \tag{2.62}$$

This solution is sometimes called the retarded solution. It is an integration over the sources evaluated not at t but at a time $|\mathbf{r} - \mathbf{r}'|/c$ seconds earlier. The time $t - |\mathbf{r} - \mathbf{r}'|/c$ is called the retarded time. The dependence on the retarded time reflects the finite velocity of propagation. Any change in the source at $\mathbf{r}'$ is not felt at $\mathbf{r}$ until a time $|\mathbf{r} - \mathbf{r}'|/c$ later.

For sinusoidal, $\exp(\mathrm{j}\omega t)$, time variation, each Fourier component of the source, $F(\mathbf{r}')\exp(\mathrm{j}\omega t')$, would yield a Fourier component of the response $\psi(\mathbf{r})\exp(\mathrm{j}\omega t')$. Thus

$$\psi(\mathbf{r})\exp(\mathrm{j}\omega t) = \iiiint F(\mathbf{r}')\exp(\mathrm{j}\omega t')\frac{1}{4\pi|\mathbf{r} - \mathbf{r}'|}\delta\left(t - t' - \frac{|\mathbf{r} - \mathbf{r}'|}{c}\right)\mathrm{d}t'\,\mathrm{d}\tau' \tag{2.63}$$

Evaluating the t' integration, (2.63) becomes

$$\psi(\mathbf{r})\exp(\mathrm{j}\omega t) = \exp(\mathrm{j}\omega t)\iiint \frac{\exp(-\mathrm{j}k|\mathbf{r} - \mathbf{r}'|)}{4\pi|\mathbf{r} - \mathbf{r}'|}F(\mathbf{r}')\,\mathrm{d}\tau'$$

or

$$\psi(\mathbf{r}) = \iiint \frac{\exp(-\mathrm{j}k|\mathbf{r} - \mathbf{r}'|)}{4\pi|\mathbf{r} - \mathbf{r}'|}F(\mathbf{r}')\,\mathrm{d}\tau' \tag{2.64}$$

where $k = \omega/c$.

Equation (2.64) is the solution to the Helmholtz equation (2.26). Applying (2.64) to each of the components of (2.21) and to (2.22), we have the following solutions to the vector potential $\mathbf{A}(\mathbf{r})$ and the scalar potential $\Phi(\mathbf{r})$ in terms of the current density $\mathbf{J}(\mathbf{r}')$ and the charge density $\rho(\mathbf{r}')$ (equations (2.42) and (2.40)):

$$\mathbf{A}(\mathbf{r}) = \frac{\mu}{4\pi} \iiint \frac{\mathbf{J}(\mathbf{r}') \exp(-jk|\mathbf{r}-\mathbf{r}'|)}{|\mathbf{r}-\mathbf{r}'|} d\tau'$$

$$\Phi(\mathbf{r}) = \frac{1}{4\pi\varepsilon} \iiint \frac{\rho(\mathbf{r}') \exp(-jk|\mathbf{r}-\mathbf{r}'|)}{|\mathbf{r}-\mathbf{r}'|} d\tau'$$

The theory of radiation from wire antennas, for which the distribution of current density is known quite well, is primarily concerned with the evaluation of the integral in (2.42). Different antennas have different current distributions, corresponding to different functions for $\mathbf{J}(\mathbf{r}')$. Once $\mathbf{A}(\mathbf{r})$ has been determined for a particular antenna, the electromagnetic fields can be obtained by carrying out the differential operations indicated by (2.23) and (2.24).

As mentioned at the end of Chapter 1, there is another class of antennas, known as aperture antennas, for which the actual current distribution is complicated and/or difficult to obtain, but for which the electromagnetic field configuration over a specific closed surface surrounding the antenna can be postulated with reasonable accuracy. To analyse aperture antennas, it is necessary to obtain a relationship between the fields existing on a closed surface and the radiation fields outside this surface. This will be the subject of Chapter 10.

2.6 WORKED EXAMPLE

Consider a localized source ($\mathbf{J}$, ρ) in a medium characterized by permeability μ, permittivity ε and conductivity σ. Beginning with Maxwell's equations in phasor form, derive the differential equations governing the vector potential $\mathbf{A}$ and the scalar potential Φ. For what choice of $\nabla \cdot \mathbf{A}$ will the two equations for $\mathbf{A}$ and Φ be decoupled? Write down the solutions for $\mathbf{A}$ and Φ and express $\mathbf{E}$ and $\mathbf{H}$ in terms of $\mathbf{A}$ at points outside the source region.

Solution

Since $\sigma \neq 0$, the radiation field from the source sets up a current $\sigma\mathbf{E}$ in the conductor. Hence we have

$$\nabla \times \mathbf{E}(\mathbf{r}) = -j\omega\mu\mathbf{H}(\mathbf{r}) \tag{2.65}$$

$$\nabla \times \mathbf{H}(\mathbf{r}) = j\omega\varepsilon\mathbf{E}(\mathbf{r}) + \sigma\mathbf{E}(\mathbf{r}) + \mathbf{J}(\mathbf{r}') \tag{2.66}$$

$$\nabla \cdot \mathbf{E} = \rho(\mathbf{r}')/\varepsilon \tag{2.67}$$

$$\nabla \cdot \mathbf{H} = 0 \tag{2.68}$$

From (2.68) and (2.65), we define $\mathbf{A}$ and Φ by the relations

$$\mathbf{H} = \mu^{-1}(\nabla \times \mathbf{A}) \tag{2.69}$$

$$\mathbf{E} + \mathrm{j}\omega\mathbf{A} = -\nabla\Phi \tag{2.70}$$

In terms of $\mathbf{A}$ and Φ, (2.66) and (2.67) become

$$\nabla^2\mathbf{A} + (\omega^2\mu\varepsilon - \mathrm{j}\omega\mu\sigma)\mathbf{A} - \nabla(\nabla\cdot\mathbf{A} + \mathrm{j}\omega\mu\varepsilon\Phi + \mu\sigma\Phi) = -\mu\mathbf{J} \tag{2.71}$$

$$\nabla^2\Phi + \mathrm{j}\omega\varepsilon\nabla\cdot\mathbf{A} = -\rho/\varepsilon \tag{2.72}$$

If we choose

$$\nabla\cdot\mathbf{A} = -(\mathrm{j}\omega\mu\varepsilon + \mu\sigma)\Phi \tag{2.73}$$

then (2.71) and (2.72) decouple, resulting in the following equations for $\mathbf{A}$ and Φ:

$$\nabla^2\mathbf{A} + \gamma^2\mathbf{A} = -\mu\mathbf{J} \tag{2.74}$$

$$\nabla^2\Phi + \gamma^2\Phi = -\rho/\varepsilon \tag{2.75}$$

where

$$\gamma^2 = (\omega^2/c^2) - \mathrm{j}\omega\mu\sigma = k^2 - \mathrm{j}\omega\mu\sigma \tag{2.76}$$

The solutions to (2.74) and (2.75) are given by

$$\mathbf{A}(\mathbf{r}) = \frac{\mu}{4\pi}\iiint \frac{\mathbf{J}(\mathbf{r}')\exp(-\mathrm{j}\gamma|\mathbf{r}-\mathbf{r}'|)}{|\mathbf{r}-\mathbf{r}'|}\mathrm{d}\tau' \tag{2.77}$$

$$\Phi(\mathbf{r}) = \frac{1}{4\pi\varepsilon}\iiint \frac{\rho(\mathbf{r}')\exp(-\mathrm{j}\gamma|\mathbf{r}-\mathbf{r}'|)}{|\mathbf{r}-\mathbf{r}'|}\mathrm{d}\tau' \tag{2.78}$$

In terms of $\mathbf{A}$, $\mathbf{H}$ is given by (2.69). At points outside the source region, $\mathbf{J}(\mathbf{r}') = 0$, and (2.66) yields the following relation between $\mathbf{E}$ and $\mathbf{A}$:

$$\mathbf{E}(\mathbf{r}) = \frac{\nabla\times(\nabla\times\mathbf{A})}{(\sigma + \mathrm{j}\omega\varepsilon)\mu} \tag{2.79}$$

2.7 PROBLEMS

1. Obtain the continuity equation relating the current density $\mathbf{J}(\mathbf{r}, t)$ and the charge density $\rho(\mathbf{r}, t)$ from Maxwell's equations.

2. Show that the magnetic field intensity $\mathbf{H}(\mathbf{r}, t)$ satisfies the following time-dependent inhomogeneous wave equation:

$$\left(\nabla^2 - \frac{1}{c^2}\frac{\partial^2}{\partial t^2}\right)\mathbf{H}(\mathbf{r}, t) = -\nabla\times\mathbf{J}(\mathbf{r}, t) \tag{2.80}$$

3. Show that the electric field intensity $\mathbf{E}(\mathbf{r}, t)$ satisfies the following time-dependent inhomogeneous wave equation:

$$\left(\nabla^2 - \frac{1}{c^2}\frac{\partial^2}{\partial t^2}\right)\mathbf{E}(\mathbf{r}, t) = \nabla\left(\frac{\rho}{\varepsilon}\right) + \mu\frac{\partial\mathbf{J}}{\partial t} \tag{2.81}$$

4. Verify that (2.31) is indeed the solution to (2.29).

5. Solve (2.29) for the case when the time variation is sinusoidal.

6. Fill out the steps that lead to (2.40).

7. Write down the solutions to equations (2.80) and (2.81). Comment on why these solutions are seldom used.

8. Verify (2.52) and (2.56).

9. Obtain the Green's function for the diffusion equation

$$\left(\nabla^2 - \frac{1}{\alpha^2}\frac{\partial}{\partial t}\right)\psi(\mathbf{r}, t) = -F(\mathbf{r}, t) \tag{2.82}$$

operating in an infinite region by applying Fourier transforms to both the spatial and time variables.

CHAPTER 3
Linear Antennas

3.1 INTRODUCTION

In this chapter, we apply the results of Chapter 2 to study the radiation from linear antennas. By a linear antenna, we mean a radiating structure consisting of straight (rather than curved) conductors. A two-wire transmission line which is fed by an alternating source at one end and open-circuited at the other end (Figure 3.1a) can be considered a linear antenna. Such a structure, however, is not an efficient radiator because the current in one wire is out of phase with the current in the opposite wire. If the separation is small compared to the wave length, the radiations from the two wires tend to cancel each other (see worked example 3 at the end of this chapter). This tendency to cancel can be reduced if we bend a section of the line of length $L/2$ near the open-circuited end outwards, in the manner indicated in Figure 3.1(b). When the two conductors that make up the flared-out section become vertical (Figure 3.1c), the currents in them flow in the same direction and the structure becomes an efficient radiator. Such an arrangement of two conductors lying in a straight line and separated by a narrow gap at the centre is called a dipole antenna. It is the most basic type of antenna.

Since an open-circuited transmission line has a sinusoidal current distribution along its length, we expect that the current also varies along the dipole antenna. As will be discussed later, this variation is, to a first approximation, also sinusoidal. We can regard the current distribution along the antenna to be a superposition of infinitesimal current elements of various strengths. Hence to find the radiation from a dipole, we must first calculate the radiation from an elemental (infinitesimal) current source. This result will be basic not only to linear antennas but also to wire antennas of any shape.

In subsequent discussions, we shall be concerned with sinusoidal time variation both because of its practical importance and because any arbitrary time variation can be considered as a superposition of sinusoidal components by means of Fourier analysis.

3.2 FIELDS OF AN INFINITESIMAL CURRENT ELEMENT (HERTZIAN DIPOLE)

Let us consider a conductor of infinitesimal area Δa and length Δl carrying a current $\mathbf{I} = I\hat{\mathbf{z}}$ of angular frequency ω, where $\hat{\mathbf{z}}$ is the unit vector along the z-axis.

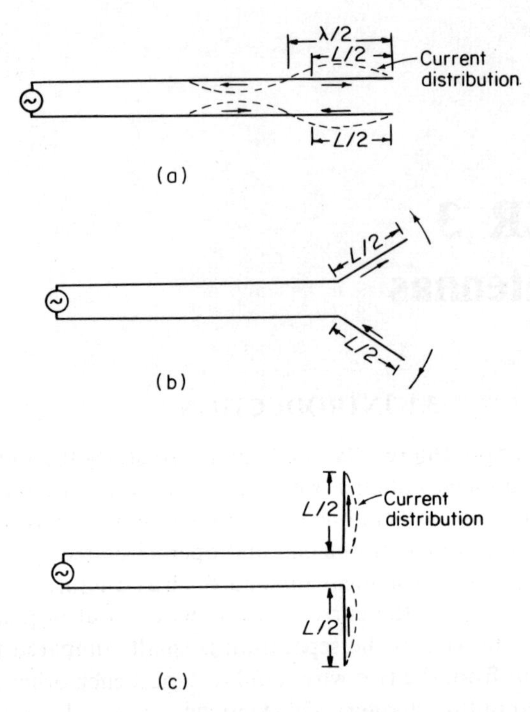

Figure 3.1 Development of a dipole antenna of length L from an open-circuited two-wire transmission line

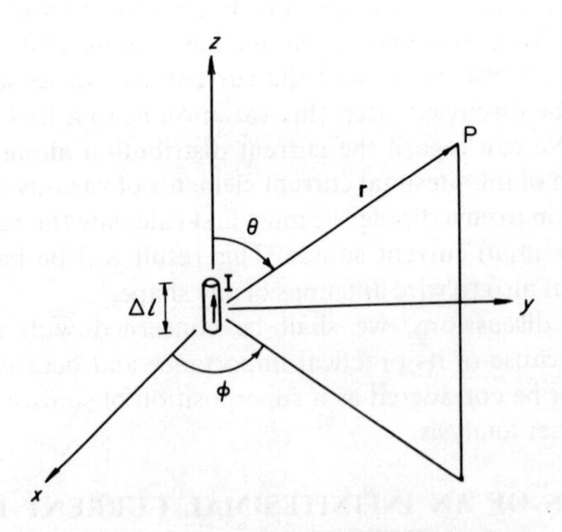

Figure 3.2 Geometry for evaluating the fields of an infinitesimal current element (Hertzian dipole)

The length Δl is also taken to be infinitesimal so that I can be regarded as constant along Δl. Such a current element is known as a Hertzian dipole. The current density $\mathbf{J} = J\hat{\mathbf{z}}$, where J is related to I by the integral

$$I = \iint_{\Delta a} J \, \mathrm{d}\sigma' \tag{3.1}$$

where $\mathrm{d}\sigma'$ is an elemental area perpendicular to z. Let the centre of the current element be situated at the origin of a coordinate system, as shown in Figure 3.2. Consider an arbitrary point P with spherical coordinates r, θ, ϕ. Using (2.42), the vector potential at P is

$$\mathbf{A}(\mathbf{r}) = \frac{\mu}{4\pi} \iiint \frac{\mathbf{J} \exp(-jk|\mathbf{r} - \mathbf{r}'|)}{|\mathbf{r} - \mathbf{r}'|} \, \mathrm{d}\tau' \tag{3.2}$$

Since both Δa and Δl are infinitesimal, we can write

$$|\mathbf{r} - \mathbf{r}'| \simeq |\mathbf{r}| = r$$

and (3.2) becomes

$$\mathbf{A}(\mathbf{r}) = \frac{\exp(-jkr)}{4\pi r} \iiint \mathbf{J} \, \mathrm{d}\tau$$

Since the axis of the conductor is along z, the volume element $\mathrm{d}\tau = \mathrm{d}\sigma \, \mathrm{d}z'$. Hence

$$\begin{aligned} \mathbf{A}(\mathbf{r}) &= \frac{\mu}{4\pi r} \exp(-jkr) \int_{-\Delta l/2}^{\Delta l/2} \iint_{\Delta a} J\hat{\mathbf{z}} \, \mathrm{d}\sigma \, \mathrm{d}z' \\ &= \frac{\mu \exp(-jkr)}{4\pi r} I \Delta l \hat{\mathbf{z}} \end{aligned} \tag{3.3}$$

The vector potential $\mathbf{A}$ is therefore along the same direction as $\mathbf{J}$, i.e. the z-axis.

Using the equation for the curl in spherical coordinates, we have, for the magnetic field $\mathbf{H}$,

$$\begin{aligned} \mu \mathbf{H} = \nabla \times \mathbf{A} = {} & \frac{\hat{\mathbf{r}}}{\sin\theta} \left(\frac{\partial}{\partial \theta} (\sin\theta \, A_\phi) - \frac{\partial A_\theta}{\partial \phi} \right) + \hat{\boldsymbol{\theta}} \left(\frac{1}{r \sin\theta} \frac{\partial A_r}{\partial \phi} - \frac{1}{r} \frac{\partial (r A_\phi)}{\partial r} \right) \\ & + \frac{\hat{\boldsymbol{\phi}}}{r} \left(\frac{\partial}{\partial r} (r A_\theta) - \frac{\partial A_r}{\partial \theta} \right) \end{aligned} \tag{3.4}$$

Since $\mathbf{A} = A_z \hat{\mathbf{z}}$, where

$$A_z = \frac{\mu}{4\pi r} \exp(-jkr) I \Delta l \tag{3.5}$$

we have

$$A_r = A_z \cos\theta \tag{3.6}$$

$$A_\theta = -A_z \sin\theta \tag{3.7}$$

$$A_\phi = 0 \tag{3.8}$$

On substituting (3.5)–(3.8) into (3.4), we find that the only non-zero component of **H** is H_ϕ, i.e.

$$\mathbf{H} = H_\phi \hat{\boldsymbol{\phi}}$$

where

$$H_\phi = -\frac{I\Delta l k^2 \sin\theta}{4\pi}\exp(-jkr)\left[\frac{1}{jkr} + \left(\frac{1}{jkr}\right)^2\right] \tag{3.9}$$

The electric field **E** is given by (2.24), namely

$$\mathbf{E} = \frac{1}{j\omega\varepsilon}(\nabla \times \mathbf{H})$$

On using (3.9) in (2.24) and performing some straightforward algebra, we find that **E** has no ϕ-component and its r- and θ-components are:

$$E_r = -\frac{\eta I\Delta l k^2 \cos\theta}{2\pi}\exp(-jkr)\left[\left(\frac{1}{jkr}\right)^2 + \left(\frac{1}{jkr}\right)^3\right] \tag{3.10}$$

$$E_\theta = -\frac{\eta I\Delta l k^2 \sin\theta}{4\pi}\exp(-jkr)\left[\frac{1}{jkr} + \left(\frac{1}{jkr}\right)^2 + \left(\frac{1}{jkr}\right)^3\right] \tag{3.11}$$

where η is the intrinsic impedance of the medium given by

$$\eta = \sqrt{\frac{\mu}{\varepsilon}} = \frac{k}{\omega\varepsilon}$$

In free space, $\eta = 120\pi$ or 377 ohm.

Equations (3.9)–(3.11) show that the fields are functions of θ and are independent of ϕ. This is a consequence of the fact that we have chosen our coordinate system such that the z-axis coincides with the direction of the current. It should be noted that in an arbitrarily oriented coordinate system, the expressions for the fields will be a function of both angular coordinates (see worked example 2 in section 3.16).

The expressions for the electromagnetic fields produced by an infinitesimal source as given by (3.9)–(3.11) are valid for any point in space. It is, however, useful to simplify them under two approximations. The first approximation applies to points satisfying the inequality $kr \ll 1$ ($r \ll \lambda/2\pi$) and is known as the near-field approximation. The second approximation applied to points satisfying the inequality $kr \gg 1$ ($r \gg \lambda/2\pi$) and is known as the far-field approximation.

Near-Field Approximation

For $kr \ll 1$, $\exp(-jkr) \simeq 1$ and the retardation effect is negligible. The phase of the fields at the point P is approximately the same as the phase of the current at the source point. Equations (3.9)–(3.11) simplify to

$$E_r \simeq \frac{I\eta\Delta l \cos\theta}{2\pi jkr^3} \tag{3.12}$$

$$E_\theta \simeq \frac{I\eta\Delta l \sin\theta}{4\pi jkr^3} \tag{3.13}$$

$$H_\phi \simeq \frac{I\Delta l \sin\theta}{4\pi r^2} \tag{3.14}$$

Note that **E** and **H** are in phase quadrature. Hence there is no power flow but merely a storage of energy in the fields which is returned every cycle. For this reason, the near field is sometimes referred to as the induction field.

Far-Field Approximation

For $kr \gg 1$, we have

$$H_\phi \simeq \frac{jkI\Delta l}{4\pi r}\exp(-jkr)\sin\theta \tag{3.15}$$

$$E_\theta = \eta H_\phi \tag{3.16}$$

The radial component E_r is much less than E_θ and can be neglected. Thus in the far zone, the radiation is linearly polarized, since the electric field vector is always directed along $\hat{\boldsymbol{\theta}}$. Moreover, **E** and **H** are mutually perpendicular and are in phase. The ratio E_θ/H_ϕ is a real number and is equal to η, the intrinsic impedance of the medium.

The far-field component varies with θ according to the factor $\sin\theta$ and is independent of the azimuthal angle ϕ. It is maximum in the direction perpendicular to the axis of the current element ($\theta = 90^\circ$) and is zero along the axis ($\theta = 0^\circ$). Thus the radiation is directive and not isotropic even for a source which is infinitesimal in size.

The expressions we have obtained for the electromagnetic fields are phasor quantities. To find the time-dependent fields, we multiply the phasors by the factor $\exp(j\omega t)$ and take the real part. For example, the electric field in the far zone is

$$\begin{aligned} E_\theta(\mathbf{r}, t) &= \mathrm{Re}\left(\frac{\eta jkI\Delta l}{4\pi r}\sin\theta\exp[j(\omega t - kr)]\right) \\ &= -\frac{kI\Delta l\eta}{4\pi r}\sin\theta\sin(\omega t - kr) \end{aligned} \tag{3.17}$$

Equation (3.17) is in the form of a spherical travelling wave. The velocity of propagation is $\omega/k = c$ and the amplitude varies inversely as the distance from the source.

In the next three sections, we define several important parameters that are used to describe the radiation characteristics of any antenna and evaluate these parameters for the Hertzian dipole.

3.3 POWER RADIATED AND RADIATION RESISTANCE

The time-averaged (over a cycle) power density (in Wm^{-2}) carried by a time-harmonic electromagnetic field is given by the real part of the complex Poynting vector $\mathbf{S}$ where

$$\mathbf{S} = \tfrac{1}{2}\mathbf{E} \times \mathbf{H}^* \tag{3.18}$$

where $*$ denotes complex conjugate. For the Hertzian dipole under consideration,

$$\mathbf{S} = \tfrac{1}{2}\begin{vmatrix} \hat{\mathbf{r}} & \hat{\boldsymbol{\theta}} & \hat{\boldsymbol{\phi}} \\ E_r & E_\theta & 0 \\ 0 & 0 & H_\phi^* \end{vmatrix} \tag{3.19}$$

where E_r, E_θ, and H_ϕ are given by (3.9)–(3.11). Expanding the determinant, we obtain

$$S_\phi = 0 \tag{3.20}$$

$$S_\theta = -\tfrac{1}{2}E_r H_\phi^* = \mathrm{j}\frac{I^2\eta(\Delta l)^2}{8\pi\lambda r^3}\sin\theta\cos\theta\left(1 + \frac{1}{k^2r^2}\right) \tag{3.21}$$

$$S_r = \tfrac{1}{2}E_\theta H_\phi^* = \frac{I^2\eta(\Delta l)^2\sin^2\theta}{8\lambda^2 r^2}\left(1 - \frac{\mathrm{j}}{k^3r^3}\right) \tag{3.22}$$

Equation (3.20) shows that there is no instantaneous flow of power in the ϕ-direction. Equation (3.21) shows that S_θ is purely imaginary. Since $\mathrm{Re}\, S_\theta = 0$, this means that there is no average power flowing in the θ-direction. There is, however, instantaneous power which oscillates back and forth during a cycle.

The only non-zero component of $\mathrm{Re}\,\mathbf{S}$ is in the radial direction. From (3.22),

$$\mathrm{Re}\,\mathbf{S} = S_r\hat{\mathbf{r}} = \frac{I^2}{8}\eta\left(\frac{\Delta l}{\lambda}\right)^2\frac{\sin^2\theta}{r^2} \tag{3.23}$$

Equation (3.23) gives the average power density carried in the electromagnetic field produced by the Hertzian dipole. It is obtained by using the full solutions for $\mathbf{E}$ and $\mathbf{H}$ as given by (3.9)–(3.11) in (3.18). We note that the same result is obtained if we use the far-field approximations to the fields, namely (3.15) and (3.16), in computing $\mathbf{S}$. This means that only the far-field components contribute to the average power density. The contributions due to the other components vanish upon being averaged over a cycle.

To compute the total power radiated by the source, we integrate the average power density over the surface of a sphere of radius r. The average power through an area $\mathrm{d}\sigma$ is

$$\mathrm{d}W = (\mathrm{Re}\, S_r)\,\mathrm{d}\sigma = (\mathrm{Re}\, S_r)(r^2\sin\theta\,\mathrm{d}\theta\,\mathrm{d}\phi) = \frac{I^2}{8}\eta\left(\frac{\Delta l}{\lambda}\right)^2\sin^3\theta\,\mathrm{d}\theta\,\mathrm{d}\phi \tag{3.24}$$

The total power (in watts) is

$$W = \frac{I^2}{8}\eta\left(\frac{\Delta l}{\lambda}\right)^2 \int_0^{2\pi}\int_0^{\pi} \sin^3\theta \, d\theta \, d\phi = \frac{I^2}{3}\pi\eta\left(\frac{\Delta l}{\lambda}\right)^2 \tag{3.25}$$

The radiation resistance R_r is defined as the value of a hypothetical resistor which dissipates a power equal to the power radiated by the antenna when fed by the same current I. Thus

$$\tfrac{1}{2}I^2 R_r \equiv W \tag{3.26}$$

For the Hertzian dipole, use of (3.25) in (3.26) yields R_r (in ohms) as

$$R_r = \frac{2\pi}{3}\eta\left(\frac{\Delta l}{\lambda}\right)^2 \tag{3.27}$$

In free space, $\eta = 120\pi$ ohm and R_r (in ohms) is

$$R_r = 80\pi^2\left(\frac{\Delta l}{\lambda}\right)^2 = 789\left(\frac{\Delta l}{\lambda}\right)^2 \tag{3.28}$$

Equation (3.25) shows that the radiated power is proportional to the square of the length-to-wavelength ratio. For the radiator to be efficient, its size must be comparable to the wavelength. This is why radiation effects can be ignored in the analysis of low-frequency circuits but must be taken into account when the size of the apparatus is comparable to the wavelength. In the limit $\lambda \to \infty$, $W \to 0$ as expected, since a source of direct current does not radiate.

3.4 RADIATION INTENSITY, DIRECTIVE GAIN, AND DIRECTIVITY

The time-averaged power per unit area, as given by $\mathrm{Re}\, S_r$, is inversely proportional to the distance squared. The time-averaged power per unit solid angle, on the other hand, has the advantage of being independent of distance. This quantity is called the radiation intensity and is denoted by U. Since

$$dW = (\mathrm{Re}\, S_r)d\sigma = (\mathrm{Re}\, S_r)r^2 \, d\Omega \tag{3.29}$$

where $d\Omega$ is the solid angle subtended by $d\sigma$, we have

$$U \equiv dW/d\Omega = (\mathrm{Re}\, S_r)r^2 \tag{3.30}$$

For the Hertzian dipole, use of (3.23) in (3.30) yields

$$U = \frac{\eta}{8}I^2\left(\frac{\Delta l}{\lambda}\right)^2 \sin^2\theta \tag{3.31}$$

The variations of E_θ or H_ϕ and U with θ are shown in Figure 3.3. The maximum occurs in the broadside direction ($\theta = 90^\circ$). The half-power points occur at $\theta = 45^\circ$ and $\theta = 135^\circ$. The angle subtended by the half-power points, which will be referred to as the half-power beamwidth, is thus 90°.

Let us define a hypothetical isotropic radiator as one that radiates the same amount of power in all directions. Then the radiation intensity is a constant,

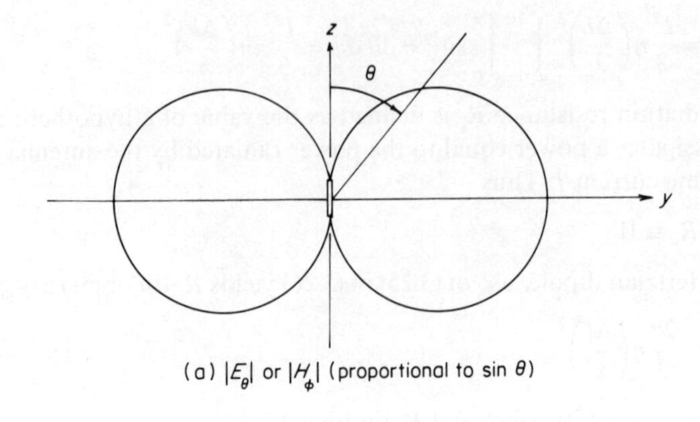

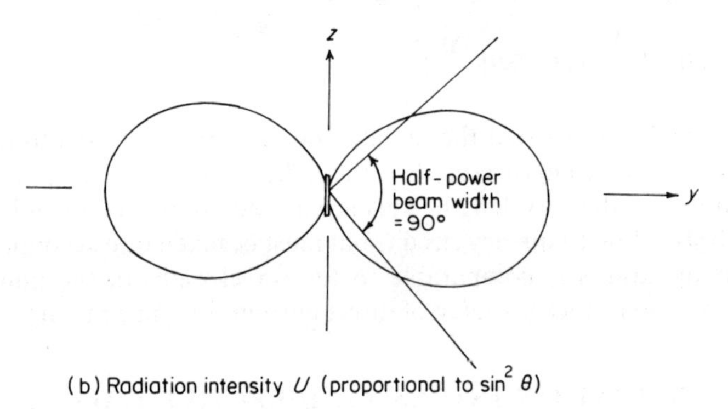

Figure 3.3 Variation of (a) the far field and (b) the radiation intensity with θ for the Hertzian dipole

say U_0. Since the solid angle subtended by the surface of a sphere is 4π radians, the total power radiated by the isotropic radiator is

$$W = 4\pi U_0 \tag{3.32}$$

To give a quantitative measure of the directional property of an antenna, we define its directive gain g as the ratio of the radiation intensity of the antenna to that of an isotropic radiator radiating the same amount of power. Thus

$$g = U/U_0 = 4\pi U/W \tag{3.33}$$

The quantity g is a function of direction. Let its maximum value be denoted by U_m. The directivity of an antenna is defined as

$$D \equiv U_m/U_0 = 4\pi U_m/W \tag{3.34}$$

For the Hertzian dipole, one readily obtains

$$U_m = \frac{\eta}{8} I^2 \left(\frac{\Delta l}{\lambda}\right)^2 \tag{3.35}$$

$$D = 1.5 \tag{3.36}$$

3.5 RADIATION EFFICIENCY AND POWER GAIN

In (3.33) and (3.34), W is the power radiated by the antenna. This is equal to the power accepted by the antenna from the source, denoted by W_{in}, minus the ohmic loss incurred by the finite conductivity of the antenna, denoted by W_{L}. The radiation efficiency e is defined as

$$e = W/W_{\text{in}} = W/(W + W_{\text{L}}) \tag{3.37}$$

If W_{in} instead of W is used in (3.33), the resultant quantity is called the power gain:

$$G = 4\pi U/W_{\text{in}} \tag{3.38}$$

In words, the power gain of an antenna in a specified direction is the ratio of its radiation intensity to that of an isotropic radiator radiating an amount of power equal to the power accepted by the antenna from its source. The maximum value of the power gain is

$$G_{\text{m}} = 4\pi U_{\text{m}}/W_{\text{in}} \tag{3.39}$$

On using (3.37), we have

$$e = G/g = G_{\text{m}}/D \tag{3.40}$$

In the literature, the power gain (also abbreviated as gain) is often given without reference to a direction, in which case it is understood to mean the maximum value.

For the Hertzian dipole of length Δl and radius a, the current is uniform along Δl and flows in an area $2\pi a\delta$ where δ is the skin depth given by

$$\delta = \sqrt{\frac{2}{\omega\mu\sigma}} \tag{3.41}$$

The loss resistance is

$$R_{\text{L}} = \frac{1}{\sigma}\frac{\Delta l}{2\pi a\delta} = \frac{R_{\text{s}}\Delta l}{2\pi a} \tag{3.42}$$

where

$$R_{\text{s}} = \sqrt{\frac{\omega\mu}{2\sigma}} \tag{3.43}$$

is the surface resistance. The radiation efficiency is therefore

$$e = \frac{R_{\text{r}}}{R_{\text{r}} + R_{\text{L}}} = 1\bigg/\left[1 + \frac{R_{\text{s}}}{1578\pi}\left(\frac{\lambda}{a}\right)\left(\frac{\lambda}{\Delta l}\right)\right] \tag{3.44}$$

The definitions for the radiation resistance R_{r}, the radiation intensity U, the directive gain g, the directivity D, the power gain G, and the radiation efficiency e are general and are applicable to all antennas. For the Hertzian dipole with its axis directed along the z-axis of a coordinate system, we have seen that U, g, and G are functions of θ and are independent of ϕ. This is also true for a finite-length dipole antenna, which can be regarded as consisting of a continuous distribution of Hertzian dipoles. However, for an arbitrarily oriented coordinate system or for more complicated radiating structures, for example, two dipoles arranged in the form of a cross, U, g, and G are in general functions of both angular coordinates.

3.6 EFFECTIVE LENGTH OF FINITE-LENGTH DIPOLE ANTENNA

Let us consider a dipole antenna of length L consisting of two conductors lying in a straight line separated by a narrow gap of width Δ, as shown in Figure 3.4. Across the gap, there exists an alternating voltage source of angular frequency ω. Let the resulting current distribution along the antenna be $I(z')$. We shall assume the gap width Δ to be infinitesimally small so that the current $I(z')$ can be regarded as a continuous function of z'. This assumption turns out to be justified if the inequality $(\Delta/\lambda) \ll 1$ is satisfied. The effect of a finite gap width will be considered in section 3.12.

In addition to the current being a function of z', the radiation from different portions of the antenna will arrive at the field point out of phase. This makes the integral for **A** difficult to evaluate. The problem is considerably simplified if we restrict ourselves to calculating the fields at points far from the antenna satisfying the inequalities

$$r \gg L \tag{3.45}$$

and

$$r \gg \lambda \tag{3.46}$$

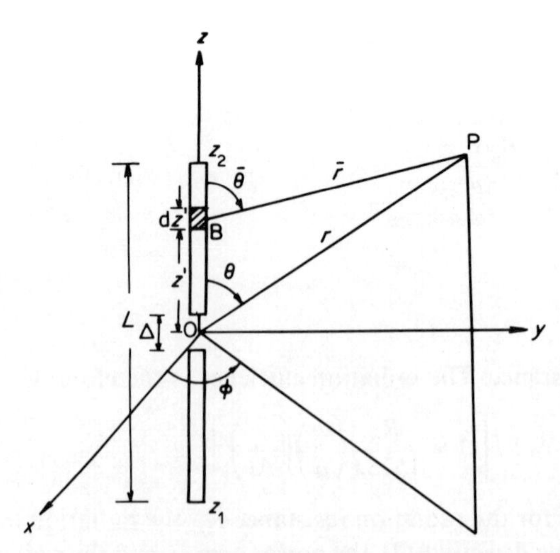

Figure 3.4 Geometry for the evaluation of the far field of a finite-length dipole antenna

Consider an element of length $\mathrm{d}z'$ centred at point B which is at a distance z' from the origin. This element carries a current $I(z')$. Let $\bar{r}$ be the distance from $\mathrm{d}z'$ to the point P. By virtue of (3.46), P will be in the far zone of the element and the field due to $I(z')$ flowing in $\mathrm{d}z'$ is, from (3.15) and (3.16),

$$\mathrm{d}H_\phi = \frac{\mathrm{j}kI(z')\mathrm{d}z'}{4\pi\bar{r}}\exp(-\mathrm{j}k\bar{r})\sin\bar{\theta} \tag{3.47}$$

$$\mathrm{dE}_\theta = \eta\,\mathrm{d}H_\phi \tag{3.48}$$

If P is sufficiently far away, the lines BP and OP can be regarded as parallel and

$$\bar{r} \simeq r - z'\cos\theta \tag{3.49}$$

Hence

$$\mathrm{d}H_\phi = \frac{\mathrm{j}kI(z')\mathrm{d}z'}{4\pi(r - z'\cos\theta)}\exp[-\mathrm{j}k(r - z'\cos\theta)]\sin\bar{\theta} \tag{3.50}$$

By (3.45), $z'\cos\theta \ll r$ for all values of z' on the antenna. In the denominator of (3.50), we can approximate $(r - z'\cos\theta)$ by r. Note that we cannot do so in the exponent since

$$\exp[-\mathrm{j}k(r - z'\cos\theta)] = \exp(-\mathrm{j}kr)\exp(\mathrm{j}kz'\cos\theta)$$

and the factor $\exp(\mathrm{j}kz'\cos\theta)$ is approximately unity only if $kz'\cos\theta \ll 1$ for all values of z'. For this to be satisfied, we must have $kL \ll 1$, implying that $L \ll \lambda$. Thus for an antenna with length comparable to wavelength, the term $z'\cos\theta$ cannot be neglected compared to r in the exponent of (3.50) even though we are considering points satisfying $r \gg L$. Hence

$$\mathrm{d}H_\phi = \frac{\mathrm{j}k\sin\bar{\theta}}{4\pi r}\exp(-\mathrm{j}kr)I(z')\exp(\mathrm{j}kz'\cos\theta)\mathrm{d}z' \tag{3.51}$$

The total magnetic field at point P is obtained by summing over the current elements of the antenna, yielding

$$H_\phi = \frac{\mathrm{j}k\exp(-\mathrm{j}kr)}{4\pi r}\int_{z_1}^{z_2}\sin\theta\, I(z')\exp(\mathrm{j}kz'\cos\theta)\mathrm{d}z' \tag{3.52}$$

where we have replaced $\sin\bar{\theta}$ by $\sin\theta$ since point P is far away. The integral in (3.52) is a weighted current moment which has the dimension of a current multiplied by a length. It is convenient to normalize the weighted current moment by the current at the feed point of the antenna, yielding a quantity that has the dimension of length. This quantity is called the effective length and is denoted by $h_\mathrm{e}(\theta)$. Thus

$$h_\mathrm{e}(\theta) = \frac{\sin\theta}{I(0)}\int_{z_1}^{z_2} I(z')\exp(\mathrm{j}kz'\cos\theta)\mathrm{d}z' \tag{3.53}$$

where $I(0)$ is the current at the feed point. In terms of $h_\mathrm{e}(\theta)$, the far fields can be written as

$$H_\phi = \frac{jkI(0)}{4\pi r}\exp(-jkr)h_e(\theta) \tag{3.54}$$

$$E_\theta = \eta H_\phi \tag{3.55}$$

The problem of calculating the far field of a finite-length dipole antenna with a given current distribution therefore reduces to that of evaluating the integral defining the effective length, which is the quantity that distinguishes one antenna from another and contains all the information about the directional properties of an antenna.

We now return to examine the condition for the validity of the parallel ray approximation (3.49). To do this, it is convenient to choose the point P to be in the y–z plane. No generality is lost because the problem has symmetry about the z-axis. From Figure 3.4, we have

$$\begin{aligned}\bar{r} &= [y^2 + (z - z')^2]^{1/2} = [r^2 - 2r(\cos\theta)z' + (z')^2]^{1/2} \\ &= r + \tfrac{1}{2}(r^2)^{-1/2}[-2r(\cos\theta)z' + (z')^2] \\ &\quad + \frac{\frac{1}{2}(-\frac{1}{2})}{2}(r^2)^{-3/2}[-2r(\cos\theta)z' + (z')^2]^2 + \cdots \\ &= r - z'\cos\theta + \frac{(z')^2\sin^2\theta}{2r} + O(1/r^2) + \cdots\end{aligned} \tag{3.56}$$

Equation (3.56) reduces to (3.49) if only the first two terms are kept. Reference to the arguments leading to (3.51) shows that the neglect of the third term introduces a phase error of the amount $k(z')^2(\sin^2\theta)/2r$. It is commonly accepted that (3.49) is a good approximation if the maximum path deviation due to the neglect of the third term is less than $\lambda/16$, corresponding to a phase error of $(2\pi/\lambda)(\lambda/16) = \pi/8$ radians or 22.5°. For an antenna of length L, the requirement is, on setting $z' = L/2$ and $\theta = 90°$, expressed as

$$(L/2)^2/2r < \lambda/16$$

or

$$r > 2L^2/\lambda \tag{3.57}$$

The far-field conditions (3.45), (3.46) and (3.57) for a finite-length antenna, upon which the simple results (3.53)–(3.55) are based, are summarized below:

$$r \gg L,$$
$$r \gg \lambda,$$
$$r > 2L^2/\lambda$$

3.7 CURRENT DISTRIBUTION BASED ON ANALOGY WITH TRANSMISSION-LINE THEORY

As indicated in section 3.1, a dipole antenna of length L can be regarded as being evolved from a two-wire open-circuited transmission line by bending a segment of the line a distance $L/2$ from the end. The two conductors of this segment are bent outwards in opposite directions, until they are perpendicular to the main transmission line, as shown in Figure 3.1. The current distribution along the line before it is bent is a standing wave whose amplitude varies sinusoidally, being zero at the end of the line and maximum at a distance $\lambda/2$ from the end. If we assume that the current distribution does not change after the line is bent, we would have the current distribution along a dipole antenna of length L to be

$$I(z') = I_m \sin\left[k(\tfrac{1}{2}L - |z'|)\right] \tag{3.58}$$

where the maximum value of the current I_m is related to the current at the feed point $I(0)$ by the formula

$$I_m = I(0)/\sin(\tfrac{1}{2}kL) \tag{3.59}$$

The assumption that the current distribution remains unchanged after the line is bent is not rigorously justified since the opened-up line is no longer a uniform transmission line, because the distributed capacitance and inductance per unit length would vary along the vertical part of the line. More fundamentally, transmission-line theory is circuit theory that neglects radiation. The current distribution of an antenna, on the other hand, is basically a problem in electromagnetic field theory and should be solved as such. This approach was first considered by Hallén (1938) who obtained an integral equation for the current distribution. The derivation of this equation will be sketched in the following section.

3.8 HALLÉN'S INTEGRAL EQUATION FOR THE CURRENT DISTRIBUTION

Consider a linear antenna in the form of a cylindrical perfect conductor of length L and radius a (Figure 3.5). It is driven by a voltage V across a narrow gap at the centre, which extends from $z = -\frac{1}{2}\Delta$ to $z = \frac{1}{2}\Delta$. As a result, a current $I(z')$ flows along the antenna, producing electromagnetic fields **E** and **H** in the space surrounding the antenna. The current and the fields must be such that they satisfy certain boundary conditions. The first is that the current must vanish at the ends of the antenna. The second is that the tangential electric field must vanish on the surface of the antenna, since we assume its conductivity to be infinite. The third is that the integral of the electric field across the narrow gap must be equal to the voltage applied across the gap. Let us now express these physical conditions in mathematical form.

In Chapter 2, it was shown that the electric field **E** outside the antenna is derivable entirely from the vector potential **A** as follows:

$$\mathbf{E} = -\nabla\Phi - j\omega\mathbf{A} \tag{3.60}$$

Since

$$\nabla\cdot\mathbf{A} = -\mu\varepsilon j\omega\Phi \tag{3.61}$$

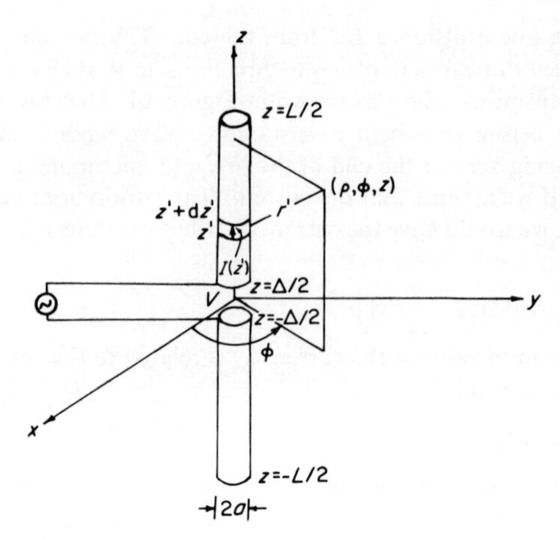

Figure 3.5 Symmetrical centre-fed cylindrical antenna

we have

$$\mathbf{E} = \frac{1}{\mathrm{j}\omega\mu\varepsilon}[\nabla(\nabla\cdot\mathbf{A})] - \mathrm{j}\omega\mathbf{A} = -\mathrm{j}\omega[(c/\omega)^2\nabla(\nabla\cdot\mathbf{A}) + \mathbf{A}] \tag{3.62}$$

As **I** has only a z-component, **A** has only a z-component. The z-component of (3.62) reads

$$-\mathrm{j}\frac{\omega}{k^2}\left(\frac{\partial^2 A_z}{\partial z^2} + k^2 A_z\right) = E_z \tag{3.63}$$

The quantity E_z is equal to zero except at the gap. Across the gap, which we assume to be infinitesimally narrow, we can write

$$E_z = -V\delta(z) \tag{3.64}$$

Hence

$$\frac{\partial^2 A_z}{\partial z^2} + k^2 A_z = \mathrm{j}\omega\mu\varepsilon V\delta(z) \tag{3.65}$$

where we have made use of $k^2 = \omega^2\mu\varepsilon$.

The right-hand side of (3.65) is equal to zero except across the gap. Thus the solution for A_z at points outside the gap is given by

$$A_z = C_1 \cos kz + C_2 \sin kz \tag{3.66}$$

For a symmetrically fed antenna,

$$I(z') = I(-z') \tag{3.67}$$

and

$$A_z(z') = A_z(-z') \tag{3.68}$$

Thus (2.56) becomes

$$A_z = C_1 \cos(kz) + C_2 \sin(k|z|). \tag{3.69}$$

To satisfy the condition at the gap, we integrate (3.65) across the gap, i.e. from $z = -\frac{1}{2}\Delta$ to $\frac{1}{2}\Delta$, obtaining

$$\frac{\partial A_z}{\partial z}\bigg|_{z=-\frac{1}{2}\Delta}^{z=\frac{1}{2}\Delta} + k^2 z A_z\bigg|_{z=-\frac{1}{2}\Delta}^{z=\frac{1}{2}\Delta} = \mathrm{j}\omega\mu\varepsilon V \tag{3.70}$$

In the limit of an infinitesimal gap, $\Delta \to 0$ and we find that, for (3.69) to satisfy (3.65), we must have

$$C_2 = \frac{\mathrm{j}\omega\mu\varepsilon V}{2k} = \frac{\mathrm{j}V}{2c} \tag{3.71}$$

Hence

$$A_z = C_1 \cos(kz) + \frac{\mathrm{j}V}{2c}\sin(k|z|). \tag{3.72}$$

From Chapter 2, we know that **A** can also be expressed in terms of an integral of the current on the antenna. At a point with cylindrical coordinates (ρ, ϕ, z),

$$A_z = \frac{\mu}{4\pi}\int_{-L/2}^{L/2} \frac{I(z')\exp(-\mathrm{j}kr)}{r}\,\mathrm{d}z' \tag{3.73}$$

where

$$r = \sqrt{(\rho^2 + (z - z')^2]} \tag{3.74}$$

On the surface of the antenna, $\rho = a$, $r = \sqrt{[a^2 + (z - z')^2]}$ and

$$A_z = \frac{\mu}{4\pi}\int_{-L/2}^{L/2} \frac{I(z')\exp\{-\mathrm{j}k\sqrt{[a^2 + (z - z')^2]}\}}{\sqrt{[a^2 + (z - z')^2]}}\,\mathrm{d}z' \tag{3.75}$$

Equating (3.72) and (3.75), there results

$$\frac{\mu}{4\pi}\int_{-L/2}^{L/2} \frac{I(z')\exp\{-\mathrm{j}k\sqrt{[a^2 + (z - z')^2]}\}}{\sqrt{[a^2 + (z - z')^2]}}\,\mathrm{d}z' = C_1 \cos(kz) + \frac{\mathrm{j}V}{2c}\sin(k|z|) \tag{3.76}$$

Equation (3.76) is Hallén's integral equation for the current distribution in the case when the conductivity of the antenna is assumed to be infinite. The problem is to find a function of z' for I such that the equation is satisfied. The remaining constant C_1 is to be determined by the boundary condition $I(z') = 0$ at $z' = \pm L/2$.

Various approximate methods have been devised to solve the integral equation of Hallén. We shall not go into the details here. One method uses the quantity

$$\frac{1}{\Omega}=\frac{1}{2\ln(L/a)} \tag{3.77}$$

as a small parameter and solves the equation in successive powers of $1/\Omega$. The result is

$$I(z')=\frac{\mathrm{j}V}{60\Omega}\left(\frac{\sin(\frac{1}{2}L-|z'|)+b_1\Omega^{-1}+b_2\Omega^{-2}+\cdots}{\cos(\frac{1}{2}kL)+d_1\Omega^{-1}+d_2\Omega^{-2}+\cdots}\right) \tag{3.78}$$

where b_1, b_2, d_1, d_2, etc. are functions of L and z. For the detailed steps leading to (3.78) and the expressions for b_1, b_2, d_1, d_2, etc., the reader is referred to the paper by King and Harrison (1943).

For very thin antennas, Ω is very large and

$$I(z')=\frac{\mathrm{j}V}{60\Omega}\left(\frac{\sin[k(\frac{1}{2}L-|z'|)]}{\cos(\frac{1}{2}kL)}\right) \tag{3.79}$$

On setting

$$I_m\equiv\frac{\mathrm{j}V}{60\Omega\cos(\frac{1}{2}kL)} \tag{3.80}$$

(3.79) becomes

$$I(z')=I_m\sin[k(\tfrac{1}{2}L-|z'|)] \tag{3.81}$$

Equation (3.81) is of the same form as (3.58). It thus appears that the sinusoidal current distribution of an open-circuit transmission line is an exact solution for an antenna which is infinitesimally thin. It is also a good approximation for antennas with finite but large length-to-radius ratio (L/a). A criterion for the sinusoidal distribution to be a good approximation is arbitrarily taken to be

$$L/a\gtrsim 60 \tag{3.82}$$

The solution for $I(z')$ as given by (3.78) is a complex function and can be denoted by $|I(z')|\angle\theta$. Figure 3.6 illustrates the current amplitude $|I(z')|$ and the phase θ distributions for $\lambda/2$, λ, and $1\frac{1}{4}\lambda$ depoles with $L/a=75$ and $L/a=\infty$. The $L/a=75$ curves are obtained from (3.78), including terms up to the first power in $1/\Omega$ (King and Harrison, 1943). The most significant difference between the $L/a=75$ and the $L/a=\infty$ curves occurs at the feed point for the $L=\lambda$ dipole. It is evident that, if L is a multiple of λ, the feed-point current is zero for $L/a=\infty$ (equation (3.81)) and is non-zero for finite L/a (equation (3.78)). This difference is important when one is concerned with calculating the driving-point impedances (section 3.11). However, for antennas with $L/a\gtrsim 60$, the difference does not affect the radiation characteristics significantly. Unless otherwise stated, we shall, in the interest of simplicity, assume the current distribution on finite-length antennas to be given by the sinusoidal function (3.81).

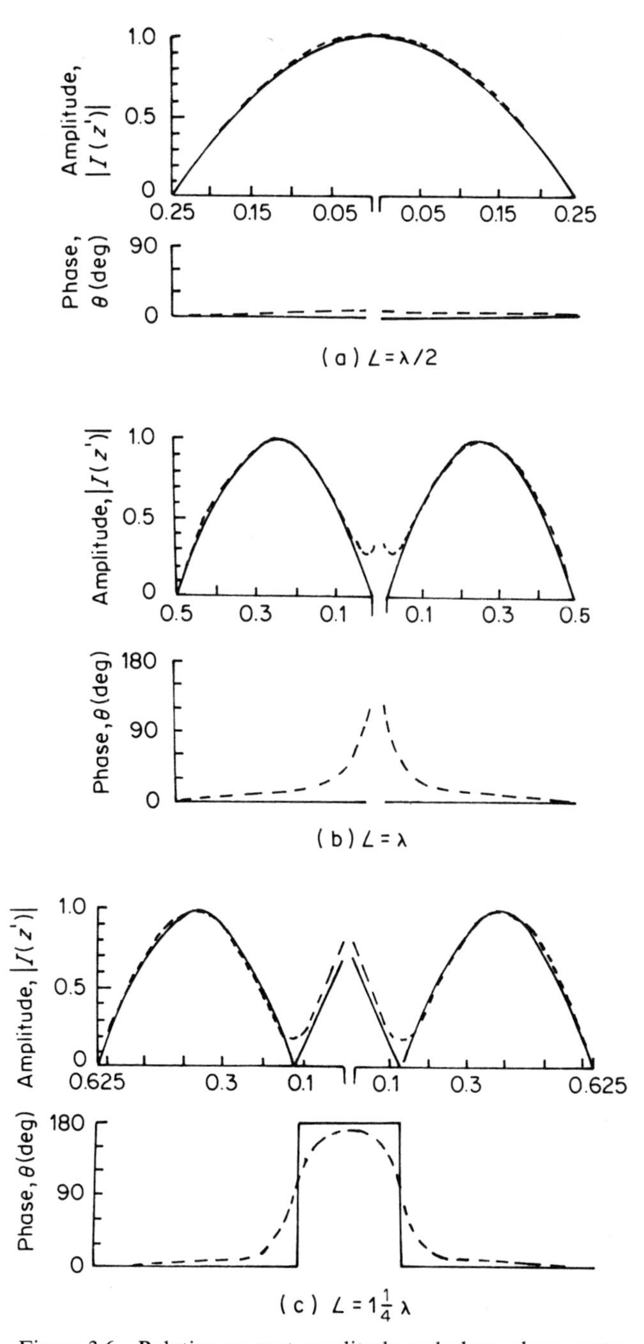

Figure 3.6 Relative current amplitude and phase along centre-fed cylindrical dipoles of lengths (a) $\lambda/2$, (b) λ, and (c) $1\frac{1}{4}\lambda$ for length-to-radius ratios (L/a) of 75 (– – –) and ∞ (——). Distance from the centre of the dipole is expressed in wavelengths (Source: King and Harrison (1943), *Proc. IRE*, **31**, 548–567. © 1943, IEEE)

3.9 PATTERN FACTOR FOR LINEAR ANTENNAS WITH SINUSOIDAL CURRENT DISTRIBUTION

The far field of a finite-length dipole antenna is given by (3.54) and (3.55):

$$H_\phi = \frac{\mathrm{j}kI(0)\exp(-\mathrm{j}kr)}{4\pi r}h_e(\theta)$$

$$E_\theta = \eta H_\phi$$

where the effective length $h_e(\theta)$ is given by (3.53):

$$h_e(\theta) = \frac{\sin\theta}{I(0)}\int_{z_1}^{z_2} I(z')\exp(\mathrm{j}kz'\cos\theta)\,\mathrm{d}z'$$

For a centre-fed antenna with a sinusoidal current distribution, $z_1 = -L/2$, $z_2 = L/2$, and

$$I(z') = I_m \sin[k(\tfrac{1}{2}L - |z'|)] = \begin{cases} I_m \sin[k(\tfrac{1}{2}L - z')] & z' > 0 \\ I_m \sin[k(\tfrac{1}{2}L + z')] & z' < 0 \end{cases} \tag{3.83}$$

The integral in (3.53) becomes

$$\int_{z_1}^{z_2} I(z')\exp(\mathrm{j}kz'\cos\theta)\,\mathrm{d}z' = I_m\int_{-L/2}^{0} \sin[k(\tfrac{1}{2}L + z')]\exp(\mathrm{j}kz'\cos\theta)\,\mathrm{d}z'$$
$$+ I_m\int_{0}^{L/2} \sin[k(\tfrac{1}{2}L - z')]\exp(\mathrm{j}kz'\cos\theta)\,\mathrm{d}z' \tag{3.84}$$

The two integrals in (3.84) can be evaluated using the formula

$$\int \exp(ax)\sin(c+bx)\,\mathrm{d}x = \frac{\exp(ax)}{a^2+b^2}[a\sin(c+bx) - b\cos(c+bx)] \tag{3.85}$$

In the first integral, $a = \mathrm{j}k\cos\theta$, $b = k$, and $c = \frac{1}{2}kL$. In the second integral, $a = \mathrm{j}k\cos\theta$, $b = -k$, and $c = \frac{1}{2}kL$.

Carrying out the integrations, adding the results, and simplifying, we have

$$h_e(\theta) = \frac{2I_m}{kI(0)}P(\theta) \tag{3.86}$$

$$H_\phi(\theta) = \frac{\mathrm{j}I_m\exp(-\mathrm{j}kr)}{2\pi r}P(\theta) \tag{3.87}$$

$$E_\theta(\theta) = \eta H_\phi(\theta) \tag{3.88}$$

where $P(\theta)$ is the pattern factor given by

$$P(\theta) = \frac{\cos(\frac{1}{2}kL\cos\theta) - \cos(\frac{1}{2}kL)}{\sin\theta} \tag{3.89}$$

The pattern factor describes how the radiation in the far zone varies with direction. Note that it is a function of the elevation angle θ only and is independent

Table 3.1 Sketches of current distributions and pattern factors for the Hertzian, half-wave, full-wave, and one-and-a-half-wave dipoles

L	Sketch of $I(z')$	$P(\theta)$	Sketch of $P(\theta)$
dl		$\sin\theta$	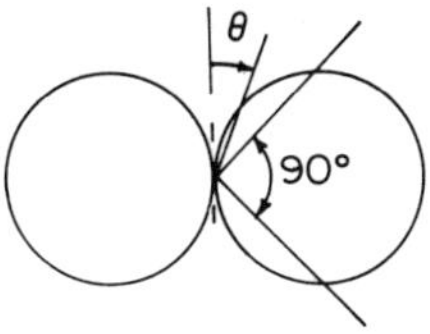 θ 90°
$\lambda/2$		$\dfrac{\cos(\frac{1}{2}\pi\cos\theta)}{\cos\theta}$	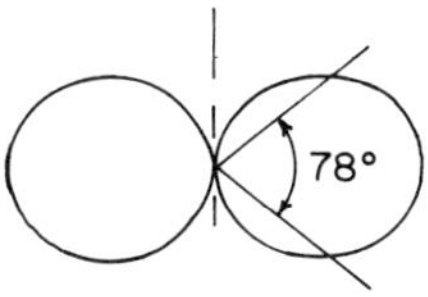 78°
λ	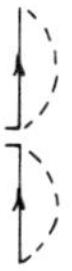	$\dfrac{\cos(\pi\cos\theta)+1}{\sin\theta}$	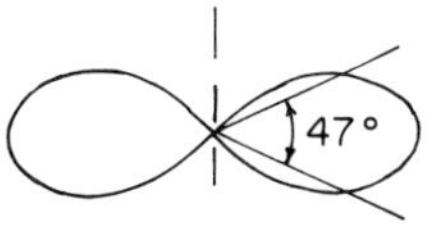 47°
$3\lambda/2$		$\dfrac{\cos(\frac{3}{2}\pi\cos\theta)}{\sin\theta}$	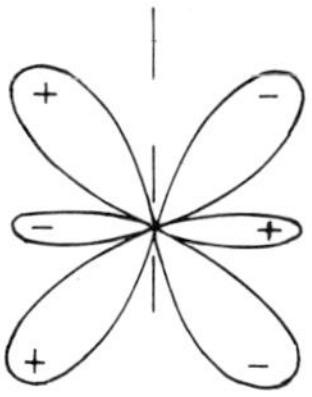 + − − + + −

of the azimuthal angle ϕ. In Table 3.1, we list the pattern factors and sketch how they vary with θ for half-wave ($L = \lambda/2$), full-wave ($L = \lambda$), and one-and-a-half-wave ($L = 3\lambda/2$) dipoles. We also include the hypothetical case of an infinitesimal current element for comparison purposes.

Note that the pattern becomes sharper (more directional) as the length increases from infinitesimal to $\lambda/2$ and λ. The half-power beamwidths for these three cases are 90°, 78°, and 47° respectively. However, when the length increases from λ to $3\lambda/2$, the radiation pattern becomes multilobed. The pattern factor also changes sign, corresponding to phase changes by 180°, as θ varies. The relative phases of the lobes are indicated by the + and − signs in the sketch for the $3\lambda/2$ dipole.

The transition from a figure-of-eight pattern to a multilobed pattern occurs at $L \simeq 1.2\lambda$. Thus one can make the dipole more directive by increasing its length only up to $L \simeq 1.2\lambda$. There is a limit to what one can achieve in directivity by using a single antenna. As we shall see in Chapter 7, an array of antennas is needed to obtain very narrow beamwidths.

In Table 3.1, we also sketch the sinusoidal current distributions of the antennas. The arrows indicate the direction of current at any instant. For the $\lambda/2$ and $3\lambda/2$ dipoles, the maximum current occurs at the feed point. For the full-wave dipole, the feed-point current is equal to zero. We must remember, however, that the sinusoidal function is only the leading term in the solution for $I(z')$ as given by (3.78). Thus the actual current at the feed point has a finite, albeit small, value.

3.10 RADIATION RESISTANCE OF DIPOLE ANTENNAS

The power radiated by a dipole of length L is

$$W = \tfrac{1}{2}\int_0^{2\pi}\int_0^{\pi} \mathrm{Re}(E_\theta H_\phi^*) r^2 \sin\theta \, d\theta \, d\phi$$

$$= \frac{\eta I_m^2}{8\pi^2}\int_0^{2\pi}\int_0^{\pi} \frac{[\cos(\frac{1}{2}kL\cos\theta) - \cos(\frac{1}{2}kL)]^2}{\sin\theta} \, d\theta \, d\phi \tag{3.90}$$

In free space, $\eta = 120\pi$ ohm. After performing the ϕ integration, (3.90) becomes

$$W = 30 I_m^2 \int_0^{\pi} \frac{[\cos(\frac{1}{2}kL\cos\theta) - \cos(\frac{1}{2}kL)]^2}{\sin\theta} \, d\theta \tag{3.91}$$

The radiation resistance can be defined in terms of either the maximum current I_m or the current at the feed point $I(0)$. For the sinusoidal current distribution, I_m and $I(0)$ are related by (3.59). In terms of $I(0)$, we have

$$W = \tfrac{1}{2} I^2(0) R_r \tag{3.92}$$

On equating (3.91) and (3.92), we obtain

$$R_r = \frac{60 I_m^2}{I^2(0)} \int_0^{\pi} \frac{[\cos(\frac{1}{2}kL\cos\theta) - \cos(\frac{1}{2}kL)]^2}{\sin\theta} \, d\theta \tag{3.93}$$

Let us proceed to evaluate (3.93) for the case of a half-wave dipole, for which $I_m = I(0)$. Putting $L = \lambda/2$, (3.93) reads

$$R_r(\lambda/2) = 60 \int_0^{\pi} \frac{\cos^2(\frac{1}{2}\pi\cos\theta)}{\sin\theta} \, d\theta \tag{3.94}$$

On letting $x = \cos\theta$, we have

$$R_r(\lambda/2) = 60 \int_{-1}^{1} \frac{\cos^2(\frac{1}{2}\pi x)}{1 - x^2} \, dx = 15 \int_{-1}^{1} \left(\frac{1 + \cos(\pi x)}{1 - x} + \frac{1 + \cos(\pi x)}{1 + x} \right) dx \tag{3.95}$$

The first and second integrals are easily shown to be equal and (3.95) becomes

$$R_r(\lambda/2) = 30 \int_{-1}^{1} \frac{1 + \cos(\pi x)}{1 - x} \mathrm{d}x \tag{3.96}$$

On letting $y = \pi(1 + x)$ in (3.96), there results

$$R_r(\lambda/2) = 30 \int_{0}^{2\pi} \frac{1 - \cos y}{y} \mathrm{d}y \tag{3.97}$$

The function $\mathrm{Cin}(x)$ is defined by the integral

$$\mathrm{Cin}(x) \equiv \int_{0}^{x} \frac{1 - \cos y}{y} \mathrm{d}y \tag{3.98}$$

and is tabulated (e.g. Abramowitz and Stegun, 1964). In particular, its value for $x = 2\pi$, i.e. $\mathrm{Cin}(2\pi)$, is equal to 2.438. Hence

$$R_r(\lambda/2) = 30\,\mathrm{Cin}(2\pi) = 73 \text{ ohm} \tag{3.99}$$

On using equations (3.30) and (3.34), the directivity of a half-wave dipole is found to be

$$D(\lambda/2) = 1.64 \tag{3.100}$$

which is to be compared with the value of 1.5 for an infinitesimal current element.

For dipoles of lengths which are multiples of $\lambda/2$, (3.93) can also be evaluated in terms of the $\mathrm{Cin}(x)$ function. For example, it can be readily shown that, for $L = 3\lambda/2$,

$$R_r(3\lambda/2) = 30\,\mathrm{Cin}(6\pi) = 30(3.51) = 105.3 \text{ ohm} \tag{3.101}$$

3.11 ANTENNA IMPEDANCE

The impedance of an antenna is defined as the ratio of the voltage at the terminals of the antenna to the resulting current flowing in the antenna. If the current is taken to be the maximum value, it is known as the impedance referred to the loop current. If the current is taken to be the value at the feed point, it is known as the impedance referred to the base current. The latter is also called the input or driving-point impedance. Unless otherwise stated, we shall define antenna impedance in terms of the base current:

$$Z_a = Z_{in} \equiv V(0)/I(0) = R_a + \mathrm{j}X_a \tag{3.102}$$

As far as the transmission line connected to the antenna is concerned, the antenna is equivalent to an impedance Z_a. The real part of Z_a, denoted by R_a, is measure of how much power is consumed by the antenna. If the antenna has no ohmic loss, this is equal to its radiation resistance $\mathbf{R}_r$. The imaginary part of Z_a, denoted by X_a, arises from the fact that the applied voltage, $V(0)$, and the resultant current at the feed point, $I(0)$, are not in phase.

An expression for Z_a can be obtained if we use the solution to Hallén's integral equation given by (3.78):

$$I(z') = \frac{\mathrm{j}V(0)}{60\Omega}\left(\frac{\sin[k(\tfrac{1}{2}L - |z'|)] + b_1\Omega^{-1} + b_2\Omega^{-2} + \cdots}{\cos(\tfrac{1}{2}kL) + d_1\Omega^{-1} + d_2\Omega^{-2} + \cdots}\right)$$

Setting $|z'| = 0$ in (3.78), there results

$$I(0) = \frac{\mathrm{j}V(0)}{60\Omega}\left(\frac{\sin(\tfrac{1}{2}kL) + b_1\Omega^{-1} + b_2\Omega^{-2} + \cdots}{\cos(\tfrac{1}{2}kL) + d_1\Omega^{-1} + d_2\Omega^{-2} + \cdots}\right) \quad (3.103)$$

It follows from (3.103) that

$$Z_a \equiv \frac{V(0)}{I(0)} = -\mathrm{j}60\Omega\left(\frac{\cos(\tfrac{1}{2}kL) + d_1\Omega^{-1} + d_2\Omega^{-2} + \cdots}{\sin(\tfrac{1}{2}kL) + b_1\Omega^{-1} + b_2\Omega^{-2} + \cdots}\right) \quad (3.104)$$

The results based on (3.104) are illustrated in Figure 3.7 for length-to-radius ratios (L/a) of 120 and 4000. The half-length $L/2$ of the antenna is given along the spirals in free-space wavelengths. The impedance variation is that which would be obtained as a function of frequency for an antenna of fixed physical dimensions. From the diagram, we obtain the important information that the variation in impedance with frequency of the thicker antenna is much less than that of the thinner antenna.

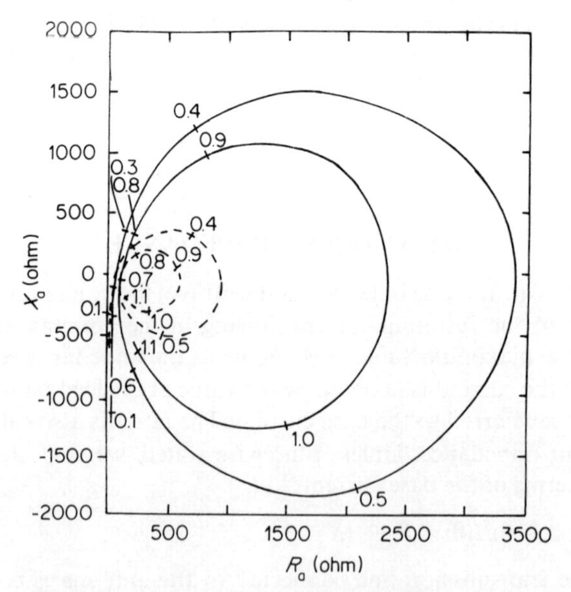

Figure 3.7 Input impedance ($R_a + \mathrm{j}X_a$) in ohm for cylindrical centre-fed dipoles with length-to-radius ratios (L/a) of 120 (----)($\Omega = 9.6$) and 4000 (——) ($\Omega = 16.6$) (Source: Hallén (1948), *Tech. Rep. No.* 46, Cruft Laboratory. Reproduced by permission of Harvard University)

In Chapter 5, the antenna impedance will be discussed again using a method known as the induced e.m.f. method.

3.12 EFFECT OF FINITE GAP WIDTH ON THE FAR FIELD OF DIPOLE ANTENNAS

In calculating the far field of dipole antennas, we have assumed the width of the gap separating the two conductors to be infinitesimally narrow so that the current distribution $I(z')$ can be assumed to be a continuous function of z'. In this section, the effect of the finite width will be investigated.

Since Maxwell's equations are linear, the principle of superposition holds. The radiation from a centre-fed dipole of length L with a gap width Δ can be considered as equal to the radiation from a continuous dipole of length $(L + \Delta)$ minus the radiation from a short dipole of length Δ, as illustrated in Figure 3.8.

Let the far-zone magnetic field due to the short dipole of length Δ be H'_ϕ. From (3.54) and (3.58), we have

$$H'_\phi = \frac{\mathrm{j}kI_m \exp(-\mathrm{j}kr)}{4\pi r} \sin\theta \int_{-\Delta/2}^{\Delta/2} \sin\left[k\left(\frac{L+\Delta}{2} - |z'|\right)\right] \exp(\mathrm{j}kz' \cos\theta)\,\mathrm{d}z' \tag{3.105}$$

In making use of (3.85), we obtain

$$H'_\phi = \frac{\mathrm{j}I_m \exp(-\mathrm{j}kr)}{2\pi r} P'(\theta, \Delta) \tag{3.106}$$

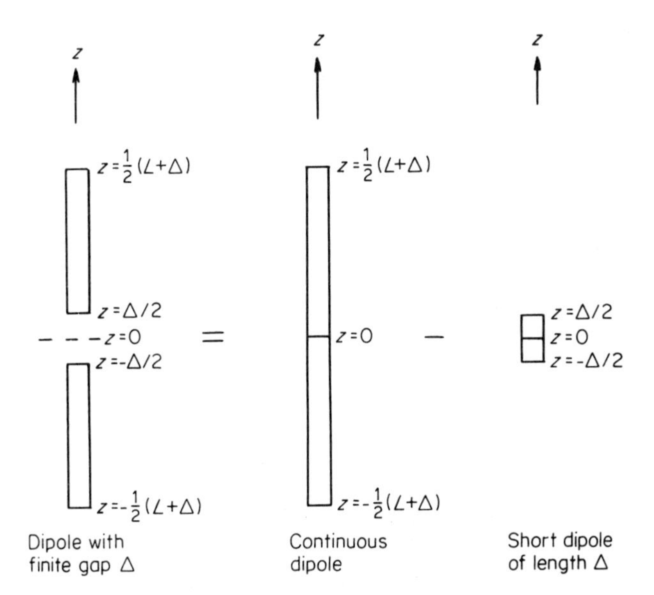

Figure 3.8 Calculation of the effect of finite gap width on the radiation pattern of a dipole antenna

where

$$P'(\theta, \Delta) = \sin^{-1}\theta\{\cos[k(\tfrac{1}{2}L - \tfrac{1}{2}\Delta)]\cos(\tfrac{1}{2}k\Delta\cos\theta) - \cos\theta\sin[k(\tfrac{1}{2}L - \tfrac{1}{2}\Delta)]\sin(\tfrac{1}{2}k\Delta\cos\theta) - \cos(\tfrac{1}{2}kL)\} \tag{3.107}$$

The far-zone magnetic field due to the continuous dipole can be obtained by replacing Δ by $(L+\Delta)$ in (3.107). Denoting it by H''_ϕ, we have

$$H''_\phi = \frac{jI_m \exp(-jkr)}{2\pi r} P''(\theta, \Delta) \tag{3.108}$$

where

$$P''(\theta, \Delta) = \sin^{-1}\theta\{\cos(\tfrac{1}{2}k\Delta)\cos[\tfrac{1}{2}k(L+\Delta)\cos\theta] - \cos\theta\sin(\tfrac{1}{2}k\Delta)\sin[\tfrac{1}{2}k(L+\Delta)\cos\theta] - \cos(\tfrac{1}{2}kL)\} \tag{3.109}$$

Substracting (3.106) from (3.108) we obtain the far-zone magnetic field of a dipole of length L with a finite gap width Δ:

$$H_\phi = H''_\phi - H'_\phi = \frac{jI_m \exp(-jkr)}{2\pi r} P(\theta, \Delta) \tag{3.110}$$

where

$$P(\theta, \Delta) = P''(\theta, \Delta) - P'(\theta, \Delta) \tag{3.111}$$

The far-zone electric field is, of course, still given by $E_\theta = \eta H_\phi$.

The effect of the finite gap width is best illustrated graphically. In Figure 3.9, we show that the factor $P(\theta, \Delta)$ for a half-wave dipole as a function of θ for

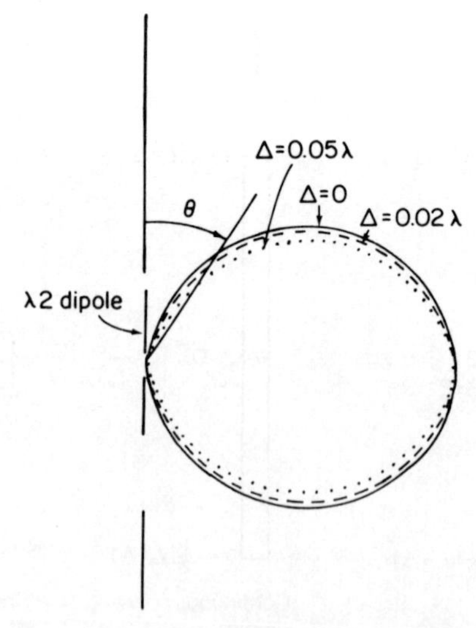

Figure 3.9 Effect of finite gap width on the radiation pattern of a half-wave dipole

$\Delta = 0, 0.02\lambda$, and 0.05λ. Only values between $\theta = 0°$ and $\theta = 180°$ are shown since the function is symmetric in θ. We see that the effect of the gap width is quite small for $\Delta = 0.02\lambda$ but is not negligible as Δ increases to 0.05λ.

3.13 FOLDED DIPOLE

A linear antenna that has a radiation pattern the same as a dipole but with a four-fold increase in radiation resistance is the folded dipole. The folded dipole consists of two straight conductors of equal length placed parallel to each other and joined together at the ends, as shown in Figure 3.10. It is excited at the centre of one of the conductors. As in the case of a dipole, a qualitative understanding of the current distribution on a folded dipole can be obtained by regarding it as being evolved from a parallel-wire transmission line. For definiteness, let us consider a $\lambda/2$ folded dipole. We begin with the short-circuited transmission line of Figure 3.11(a). We then start bending the line in the manner shown in Figure 3.11(b). When the bending process has been completed, we obtain the configuration shown in Figure 3.11(c), namely that of a folded dipole. Assuming the current distribution to be the same as the short-circuited transmission line from which it is evolved, we have an identical sinusoidal current distribution on each of the conductors. The radiation pattern produced by the folded dipole will therefore be the same as a dipole with the same terminal current, except that the strength is doubled. This results in a four-fold increase in the power radiated. For the $\lambda/2$ folded dipole, the radiation resistance is therefore $4 \times 73 = 292$ ohm. More information on the folded dipole will be given in section 6.3.2.

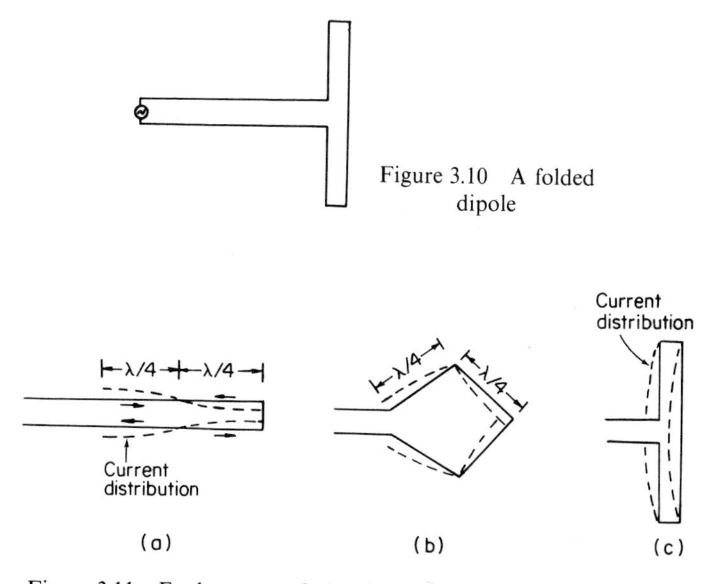

Figure 3.10 A folded dipole

Figure 3.11 Evolvement of the $\lambda/2$ folded dipole from a short-circuited transmission line

3.14 IMAGES OF ANTENNAS ABOVE A PERFECTLY CONDUCTING PLANE

All our discussion so far is for an antenna radiating in an infinite medium without boundaries. The presence of obstacles requires the electromagnetic field to satisfy the boundary conditions imposed by the obstacles, and such problems are generally too complicated to be considered here. One type of boundary value problem, however, is particularly simple and yet of considerable practical importance. This concerns an antenna radiating above a perfectly conducting plane of infinite extent, which can be solved by the method of images.

Consider a conductor carrying a current I which is parallel to a perfectly conducting plane of infinite size. By convention, the current I flowing to the right is identified with the motion of positive charges to the right. This is imaged by the motion of negative charges to the right, which is equivalent to the motion of positive charges to the left. Hence the image of a current parallel to the conducting plane is a current of equal magnitude flowing in the opposite direction, as shown in Figure 3.12(a). The boundary condition that the tangential

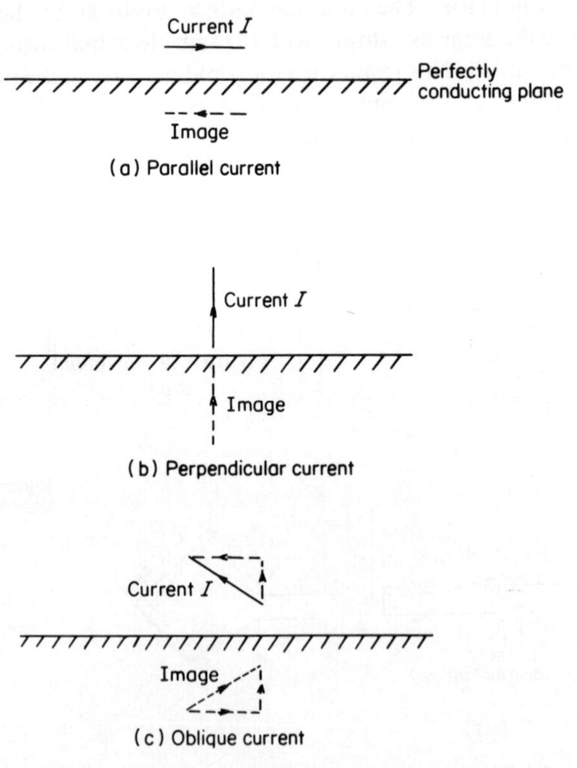

Figure 3.12 Images of currents above a perfectly conducting plane

electric field vanishes at the conducting plane will be satisfied by the actual current and its image.

Consider now a conductor carrying a current I perpendicular to a metallic ground plane, as shown in Figure 3.12(b). Using similar reasoning, it is seen that its image is a perpendicular current flowing in the same direction. Finally, the image of a current making an arbitrary angle with the metallic ground plane can be found by resolving the current into vertical and horizontal components, as shown in Figure 3.12(c).

3.15 MONOPOLE ANTENNA

The monopole antenna consists of a straight conductor above a conducting plane which is perpendicular to the conductor axis, as shown in Figure 3.13(a). It is fed by connecting the source between the end of the conductor and the conducting plane. The monopole will be imaged in the manner discussed in section 3.14. The phase of the current in the image conductor is such that the antenna plus image may be considered a single antenna in free space. Hence the monopole behaves like a dipole of twice its length. The pattern, however, is actually only half the free-space pattern, since the conducting plane cuts off the other half. For a given current at the base of the antenna, the total radiated power is only half as great as the dipole in free space since electromagnetic fields exist only in the upper hemisphere. On the other hand, because the total power radiated is effectively concentrated into half the solid angle of the free-space case, the directivity will be twice that of a free-space dipole. As an example, for a quarter-wave ($\lambda/4$) monopole, we have $R_r = 36.5$ ohm, and $D = 3.28$.

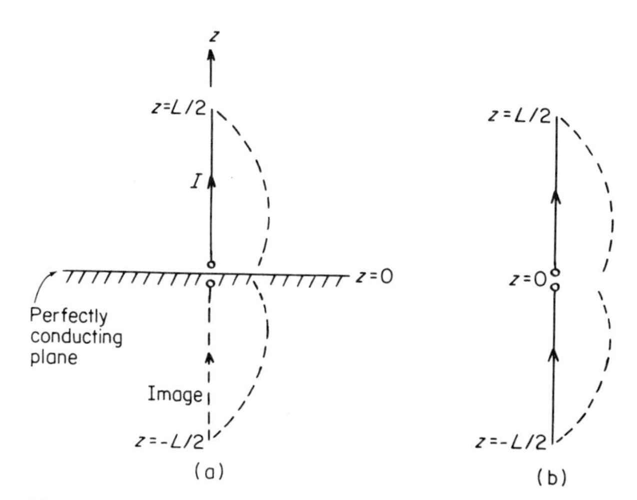

Figure 3.13 (a) A monopole of length $L/2$ with assumed sinusoidal current distribution; (b) the corresponding dipole

3.16 WORKED EXAMPLES

Example 1

For sinusoidal time variation, the electric field is, on combining (2.10) and (2.13), given by

$$\mathbf{E} = -\mathrm{j}\omega\mathbf{A} + (\mathrm{j}\omega\mu\varepsilon)^{-1}\nabla(\nabla\cdot\mathbf{A}) \tag{3.112}$$

Use (3.112) to find the electric field of a z-directed Hertzian dipole. Show that, in the far zone, **E** is given by the portion of the second term that is transverse to $\hat{\mathbf{r}}$, i.e.

$$\mathbf{E} = -\mathrm{j}\omega\mathbf{A}_\mathrm{t} \qquad \text{(far field)} \tag{3.113}$$

where

$$\mathbf{A}_\mathrm{t} = \mathbf{A} - (\mathbf{A}\cdot\hat{\mathbf{r}})\hat{\mathbf{r}} \tag{3.114}$$

Solution

The magnetic vector potential **A** of a z-directed Hertzian dipole is given by (3.3):

$$\mathbf{A} = \frac{\mu\exp(-\mathrm{j}kr)}{4\pi r} I\Delta l\hat{\mathbf{z}}$$

In spherical coordinates, **A** has two components:

$$\mathbf{A} = \hat{\mathbf{r}}A_z\cos\theta - \hat{\theta}A_z\sin\theta$$

The divergence of **A** in spherical coordinates is

$$\nabla\cdot\mathbf{A} = \frac{\mu I\Delta l}{4\pi}\left[\frac{1}{r^2}\frac{\partial}{\partial r}\left(r^2\cos\theta\frac{\exp(-\mathrm{j}kr)}{r}\right) - \frac{1}{r\sin\theta}\frac{\partial}{\partial\theta}\left(\sin^2\theta\frac{\exp(-\mathrm{j}kr)}{r}\right)\right] \tag{3.115}$$

Carrying out the differentiation, we have

$$\nabla\cdot\mathbf{A} = -\frac{\mu I\Delta l}{4\pi}\cos\theta\left(\frac{\mathrm{j}k}{r} + \frac{1}{r^2}\right)\exp(-\mathrm{j}kr) \tag{3.116}$$

The operator ∇ in spherical coordinates is

$$\nabla = \hat{\mathbf{r}}\frac{\partial}{\partial r} + \hat{\boldsymbol{\theta}}\frac{1}{r}\frac{\partial}{\partial\theta} \tag{3.117}$$

thus

$$E_r = -\mathrm{j}\omega A_r + \frac{1}{\mathrm{j}\omega\mu\varepsilon}[\nabla(\nabla\cdot\mathbf{A}]_r$$

$$= -\mathrm{j}\omega A_z\cos\theta + \frac{\mu I\Delta l\cos\theta}{4\pi}\frac{1}{\mathrm{j}\omega\mu\varepsilon}\frac{\partial}{\partial r}\left(\frac{\mathrm{j}k}{k} + \frac{1}{r^2}\right)\exp(-\mathrm{j}kr)$$

$$= -\frac{\eta I\Delta l k^2 \cos\theta}{2}\exp(-jkr)\left[\left(\frac{1}{jkr}\right)^2 + \left(\frac{1}{jkr}\right)^3\right] \quad (3.118)$$

$$E_\theta = -j\omega A_\theta + \frac{1}{j\omega\mu\varepsilon}[\nabla(\nabla\cdot\mathbf{A})]_\theta$$

$$= j\omega A_z \sin\theta - \frac{\mu I\Delta l}{j\omega\mu\varepsilon 4\pi r}\left(\frac{jk}{r} + \frac{1}{r^2}\right)\exp(-jkr)\frac{\partial}{\partial\theta}(\cos\theta)$$

$$= -\frac{\eta I\Delta l k^2 \sin\theta}{4\pi}\exp(-jkr)\left[\frac{1}{jkr} + \left(\frac{1}{jkr}\right)^2 + \left(\frac{1}{jkr}\right)^3\right] \quad (3.119)$$

The expressions for E_r and E_θ given above had been obtained in section 3.2 using equation (2.24):

$$\mathbf{E} = \frac{1}{j\omega\mu\varepsilon}\nabla\times(\nabla\times\mathbf{A})$$

In the far zone, the terms that vary as r^{-2} and r^{-3} are negligible. The r-component and the term $[\nabla(\nabla\cdot\mathbf{A})]_\theta$ in the θ-component therefore do not contribute to the far zone electric field, which is given by

$$\mathbf{E} = -j\omega A_\theta\,\hat{\boldsymbol{\theta}} = \frac{\eta jkI\Delta l}{4\pi r}\exp(-jkr)(\sin\theta)\hat{\boldsymbol{\theta}} \quad (3.120)$$

Example 2

To illustrate the mathematical manipulations associated with a rotation of axis, consider a Hertzian dipole directed along the x-axis. Obtain the far zone electromagnetic fields and show that the components of **E** and **H** are related by $E_\theta = \eta H_\phi$, $E_\phi = -\eta H_\theta$. Repeat for the case of an x-directed half-wave dipole.

Solution

The geometry of the x-directed Hertzian dipole is shown in Figure 3.14. The spherical coordinates of a point P in the far zone are (r, θ, ϕ). Let ψ be the angle

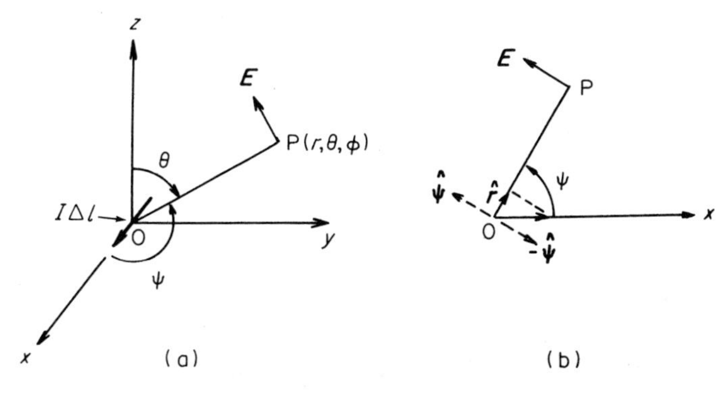

Figure 3.14 A Hertzian dipole directed along the x-axis

between OP and the x-axis and $\hat{\boldsymbol{\psi}}$ be the unit vector perpendicular to OP lying on the POX plane. Then the electric field at P is given by

$$\mathbf{E}(r) = \frac{I\Delta l}{4\pi r} \exp(-jkr)\eta\, jk(\sin\psi)\hat{\boldsymbol{\psi}} \tag{3.121}$$

We must express $\sin\psi$ and $\hat{\boldsymbol{\psi}}$ in terms of θ and ϕ. Since

$$\hat{\mathbf{r}} = \hat{\mathbf{x}} \sin\theta \cos\phi + \hat{\mathbf{y}} \sin\theta \sin\phi + \hat{\mathbf{z}} \cos\theta \tag{3.122}$$

we have

$$\hat{\mathbf{x}} \cdot \hat{\mathbf{r}} = \cos\psi = \sin\theta \cos\phi \tag{3.123}$$

Hence

$$\sin\psi = (1 - \sin^2\theta \cos^2\phi)^{1/2} \tag{3.124}$$

From Figure 3.14(b), we see that

$$\hat{\mathbf{x}} = -\hat{\boldsymbol{\psi}} \sin\psi + \hat{\mathbf{r}} \cos\psi \tag{3.125}$$

Thus

$$\hat{\boldsymbol{\psi}} = \sin^{-1}\psi(-\hat{\mathbf{x}} + \hat{\mathbf{r}} \cos\psi) \tag{3.126}$$

Substituting (3.124) and the relation

$$\hat{\mathbf{x}} = \hat{\mathbf{r}} \sin\theta \cos\phi + \hat{\boldsymbol{\theta}} \cos\theta \cos\phi - \hat{\boldsymbol{\phi}} \sin\phi \tag{3.127}$$

into (3.125), we obtain

$$\hat{\boldsymbol{\psi}} = \frac{(-\hat{\boldsymbol{\theta}} \cos\theta \cos\phi + \hat{\boldsymbol{\phi}} \sin\phi)}{(1 - \sin^2\theta \cos^2\phi)^{1/2}} \tag{3.128}$$

In terms of θ and ϕ, (3.121) becomes

$$\mathbf{E}(\mathbf{r}) = \frac{I\Delta l}{4\pi r} \exp(-jkr)\eta jk(-\hat{\boldsymbol{\theta}} \cos\theta \cos\phi + \hat{\boldsymbol{\phi}} \sin\phi) \tag{3.129}$$

Equation (3.129) shows that, for an x-directed Hertzian dipole, the far-zone electric field has both θ- and ϕ-components. The θ-component is a function of both angular variables while the ϕ-component depends on the azimuthal angle ϕ.

The magnetic field is related to the electric field by Maxwell's equation

$$\mathbf{H} = -\frac{1}{j\omega\mu} \nabla \times \mathbf{E} \tag{3.130}$$

Use of (3.129) in (3.130) yields the result

$$\mathbf{H} = \frac{I\Delta l}{4\pi r} \exp(-jkr) jk(-\hat{\boldsymbol{\theta}} \sin\phi - \hat{\boldsymbol{\phi}} \cos\theta \cos\phi) \tag{3.131}$$

From (3.129) and (3.131), we obtain the relationships

$$E_\theta = \eta H_\phi \tag{3.132}$$

$$E_\phi = -\eta H_\theta \tag{3.133}$$

Equation (3.129) can also be obtained by first finding the magnetic vector potential **A** and then using (3.113) (problem 13 in section 3.17).

If the x-directed dipole is a half-wavelength long, we have, from (3.88) and Table 3.1,

$$\mathbf{E}(\mathbf{r}) = \frac{\eta j I(0) \exp(-jkr)}{2\pi r} \frac{\cos(\frac{1}{2}\pi \cos\psi)}{\sin\psi} \tag{3.134}$$

Substituting (3.123), (3.124), and (3.127) into (3.134), we obtain

$$\mathbf{E}(\mathbf{r}) = \frac{\eta j I(0) \exp(-jkr)}{2\pi r} \frac{\cos(\frac{1}{2}\pi \sin\theta \cos\phi)}{(1 - \sin^2\theta \cos^2\phi)} (-\hat{\boldsymbol{\theta}} \cos\theta \cos\phi + \hat{\boldsymbol{\phi}} \sin\phi) \tag{3.135}$$

Use of (3.135) in (3.130) yields the result

$$\mathbf{H}(\mathbf{r}) = \hat{\theta} H_\theta + \hat{\phi} H_\phi)$$

where H_θ and H_ϕ are related to E_ϕ and E_θ by (3.132) and (3.133).

Equations (3.132) and (3.133) hold for the far field of any radiating source. This can be shown to be true by assuming the θ- and ϕ-components of **E** and **H** to be an outward-propagating wave of the form $(1/r)\exp[j(\omega t - kr)]$ and determining their relationships from Maxwell's equations (problem 12 in section 3.17).

Example 3

Two infinitesimal current elements are placed adjacent to each other and excited in antiphase as shown in Figure 3.15. Such antiphase current elements are found, for example, in a balanced two-wire transmission line.

(a) Find the far-zone **E** and **H** fields.
(b) For the case $d \ll \lambda$, find the ratio of the radiated power of the pair of current elements to that of a single current element $I\Delta l$,

Solution

(a) As shown in Figure 3.15, for a point $P(r, \theta, \phi)$ in the far zone,

$$r_1 \simeq r - \tfrac{1}{2}d \cos\alpha \tag{3.136}$$

$$r_2 \simeq r + \tfrac{1}{2}d \cos\alpha \tag{3.137}$$

Since $\cos\alpha = \hat{\mathbf{r}} \cdot \hat{\mathbf{y}}$, we have, on using (3.122),

$$\cos\alpha = \sin\theta \sin\phi \tag{3.138}$$

Hence

$$r_1 \simeq r - \tfrac{1}{2}d \sin\theta \sin\phi \tag{3.139}$$

$$r_2 \simeq r + \tfrac{1}{2}d \sin\theta \sin\phi \tag{3.140}$$

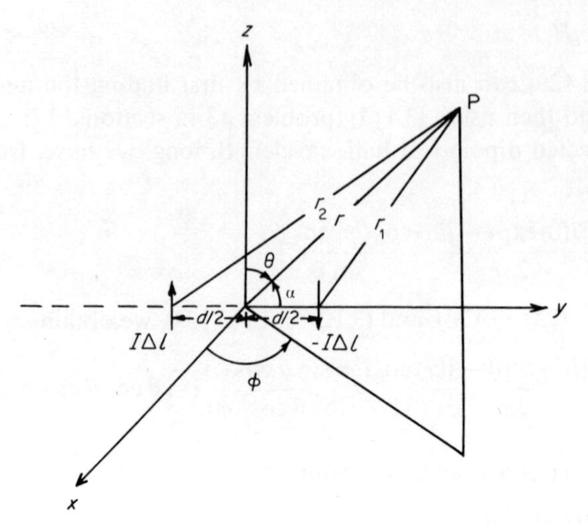

Figure 3.15 A pair of antiphase infinitesimal current elements

The far-zone electric field is given by the sum of the two elements and has only a θ-component:

$$E_\theta = \frac{I\Delta l}{4\pi r_2}\exp(-\mathrm{j}kr_2)\eta \mathrm{j}k\sin\theta - \frac{I\Delta l}{4\pi r_1}\exp(-\mathrm{j}kr_1)\eta \mathrm{j}k\sin\theta \tag{3.141}$$

In the denominators of (3.141), we can approximate r_2 and r_1 by r. Substituting (3.136) and (3.137) in the exponents, we obtain

$$E_\theta = \frac{\mathrm{j}\eta k I\Delta l}{4\pi r}\exp(-\mathrm{j}kr)\sin\theta\left[\exp(-\mathrm{j}k\tfrac{1}{2}d\sin\theta\sin\phi) - \exp(\mathrm{j}k\tfrac{1}{2}d\sin\theta\sin\phi)\right]$$

$$= \frac{2\eta k I\Delta l}{4\pi r}\exp(-\mathrm{j}kr)\sin\theta\sin(\tfrac{1}{2}kd\sin\theta\sin\phi) \tag{3.142}$$

The magnetic field in the far zone is given by $H_\phi = E_\theta/\eta$

(b) For $d \ll \lambda$, (3.142) can be simplified to

$$E_\theta \simeq E_0 kd\frac{\exp(-\mathrm{j}kr)}{r}\sin^2\theta\sin\phi \tag{3.143}$$

where

$$E_0 = \frac{\eta I k\Delta l}{4\pi} \tag{3.144}$$

Power radiated by the pair of current elements is

$$W_1 = \int_0^{2\pi}\int_0^{\pi}\tfrac{1}{2}(\mathrm{Re}\,S)r^2\sin\theta\,\mathrm{d}\theta\,\mathrm{d}\phi$$

$$= \int_0^{2\pi}\int_0^{\pi} \frac{|E_\theta|^2}{2\eta} r^2 \sin\theta \, d\theta \, d\phi$$

$$= k^2 d^2 \frac{E_0^2}{2\eta} \int_0^{2\pi}\int_0^{\pi} \sin^5\theta \sin^2\phi \, d\theta \, d\phi \tag{3.145}$$

Since

$$\int_0^{\pi} \sin^5\theta \, d\theta = 16/15 \tag{3.146}$$

$$\int_0^{2\pi} \sin^2\phi \, d\phi = \pi \tag{3.147}$$

equation (3.145) evaluates to

$$W_1 = \frac{8E_0^2\pi}{15} k^2 d^2 \tag{3.148}$$

For a single element $I\Delta l$, the power radiated is

$$W_2 = \frac{E_0^2}{2\eta} \int_0^{2\pi}\int_0^{\pi} \sin^3\theta \, d\theta \, d\phi = \frac{4E_0^2\pi}{3\eta} \tag{3.149}$$

Hence

$$\frac{W_1}{W_2} = \frac{2}{5} k^2 d^2 = \frac{8}{5}\pi^2 \left(\frac{d}{\lambda}\right)^2 = 15.78 \left(\frac{d}{\lambda}\right)^2 \tag{3.150}$$

Example 4

Find the radiation efficiencies of the following antennas made of No. 20 AWG copper wire of radius $a = 4.06 \times 10^{-4}$ m:

(a) a dipole of length 2 m operating at 1 MHz:
(b) a dipole of length 1.5 m operating at 100 MHz.

Assume $\sigma = 5.7 \times 10^7$ ohm^{-1} m^{-1} and $\mu = 4\pi \times 10^{-7}$ H m^{-1} for copper.

Solution

(a) Since the length of the dipole in units of wavelength is only 0.0066λ, we can consider it to be a Hertzian dipole. At 1 MHz.

$$R_s = \frac{4\pi \times 10^{-7} \times 2\pi \times 10^6}{2 \times 5.7 \times 10^7} = 2.63 \times 10^{-4} \text{ ohm}$$

$$R_L = \frac{R_s}{2\pi a} \Delta l = \frac{2.63 \times 10^{-4} \times 2}{2\pi \times 4.06 \times 10^{-4}} = 0.206 \text{ ohm}$$

$$R_r = 789(\Delta l/\lambda)^2 = 0.0352 \text{ ohm}$$

$$e = \frac{R_r}{R_r + R_L} = \frac{0.0352}{0.0352 + 0.206} = \frac{0.0352}{0.241} = 14.59\%$$

(b) The dipole is a half-wavelength long at 100 MHz.

$$R_s = 2.63 \times 10^{-3} \text{ ohm}$$

Let the current at the feed point be I_0.
Power radiated

$$W = \tfrac{1}{2} I_0^2 R_r = \tfrac{1}{2} I_0^2 (73) \text{ watts}$$

Power loss

$$W_L = \frac{R_s}{2\pi a} \int_{-\lambda/4}^{\lambda/4} \tfrac{1}{2} |I(z')|^2 \, dz' = \frac{R_s}{2\pi a} \int_{-\lambda/4}^{\lambda/4} \tfrac{1}{2} I_0^2 \cos^2 kz' \, dz'$$

$$= \frac{R_s}{2\pi a} \tfrac{1}{2} I_0^2 (\pi/2) = 1.62(\tfrac{1}{2} I_0^2) \text{ watts}$$

$$e = \frac{W}{W + W_L} = \frac{\tfrac{1}{2}(73) I_0^2}{\tfrac{1}{2}(73) I_0^2 + \tfrac{1}{2}(1.62) I_0^2} = 97.83\%$$

3.17 PROBLEMS

1. Verify equations (3.9)–(3.11).

2. Show that the directivity of a half-wave dipole is equal to 1.64.

3. Show that the magnitude of the far-zone electric field due to a Hertzian dipole is (in V m^{-1})

$$E = 9.49 \sqrt{W} \frac{\sin\theta}{r}$$

where W is the time-averaged radiated power in watts, r is the distance in metres from the observation point to the location of the dipole and θ is the angle made by the radius vector r with the axis of the dipole. Assume $\eta = 120\pi$ ohm.

4. A half-wave dipole radiates a time-averaged power of 146 W in free space at a frequency 300 MHz. Find the electric and magnetic fields strengths at a point $P(r, \theta, \phi)$ when $r = 100$ m, $\theta = 90°$, and $\phi = 30°$.

5. Verify that the radiation resistance of a $\tfrac{3}{2}\lambda$ dipole with sinusoidal current distribution is equal to 105.3 ohm.

6. If a dipole is relatively short so that the electrical length of each arm is less than 6°, the sinusoidal current distribution can be approximated by the triangular distribution of the form

$$I(z') = \frac{2I_m}{L}(\tfrac{1}{2}L - |z'|) \qquad L/\lambda < 1/30$$

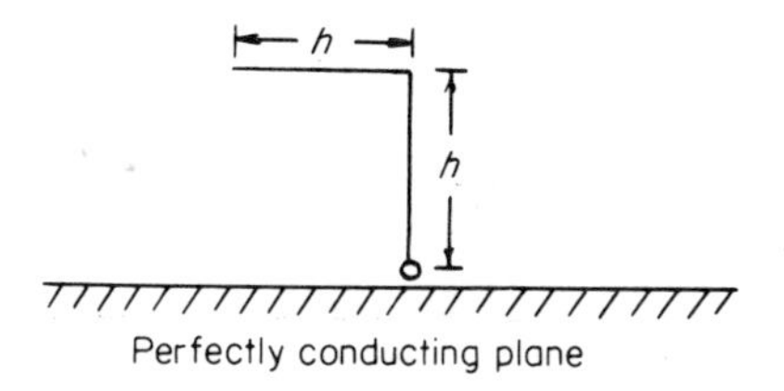

Figure 3.16 A short inverted L-antenna (problem 7)

Show that the radiation resistance of a short dipole with a triangular current distribution is given by

$$R_r = 20\pi^2(L/\lambda)^2$$

7. A short inverted L-antenna of height $L = \lambda/80$ is mounted above a perfectly conducting ground plane, as shown in Figure 3.16.

   (a) Give a resonable estimate of the current distribution of the antenna.
   (b) In calculating the radiation resistance of this antenna, considerable simplification is achieved if either the horizontal or the vertical part can be neglected. Which part will you neglect? Explain.

8. A half-wave dipole with feed-point current $I(0)$ at the centre radiates in a medium whose properties are represented by permeability μ, permittivity ε, and conductivity σ. If $\sigma/\omega\varepsilon \ll 1$, what are the electromagnetic fields in the far-zone?
   [Hint: Use the results of section 2.6.]

9. Consider a small circular loop antenna with a constant current distribution around its circumference, as shown in Figure 3.17(a). Also, consider a quarter of a small circular loop antenna placed against two perfectly conducting half-planes located perpendicular to each other as shown in Figure 3.17(b). The constant current distribution around the circumference of the quarter of a loop in Figure 3.17(b) is assumed to be the same as that for the whole circular loop shown in Figure 3.17(a). If the radiation resistance of the whole loop in (a) is R_r, what is the radiation resistance of the quarter loop in (b)? Explain your answer fully.

10. A wire of length $L = \frac{1}{2}\lambda$ supports a travelling wave of current

$$I(z') = I_0 \exp[j(\omega t - kz')]$$

(This wire may be considered as a segment of a transmission line terminated in its characteristic impedence.) Let the wire extend from $z = -\frac{1}{2}L$ to $\frac{1}{2}L$.

   (a) Determine the far-zone electric field.
   (b) Plot the pattern and discuss the main difference with the pattern of a $\frac{1}{2}\lambda$ dipole with a standing wave of current.
   (c) Determine the radiation resistance of the $\frac{1}{2}\lambda$ travelling-wave antenna.

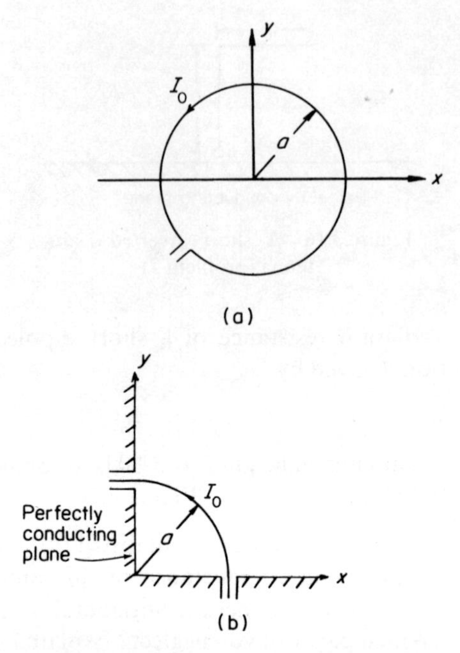

Figure 3.17 (a) Circular loop antenna and (b) quarter of a circular loop antenna with perfectly conducting planes (problem 9)

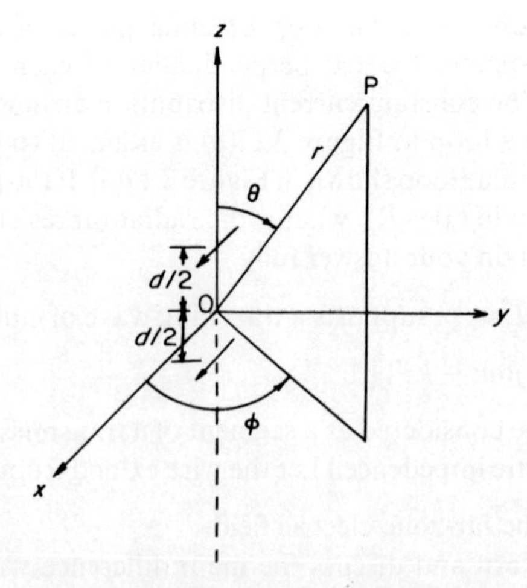

Figure 3.18 A pair of x-directed Hertzian dipoles (problem 11)

11. Figure 3.18 shows two x-directed Hertzian dipoles with centres at $(0, 0, \frac{1}{2}d)$ and $(0, 0, -\frac{1}{2}d)$ respectively. They are both of length dl and carrying the same current I (angular frequency ω). For any point P (r, θ, ϕ) in the far zone, find:

    (a) the electric and magnetic fields,
    (b) the radiation intensity.

12. By assuming the θ- and ϕ-components of the far-zone electromagnetic field to be an outward-propagating wave of the form

$$\frac{1}{r}\exp[j(\omega t - kr)],$$

show that $E_\theta = \eta H_\phi$ and $E_\phi = -\eta H_\theta$ for any radiating source.

13. Obtain equation (3.129) by first finding the magnetic vector potential **A** and then using (3.113).

BIBLIOGRAPHY

Kraus, J. (1950). *Antennas*, McGraw-Hill, New York, Chapters 5 and 9.

CHAPTER 4
The Receiving Antenna

4.1 INTRODUCTION

In Chapter 3, the linear antenna was analysed by considering it as a transmitter. In this operation, the antenna is connected to a source and the resultant radiation is calculated. Several parameters were introduced to describe the basic properties of the transmitting antenna. These are the effective length, the radiation pattern, the directive gain, and the antenna impedance. In this chapter, we consider the linear antenna as a device to receive electromagnetic waves. We begin by introducing the parameters for describing the characteristics of the receiving antenna.

4.2 CHARACTERISTICS OF THE RECEIVING ANTENNA

4.2.1 Effective Length

Consider an antenna in the radiation field of a distant transmitter. To be specific, let us take the antenna to be a symmetric dipole of length L, with an infinitesimally narrow gap at the centre, which we denote by point O. The transmitter is in the direction OP, as shown in Figure 4.1. The antenna axis

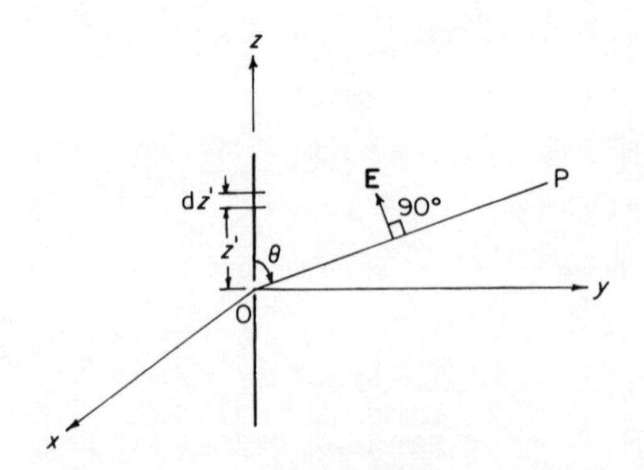

Figure 4.1 A receiving antenna in the field of a distant transmitter

and the line OP define a plane, which is referred to as the incident plane. Since the antenna is in the far zone of the transmitter, the electric vector of the incoming wave, **E**, is perpendicular to OP and we can assume its magnitude, $|\mathbf{E}| = E$, to be the same at every point along the antenna. Suppose **E** lies in the incident plane. Let V_{oc} be the open-circuit voltage induced at the antenna terminals, which we can measure with an ideal voltmeter. The ratio $-V_{oc}/E$ has the dimension of length, and is obviously a measure of the ability of the antenna to intercept electromagnetic waves. It is called the effective length of the receiving antenna. Using the superscript r to indicate 'receiving', we have

$$h_e^r(\theta) = -V_{oc}/E \tag{4.1}$$

The parameter $h_e^r(\theta)$ is clearly a function of the direction of the wave normal OP, since only the component of **E** parallel to the conductor will contribute to the induced voltage at the antenna terminals. For the case when the antenna axis is directed along z, this component depends on θ only. For example, if $\theta = 90°$, **E** is parallel to the conductors and the induced voltage will be maximum. If $\theta = 0°$, **E** is perpendicular to the conductors and no voltage will appear across the gap. For an arbitrarily oriented coordinate system, however, the induced voltage is a function of both θ and ϕ.

If the incoming electric field **E** does not lie in the incident plane, let the angle between **E** and its projection on the incident plane be ψ. Then

$$V_{oc} = -h_e^r(\theta)E\cos\psi \tag{4.2}$$

since the component of **E** perpendicular to the incident plane is perpendicular to the antenna conductors and does not contribute to V_{oc}.

4.2.2 Equivalent Circuit

In section 4.2.1, we have seen that an incoming electromagnetic field induces an open-circuit voltage V_{oc} at the terminals of an antenna. Suppose now that the antenna is connected to an external circuit, which consists of a transmission line and a load at the end of the line. As far as the external circuit is concerned, the antenna acts as a voltage source, and can be replaced by a Thevenin equivalent. The Thevenin equivalent consists of a voltage generator with an internal impedance. The size of the voltage generator is simply the open-circuit voltage V_{oc}. The internal impedance is the value of impedance that the external circuit sees when the voltage source is turned off, corresponding to the removal of the electric field **E**. This impedance will be called the impedance of the receiving antenna, and is denoted by $Z_a^r = R_a^r + jX_a^r$. The equivalent circuit of the receiving antenna is shown in Figure 4.2, where Z_L is the impedance of the external circuit, including the transmission line. We have

$$I_1 = \frac{V_{oc}}{Z_L + Z_a^r} \tag{4.3}$$

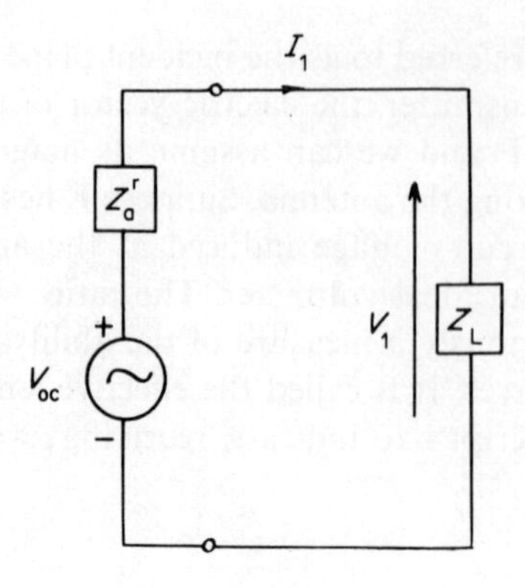

Figure 4.2 Equivalent circuit of a receiving antenna

$$V_1 = \frac{V_{oc} Z_L}{Z_L + Z_a^r} \tag{4.4}$$

If the terminals are short-circuited, $Z_L = 0$ and

$$I_{sc} = V_{oc}/Z_a^r \tag{4.5}$$

The value of the external impedance Z_L for which maximum power transfer occurs is

$$Z_L = (Z_a^r)^* = R_a^r - jX_a^r \tag{4.6}$$

The maximum time-averaged power delivered to the load Z_L is

$$W = \tfrac{1}{2} R_a^r |I_1|^2 = |V_{oc}|^2 / 8R_a^r \tag{4.7}$$

4.2.3 Effective Area

Given the magnitude of the incident electric field $|\mathbf{E}|$ and its direction of arrival as specified by OP, what is the maximum value of power which the antenna can deliver? This will depend on two factors. The first is the impedance of the external circuit, which must be adjusted so that it is the complex conjugate of Z_a^r. The second is the orientation of the vector $\mathbf{E}$ with respect to the incident plane. For $|V_{oc}|$ to be maximum, $\mathbf{E}$ must be lie on the incident plane ($\psi = 0$). When both conditions are satisfied, the power delivered to the load is called the available power, W_a. From (4.2) and (4.7), we have

$$W_a = \frac{|h_c^r(\theta)|^2}{8R_a^r} |\mathbf{E}|^2 \tag{4.8}$$

The power per unit area of the incident electromagnetic field is given by

$$S = |\mathbf{E}|^2 / 2\eta \tag{4.9}$$

Since the receiving antenna is in the incident electromagnetic field and it can be arranged to deliver a power W_a, it is as if the receiving antenna has a certain effective area in intercepting the incident electromagnetic energy.

The effective area or effective aperture is defined as a hypothetical area such that, when multiplied by the power density of the incident wave, it yields the available power. Thus

$$A_e^r(\theta)S = W_a \tag{4.10}$$

Using (4.8) and (4.9), we have

$$A_e^r(\theta) = \frac{\eta}{4R_a^r}|h_e^r(\theta)|^2 \tag{4.11}$$

4.2.4 Directional Pattern

The open-circuit voltage and the power-delivering ability of the receiving antenna are dependent on the direction of arrival of the incident wave. The voltage pattern is given by $|h_e^r(\theta)|$ and the power pattern is given by $A_e^r(\theta)$ or $|h_e^r(\theta)|^2$.

We have introduced several basic parameters that describe the characteristics of the receiving antenna. It turns out that these parameters are very much related to the parameters describing a transmitting antenna. For example, the effective length of an antenna used as a receiver. $h_e^r(\theta)$, is the same as $h_e(\theta)$, the effective length when it is used as transmitter, although the concepts of $h_e(\theta)$ and $h_e^r(\theta)$ arise in completely different contexts. Similarly, the antenna impedance and directional patterns turn out to be the same regardless of whether the antenna is used as a transmitter or as a receiver. The basis of relating the receiving properties to the transmitting properties is the reciprocity theorem, which we now discuss.

4.3 THE RECIPROCITY THEOREM

4.3.1 General Form of the Reciprocity Theorem

We begin with the reciprocity theorem of electromagnetic theory in its general form and apply it to the antenna problem in section 4.3.2. Consider a region in space that is linear and isotropic though not necessarily homogeneous. By isotropic, we mean the properties of the medium are independent of the direction of the applied electromagnetic fields, i.e. the parameters μ, ε, and σ are scalar quantities. Suppose that two sets of source currents, $\mathbf{J}'$ and $\mathbf{J}''$ can exist in this region, producing the fields $\mathbf{E}'$, $\mathbf{H}'$ and $\mathbf{E}''$, $\mathbf{H}''$. These source–field pairs satisfy Maxwell's equations:

$$\nabla \times \mathbf{E}' = -\mathrm{j}\omega\mu\mathbf{H}' \tag{4.12}$$

$$\nabla \times \mathbf{H}' = \mathrm{j}\omega\varepsilon\mathbf{E}' + \mathbf{J}' \tag{4.13}$$

$$\nabla \times \mathbf{E}'' = -\mathrm{j}\omega\mu\mathbf{H}'' \tag{4.14}$$

$$\nabla \times \mathbf{H}'' = \mathrm{j}\omega\varepsilon\mathbf{E}'' + \mathbf{J}'' \tag{4.15}$$

The two sets of sources are assumed to operate in the same medium and at the same frequency, but whether or not they operate simultaneously is unimportant.

The principle of reciprocity is the statement of an interrelationship between the two source–field pairs. It says that

$$\iiint \mathbf{E}'\cdot\mathbf{J}''\,\mathrm{d}\tau = \iiint \mathbf{E}''\cdot\mathbf{J}'\,\mathrm{d}\tau \tag{4.16}$$

where integration is over all of space.

To derive (4.16), consider the divergence of the quantity $(\mathbf{E}' \times \mathbf{H}'' - \mathbf{E}'' \times \mathbf{H}')$. Expanding the divergence of a cross-product and making use of (4.12)–(4.15), we have

$$\begin{aligned}\nabla\cdot(\mathbf{E}' \times \mathbf{H}'' - \mathbf{E}'' \times \mathbf{H}') &= \mathbf{H}''\cdot(\nabla \times \mathbf{E}') - \mathbf{E}'\cdot(\nabla \times \mathbf{H}'') \\ &\quad - \mathbf{H}'\cdot(\nabla \times \mathbf{E}'') + \mathbf{E}''\cdot(\nabla \times \mathbf{H}') \\ &= \mathbf{H}''\cdot(-\mathrm{j}\omega\mu\mathbf{H}') - \mathbf{E}'\cdot(\mathrm{j}\omega\varepsilon\mathbf{E}'' + \mathbf{J}'') \\ &\quad - \mathbf{H}'\cdot(-\mathrm{j}\omega\mu\mathbf{H}'') + \mathbf{E}''\cdot(\mathrm{j}\omega\varepsilon\mathbf{E}' + \mathbf{J}') \\ &= -\mathbf{E}'\cdot\mathbf{J}'' + \mathbf{E}''\cdot\mathbf{J}'\end{aligned} \tag{4.17}$$

If we integrate the left- and right-hand sides of (4.17) throughout the volume V enclosed by the surface S and apply the divergence theorem, there results

$$\iint_S (\mathbf{E}' \times \mathbf{H}'' - \mathbf{E}'' \times \mathbf{H}')\cdot\mathrm{d}\boldsymbol{\sigma} = \iiint_V (-\mathbf{E}'\cdot\mathbf{J}'' + \mathbf{E}''\cdot\mathbf{J}')\,\mathrm{d}\tau \tag{4.18}$$

Let S be the surface area of a sphere of radius r tending to infinity. In the limit, V is equal to all of space. If all the sources are contained within a finite volume, the fields at infinity must be outgoing spherical waves satisfying the relations

$$E_\theta = \eta H_\phi \tag{4.19a}$$

and

$$E_\phi = -\eta H_\theta \tag{4.19b}$$

Equations (4.19a) and (4.19b) are obtained by looking for solutions of the form $(1/r)\exp[\mathrm{j}(\omega t - kr)]$ in Maxwell's equations (problem 12 in section 3.17). On using (4.19a) and (4.19b), the left-hand side of (4.18) evaluates to zero since

$$\begin{aligned}\text{LHS} &= \iint_S (E'_\theta H''_\phi - E'_\phi H''_\theta - E''_\theta H'_\phi + E''_\phi H'_\theta)\,\mathrm{d}\sigma \\ &= \iint_S (\eta H'_\phi H''_\phi + \eta H'_\theta H''_\theta - \eta H''_\phi H'_\phi - \eta H''_\theta H'_\theta)\,\mathrm{d}\sigma = 0\end{aligned} \tag{4.20}$$

Hence (4.18) reduces to (4.16), which is the mathematical statement of the reciprocity theorem.

4.3.2 Application to Antennas

Let us apply the theorem to the antenna problem. Consider two dipole antennas of lengths L_1 and L_2 and radii a_1 and a_2 as shown in Figure 4.3.

In situation 1, antenna no. 1 is driven by a voltage V_1' across a narrow gap of width Δ_1. Since the gap is narrow, the corresponding electric field $\mathbf{E}_1$ is assumed to be uniform across the cross-sectional area and directed along the axis of the antenna. Antenna no. 2 is used to receive the radiation of no. 1. Suppose that, in placing an ideal ammeter across the gap (of width Δ_2), a reading I_2' is obtained. I_2' is therefore the short-circuit current at the centre of antenna no. 2 ($z_2 = 0$) when a voltage V_1' is applied at the centre of antenna no. 1 ($z_1 = 0$).

Let us now consider situation 2, which is obtained by reversing the roles played by the two antennas. Antenna no. 2 is now transmitting, and antenna no. 1 is receiving. The voltage at $z_2 = 0$ is V_2'' and the corresponding electric field $\mathbf{E}_2''$. The short-circuit current produced at the centre of antenna no. 1 ($z_1 = 0$) is I_1''.

Applying (4.16) to the two situations, we have

$$\int_0^{a_1}\int_0^{2\pi}\int_{-\Delta/2}^{\Delta/2} E_1' J_1'' \rho_1 \, dz_1 \, d\phi_1 \, d\rho_1 = \int_0^{a_2}\int_0^{2\pi}\int_{-\Delta/2}^{\Delta/2} E_2'' J_2' \rho_2 \, dz_2 \, d\phi_2 \, d\rho_2 \tag{4.21}$$

where integration over the z_1 and z_2 variables extends over the regions of the gaps only, since the antennas are assumed to be perfect conductors.

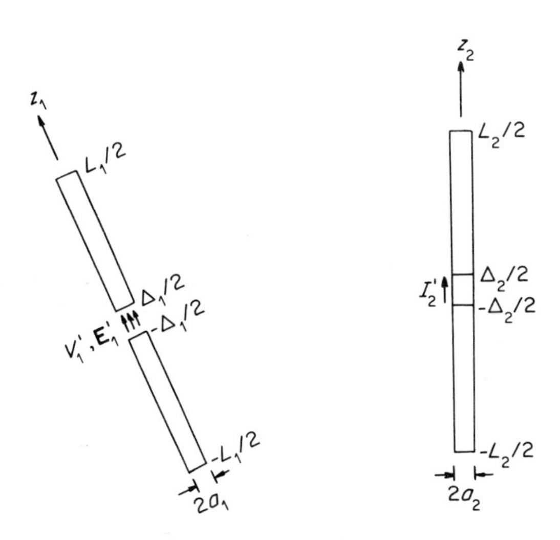

Figure 4.3 Reciprocity theorem applied to a system of transmitting and receiving antennas

Let us denote the left- and right-hand sides by I_L and I_R respectively; then

$$
\begin{aligned}
I_{\mathrm{L}} &= \int_0^{a_1}\int_0^{2\pi}\int_{-\Delta/2}^{\Delta/2} E'_1 J''_1 \rho_1 \,\mathrm{d}\rho_1 \,\mathrm{d}\phi_1 \,\mathrm{d}z_1 \\
&= \int_0^{a_1}\int_0^{2\pi} (J''_1 \rho_1 \,\mathrm{d}\rho_1 \,\mathrm{d}\phi_1) \int_{-\Delta/2}^{\Delta/2} E'_1 \,\mathrm{d}z_1
\end{aligned} \tag{4.22}
$$

since E'_1 is assumed to be uniform across the cross-sectional area of the antenna. From (2.10), we have,

$$
E'_1 = -\frac{\mathrm{d}\Phi}{\mathrm{d}z_1} - \mathrm{j}\omega A_{z_1} \tag{4.23}
$$

$$
\int_{-\Delta/2}^{\Delta/2} E'_1 \,\mathrm{d}z_1 = -[\Phi(z_1 = \tfrac{1}{2}\Delta) - \Phi(z_1 = -\tfrac{1}{2}\Delta)] - \mathrm{j}\omega \int_{-\Delta_1/2}^{\Delta_1/2} A_{z_1} \,\mathrm{d}z_1 \tag{4.24}
$$

The z_1-component of the vector potential is a smooth function near $z_1 = 0$ and therefore the integral of A_{z_1} vanishes in the limit of an infinitesimal gap ($\Delta_1 \to 0$). Let

$$
V'_1(z_1 = 0) = \Phi(z_1 = \tfrac{1}{2}\Delta_1) - \Phi(z_1 = -\tfrac{1}{2}\Delta_1)
$$

be the potential difference between the input terminals in the case of an infinitesimal narrow input region. Then

$$
\begin{aligned}
I_{\mathrm{L}} &= -V'_1(z_1 = 0) \int_0^{a_1}\int_0^{2\pi} J''_1 \rho_1 \,\mathrm{d}\rho_1 \,\mathrm{d}\phi_1 \\
&= -V'_1(z_1 = 0) I''_1(z_1 = 0)
\end{aligned} \tag{4.25}
$$

Similarly,

$$
I_{\mathrm{R}} = -V''_2(z_2 = 0) I'_2(z_2 = 0) \tag{4.26}
$$

Hence (4.21) is written

$$
\frac{V'_1(z_1 = 0)}{I'_2(z_2 = 0)} = \frac{V''_2(z_2 = 0)}{I''_1(z_1 = 0)} \tag{4.27}
$$

Equation (4.27) states that the ratio of the open-circuit voltage at $z_1 = 0$ to the resulting short-circuit current at $z_2 = 0$ is equal to the ratio of the open-circuit voltage at $z_2 = 0$ to the resulting short-circuit current at $z_1 = 0$. This form of reciprocity theorem is the same as that encountered in the study of circuit theory.

In the discussion leading to (4.27), we have for convenience assumed that the input terminals (where the voltage is applied) and the output terminals (where the short-circuit current is measured) are on two different antennas. However, the reciprocity theorem applies also to the case when the input and output terminals refer to two different locations on the same antenna. This is illustrated in Figure 4.4.

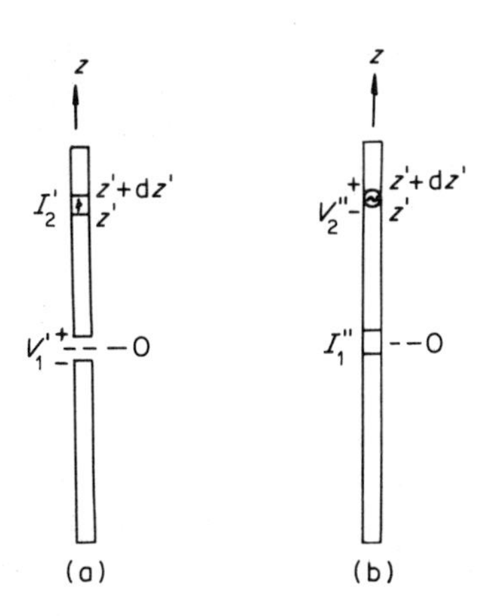

Figure 4.4 The reciprocity theorem applied to the case when the input and output terminals refer to two different locations on the same antenna. (a) A voltage V'_1 is applied to the input terminals at $z = 0$, and the current at the element dz' is measured as I'_2. (b) A voltage V''_2 is applied across dz', and the resulting short-circuit current at $z = 0$ is measured as I''_1. The reciprocity theorem reads

$$V'_1(z = 0)/I'_2(z = z') = V''_2(z = z')/I''_1(z = 0)$$

In section 4.4, we show that the receiving characteristics of an antenna can be obtained from its transmitting characteristics by means of the reciprocity theorem.

4.4 RELATION BETWEEN RECEIVING AND TRANSMITTING ANTENNA PARAMETERS

4.4.1 Impedance

The impedance of a transmitting antenna is defined as ratio of the voltage applied at the terminals of the antenna to the resulting current at the terminals, i.e.

$$Z_a = V(0)/I(0)$$

The concept of the impedance of a receiving antenna Z^r_a arises when we attempt to replace the antenna in the field of an incident wave by a Thevenin equivalent circuit consisting of a voltage generator V_{oc} with an internal impedance Z^r_a. The internal impedance is the impedance that the external circuit sees when the source is turned off, i.e. when there is no incident electric field. To measure this impedance, we connect the antenna terminals to a voltage source and measure the resulting current. The ratio is Z^r_a, which, of course, is the same as the definition of Z_a. Thus the impedance of an antenna is the same regardless of whether it is operating as a transmitter or as a receiver, and the superscript r will hence be removed.

4.4.2 Effective Length

The effective length of a receiving antenna is defined by (4.1):

$$h^r_e(\theta) = -V_{oc}/E$$

where V_{oc} is the open-circuit voltage induced at the terminals of the antenna and E is the magnitude of the electric field of a plane wave arriving from the direction specified by θ. On the other hand, the concept of the effective length of the antenna used as a transmitter arises when we calculate the far field by adding the contributions due to all the current elements that make up the current distribution along the antenna. It is defined by (3.53):

$$h_e(\theta) = \frac{\sin\theta}{I(0)} \int_{z_1}^{z_2} I(z') \exp(jkz' \cos\theta)\, dz'$$

It is remarkable that, although arising from quite different contexts, the effective length of the receiving antenna is the same as the effective length of the transmitting antenna. To show this, consider the antenna in the electromagnetic field of an incident plane wave, as shown in Figure 4.1. If we take the reference phase of the incident electric field to be at the centre of the antenna, the component of the electric field parallel to the element dz' is

$$E_{\parallel} = E \sin\theta \exp(jkz' \cos\theta) \tag{4.28}$$

This electric field will induce a field $E_{\parallel i}$ in the antenna so that the total electric field is equal to zero, in accordance with the boundary condition on a perfect conductor:

$$E_{\parallel t} = E_{\parallel} + E_{\parallel i} = 0$$

or

$$E_{\parallel i} = -E_{\parallel} \tag{4.29}$$

The voltage (electromotive force) induced in dz' is

$$E_{\parallel i}\, dz' = -E \sin\theta \exp(jkz' \cos\theta)\, dz' \tag{4.30}$$

producing a short-circuit current dI_{sc} at the terminals of the antenna.

Let us now consider another situation in which the antenna operates as a transmitter. Let a voltage V be applied to the terminals and a current distribution $I(z')$ results. At the element dz', its value is $I(z')$. According to the reciprocity theorem, we have

$$\frac{-E \sin\theta \exp(jkz' \cos\theta)\, dz'}{dI_{sc}} = \frac{V}{I(z')}$$

or

$$dI_{sc} = -\frac{E \sin\theta}{V} I(z') \exp(jkz' \cos\theta)\, dz' \tag{4.31}$$

It follows from (4.31) that the total short-circuit current is given by

$$I_{sc} = -\frac{E \sin\theta}{V} \int_{z_1}^{z_2} I(z') \exp(jkz' \cos\theta)\, dz' \tag{4.32}$$

Using the definition of $h_e(\theta)$ given by (3.53), (4.32) can be written as

$$h_e(\theta) = -\frac{I_{sc}}{E}\frac{V}{I(0)} \tag{4.33}$$

From (4.5) and the fact that $Z_a^r = Z_a$, we have

$$I_{sc}\frac{V}{I(0)} = I_{sc}Z_a = V_{oc}$$

Hence

$$h_e(\theta) = -\frac{V_{oc}}{E} = h_e^r(\theta) \tag{4.34}$$

We have thus shown that the effective lengths are the same regardless of whether the antenna is operating as a transmitter or as a receiver. The superscript r will hence be dropped in both $h_e^r(\theta)$ and $A_e^r(\theta)$.

4.4.3 Directional Pattern

The directional pattern of a transmitting antenna indicates the strength of the far-zone electric field at a fixed distance from the centre of the transmitting antenna in all different directions in space. From (3.54), it is described by the effective height $h_e(\theta)$. The directional pattern of a receiving antenna indicates the response of the antenna, namely the open-circuit voltage V_{oc} induced at its terminals, to an electromagnetic wave of specified strength arriving from all different directions in space. From (4.1), it is described by $h_e^r(\theta)$. Since we have shown in section 4.4.2 that $h_e^r(\theta) = h_e(\theta)$, it follows that the transmitting and receiving directional patterns are the same.

The identity of the directional patterns can also be proved in a more fundamental way as follows. Let the antenna for which we want to obtain its directional patterns be placed with its centre at the origin of a coordinate system and directed along the z-axis, and let it be referred to as the test antenna. Another antenna, called the probe antenna, is placed at a distance r, which is in the far zone of the test antenna. When the test antenna is being driven as a transmitter, its electric field in the far zone will be in the plane containing the axis of the antenna and the line joining the centres of the two antennas. Also, the electric vector is in the θ-direction, as shown in Figure 4.5. The probe antenna is adjusted so that it is always parallel to the electric vector. We perform the following experiments.

(a) Apply a voltage V to the input terminals of the test antenna and measure the short-circuit current I_p at the terminals of the probe antenna. I_p is a measure of the strength of $\mathbf{E}$ at the location of the probe antenna and hence provides data for the directional pattern of the test antenna as a transmitter.

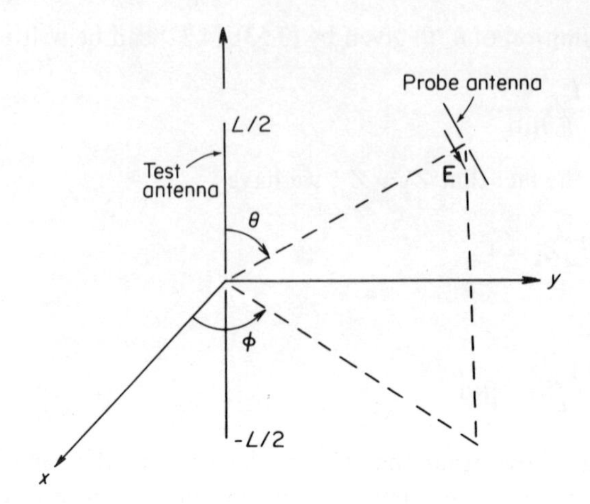

Figure 4.5 Measurement of the directional patterns of an antenna

(b) Apply a voltage V to the input terminals of the probe antenna and measure the short-circuit current I_t at the terminals of the test antenna. This short-circuit current I_t is a measure of the response of the test antenna to the EM field arriving in the direction specified by θ and hence provides data for the directional pattern of the test antenna used as a receiver.

(c) Move the probe antenna to every point on the surface of a large sphere of radius r, making sure that the probe antenna is always parallel to the θ-direction. Repeat measurements (a) and (b).

By the reciprocity theorem, $I_t = I_p$ for every location of the probe antenna. Hence the directional patterns of the test antenna are the same irrespective of whether we use it as a transmitter or as a receiver.

4.4.4 Ratio of Effective Area to Directive Gain

As discussed in section 4.2.3, the effective area is a parameter that indicates the power-receiving capacity of an antenna. It can be computed according to (4.11). In this section, we show that there is a relation between the directive gain and the effective area which holds for any antenna, namely

$$\frac{A_e(\theta, \phi)}{g(\theta, \phi)} = \frac{\lambda^2}{4\pi} \tag{4.35}$$

To be general, we have allowed for the possibility of A_e and g to be dependent not only on θ but also on ϕ.

To prove (4.35), consider two antennas, no. 1 and no. 2. The centre of antenna no. 1 is at the origin of coordinate system 1 and the centre of antenna no. 2 is at the origin of coordinate system 2. The centres O_1 and O_2 are separated by a distance r. Let the vector $\overrightarrow{O_1O_2}$ be specified by (θ_1, ϕ_1) with respect to coordinate system 1 and the vector $\overrightarrow{O_2O_1}$ be specified by (θ_2, ϕ_2) with respect to coordinate system 2.

Consider antenna no. 1 to be driven as a transmitter with voltage V_1 at the terminals. The power radiated is given by

$$W_t = \tfrac{1}{2}|I_1|^2 R_{a1} = \tfrac{1}{2}\left|\frac{V_1}{Z_{a1}}\right|^2 R_{a1} \tag{4.36}$$

where Z_{a1} is the antenna impedance of antenna no. 1 and R_{a1} is its radiation resistance (real part of Z_{a1}). If $g_1(\theta, \phi)$ is its directive gain, the radiation intensity $U_1(\theta_1, \phi_1)$ and the power density $S_1(\theta_1, \phi_1)$ are given by

$$U_1(\theta_1, \phi_1) = \frac{W_t}{4\pi} g_1(\theta_1, \phi_1) \tag{4.37}$$

$$S_1(\theta_1, \phi_1) = \frac{W_t}{4\pi r^2} g_1(\theta_1, \phi_1) \tag{4.38}$$

If $A_{e2}(\theta_2, \phi_2)$ is the effective area of antenna no. 2 in the direction (θ_2, ϕ_2) of the first antenna, we obtain the available power at the terminals of the second antenna as

$$\begin{aligned} W_a &= \frac{W_t}{4\pi r^2} g_1(\theta_1, \phi_1) A_{e2}(\theta_2, \phi_2) \\ &= \frac{R_{a1}|V_1|^2}{8\pi r^2 |Z_{a1}|^2} g_1(\theta_1, \phi_1) A_{e2}(\theta_2, \phi_2) \end{aligned} \tag{4.39}$$

From (4.7), the available power can also be expressed as

$$W_a = \frac{|V_{oc2}|^2}{8R_{a2}} = \frac{|I_{sc2} Z_{a2}|^2}{8R_{a2}} \tag{4.40}$$

where V_{oc2} is the open-circuit voltage, I_{sc2} is the short-circuit current, Z_{a2} is the antenna impedance, and R_{a2} is the radiation resistance (real part of Z_{a2}) of antenna no. 2. It follows from (4.39) and (4.40) that

$$\frac{|I_{sc2}|^2}{|V_1|^2} = \frac{R_{a1} R_{a2}}{\pi r^2 |Z_{a1}|^2 |Z_{a2}|^2} g_1(\theta_1, \phi_1) A_{e2}(\theta_2, \phi_2) \tag{4.41}$$

We now use antenna no. 2 as transmitter and antenna no. 1 as receiver. Using the same reasoning as before, we obtain

$$\frac{|I_{sc1}|^2}{|V_2|^2} = \frac{R_{a2} R_{a1}}{\pi r^2 |Z_{a1}|^2 |Z_{a2}|^2} g_2(\theta_1, \phi_1) A_{e1}(\theta_2, \phi_2) \tag{4.42}$$

By the reciprocity theorem,

$$\frac{I_{sc2}}{V_1} = \frac{I_{sc1}}{V_2} \tag{4.43}$$

Hence we obtain

$$\frac{A_{e1}(\theta_1, \phi_1)}{g_1(\theta_1, \phi_1)} = \frac{A_{e2}(\theta_2, \phi_2)}{g_2(\theta_2, \phi_2)} \tag{4.44}$$

Note that no assumptions have been made on the relative directions and orientations, nor on the types of the two antennas. Consequently, the ratio $A_e(\theta, \phi)/g(\theta, \phi)$ must be the same for all types of antennas:

$$\frac{A_e(\theta, \phi)}{g(\theta, \phi)} = \text{constant}$$

Once the constant has been evaluated for one particular antenna, it would hold for any other antenna. Let us therefore use the simplest antenna possible, namely, the infinitesimal current element of length dl. For this antenna,

$$h_e(\theta, \phi) = \sin\theta \, \mathrm{d}l \tag{4.45}$$

$$R_a = \frac{2\pi}{3}\eta\left(\frac{\mathrm{d}l}{\lambda}\right)^2 \tag{4.46}$$

$$g(\theta, \phi) = \tfrac{3}{2}\sin^2\theta \tag{4.47}$$

$$A_e(\theta, \phi) = \frac{\eta}{4R_a}|h_e(\theta, \phi)|^2 = \frac{3\lambda^2}{8\pi}\sin^2\theta \tag{4.48}$$

From (4.47) and (4.48), we obtain the ratio A_e/g to be $\lambda^2/4\pi$. This completes the proof of (4.35).

From (4.36) and (4.39), the ratio of the available power W_a at the terminals of the receiving antenna in the direction (θ_2, ϕ_2) to the power W_t radiated by the transmitting antenna is

$$\frac{W_a}{W_t} = \frac{g_1(\theta_1, \phi_1)A_{e2}(\theta_2, \phi_2)}{4\pi r^2}$$

$$= \frac{A_{e1}(\theta_1, \phi_1)A_{e2}(\theta_2, \phi_2)}{(\lambda r)^2} \tag{4.49}$$

Equation (4.49) is known as the Friss transmission formula.

4.5 MEASUREMENTS OF PATTERN, DIRECTIVITY, AND IMPEDANCE

In this section, we make use of the theory developed in this chapter to describe briefly how the three important parameters of an antenna, namely, pattern, directivity, and impedance, can be measured. For detailed information on antenna measurements, the reader is referred to Hollis *et al.* (1970) and Kummer and Gillespie (1978).

4.5.1 Pattern Measurement

The pattern of an antenna refers to the variation of the amplitude of the far field as a function of direction. The far-field criterion was discussed in section

3.6. In the far zone, the electric field has only transverse components and the pattern is independent of the distance at which it is taken. For an antenna located at the origin of a coordinate system, the far-zone electric field will, in general, consist of a θ-component, E_θ, and a ϕ-component, E_ϕ. A complete description of the pattern, for a given orientation of the antenna, will require the measurement of $|E_\theta|$ and $|E_\phi|$ in all directions. Such detailed information, however, is seldom required in practice. For many purposes, it is necessary only to obtain the patterns in two mutually perpendicular planes. If the antennas have symmetry properties, the complete pattern can often be inferred from the patterns in these planes.

To illustrate the procedures involved, consider first the specific example of a dipole antenna with its axis along the z-direction. For this orientation, the electric field has only the θ-component. Also, on account of symmetry, it is sufficient to measure the patterns in the x–z plane, $|E_\theta|(\theta, \phi = 0)$, and the x–y plane, $|E_\theta|(\theta = 90°, \phi)$. The former is called the E-plane pattern because the $\phi = 0$ plane contains the electric field vector. The latter is called the H-plane pattern because the $\theta = 90°$ plane contains the magnetic field vector.

There are two ways to measure $|E_\theta|(\theta, \phi = 0)$ and $|E_\theta|(\theta = 90°, \phi)$ patterns. In the first method, the antenna whose pattern is desired (the test antenna) is fixed in space and is used as the transmitting antenna. A second antenna, called the probe antenna, is placed at a distance $\mathbf{r}$ in the far zone to receive the signals from the test antenna. To measure $|E_\theta|(\theta, \phi = 0°)$, the probe antenna is made to lie in the x–z plane, with its axis oriented perpendicular to $\mathbf{r}$. The open-circuit voltage induced in the terminals will be proportional to $|E_\theta|$. If the probe antenna is moved to different positions along the circle of constant r in the x–z plane, with the antenna always lying in the x–z plane and oriented perpendicular to $\mathbf{r}$, the variation of the open-circuit voltage with θ will result in the E-plane pattern. Similarly, if the probe antenna is placed at a distance $\mathbf{r}$ in the x–y plane and oriented perpendicular to it, the received signal will be proportional to $|E_\theta|(\theta = 90°, \phi)$. By moving the probe antenna around a circle of constant r in the x–y plane, the H-plane pattern is obtained.

In practice, the probe antenna is connected to a receiver so that the voltage induced at its terminals is detected and amplified. The meter in the receiver may be calibrated in terms of either voltage or power. In the latter case, the relative power pattern is obtained.

In the second method, the probe antenna is fixed in space and is used as the transmitting antenna. The test antenna is used as the receiving antenna. It is to be rotated, usually by mounting it on a turntable. According to the reciprocity theorem, the pattern of the antenna is the same regardless of whether it is used as a transmitter or as a receiver. Suppose the probe antenna is located at the position $(x, 0, 0)$ and directed along z. To obtain the $|E_\theta|(\theta, \phi = 0)$ pattern, the test antenna, located at the origin, is rotated in the x–z plane. To obtain the $|E_\theta|(\theta = 90°, \phi)$ pattern, it is rotated with respect to the z-axis. The output of the receiver connected to the test antenna is usually fed to a recorder. In an automatic pattern recording system, a turntable beneath the pen of the recorder is synchronized in angle with the turntable on which the test antenna is mounted.

A sheet of polar coordinate paper attached to the turntable is then marked by the moving pen as the antenna rotates.

If the far field of the test antenna has an E_ϕ-component in addition to an E_θ-component (e.g. a cross-dipole), the patterns $|E_\phi|(\theta, \phi = 0)$ and $|E_\phi|(\theta = 90°, \phi = 0)$ should also be measured. These can be obtained simply by orienting the probe antenna so that its axis is parallel to the ϕ-direction and repeating the procedures described above.

The first method (fixed test antenna) is convenient when the test antenna is large and cannot be easily rotated. The second method (fixed probe antenna) is preferred if the test antenna is small and can be rotated easily.

4.5.2 Directivity Measurement

The directivity of an antenna can be determined by either the comparison method or the absolute method.

In the comparison method, it is necessary to make use of a standard antenna whose directivity is known accurately. At microwave frequencies, the standard antenna is usually a pyramidal horn (described in section 10.3.5) and at the lower frequencies it is usually a half-wave dipole. Let a source antenna be driven by a constant-power transmitter. First the standard antenna is placed in the far zone of the source antenna and is connected to a matched receiving system. The standard antenna is oriented such that the received power output is maximum. This power is equal to

$$W_s = A_{ems} S \tag{4.50}$$

where A_{ems} is the maximum value of the effective area of the standard antenna and S is the power density of the incident wave. Next the standard is replaced by the test antenna. The receiving system connected to it is again adjusted for conjugate match and the test antenna is oriented for maximum power output. This power is equal to

$$W_t = A_{emt} S \tag{4.51}$$

where A_{emt} is the maximum value of the effective area of the test antenna. Dividing (4.50) by (4.51), we obtain

$$W_s/W_t = A_{ems}/A_{emt} = D_s/D_t \tag{4.52}$$

where D_t and D_s are the directivities of the test and standard antennas respectively. Hence

$$D_t = D_s W_t/W_s \tag{4.53}$$

Since D_s is known, the directivity of the test antenna can be obtained from the ratio of the received powers. If losses are negligible, we also have

$$G_{mt} = G_{ms} W_t/W_s \tag{4.54}$$

where G_{mt} and G_{ms} are the maximum power gains of the test and standard antennas respectively.

The directivity can also be measured by the absolute method if two identical test antennas are available. In this method, one antenna is used as a transmitter and the other as a receiver. From (4.49), if the receiving antenna is oriented for maximum power output, the ratio of the received power to the transmitted power is

$$\frac{W_r}{W_t} = \frac{D^2\lambda^2}{(4\pi)^2 r^2}$$

or

$$D = \frac{4\pi r}{\lambda}\sqrt{\frac{W_r}{W_t}} \tag{4.55}$$

Thus, by measuring the distance r, wavelength λ, and the received and transmitted powers, the directivity can be calculated. If the two antennas are not identical, the method can be modified in the manner described in problem 10 of section 4.7.

4.5.3 Impedance Measurement

A frequencies up to about 1000 MHz, the impedance of an antenna can be measured directly on commercially available frequency bridges. Examples are the General Radio (GR)-1602-A and the Hewlett-Packard (HP)-803-A. The frequency range of the former is 41–1000 MHz and that of the latter is 52–500 MHz.

At VHF and UHF frequencies, the impedance of an antenna can also be measured by means of a slotted section of a transmission line. The antenna impedance, unless it is equal to the characteristic impedance Z_0 of the transmission line connected to it, will reflect some of the energy back towards the generator, creating a standing wave on the line. The slotted section enables one to measure the voltage standing wave ratio (VSWR) and the distance of the voltage minimum nearest the antenna terminals (d_{min}). The impedance at the end of the line, in this case the antenna impedance Z_a, can be calculated from the VSWR and d_{min} by the following formula:

$$Z_a = Z_0 \frac{1 - Sj\tan(kd_{min})}{S - j\tan(kd_{min})} \tag{4.56}$$

Alternatively, Z_a can be obtained simply from a Smith chart, the use of which is explained in textbooks on electromagnetic theory (e.g. Jordan and Balmain, 1968).

In practice, it is not necessary to measure d_{min} directly, which may be difficult or impossible at short wavelengths since this minimum may occur at a point in the line ahead of the beginning of the slotted section. An indirect measurement of d_{min} is as follows. First, the antenna terminals are short-circuited. The standing wave pattern will then exhibit voltage minima at integral multiples of half-wavelengths from the short. The location of one of these minimum positions is recorded. Then the short-circuit is removed. This will cause the voltage minimum to move to a new position. The distance that it has moved, in the direction of the generator, corresponds to the parameter d_{min}.

4.6 WORKED EXAMPLES

Example 1

A half-wave dipole situated with its centre at the origin radiates a time-averaged power of 584 W in free space at a frequency 300 MHz. A second half-wave dipole is placed with its centre at a point $P(r, \theta, \phi)$ where $r = 100$ m, $\theta = 90°$, $\phi = 30°$. It is oriented so that its axis is parallel to that of the transmitting half-wave dipole. What is the available power at the terminals of the second dipole. Assume the impedance of half-wave dipoles to be $73 + j42.5$ ohm.

Solution

The available power at point P is given by (4.8):

$$W_a = \frac{|h_e(\theta)|^2}{8R_a}|\mathbf{E}|^2$$

For a half-wave dipole,

$$h_e(\theta) = \sin\theta \int_{-\lambda/4}^{\lambda/4} \cos(kz')\exp(jkz'\cos\theta)\mathrm{d}z' = \frac{2\cos(\frac{1}{2}\pi\cos\theta)}{k\sin\theta}$$

At $\theta = 90°$, $h_e(90°) = \lambda/\pi$. Since $\lambda = c/f = 3 \times 10^8/3 \times 10^8$ m $= 1$ m, $h_e(90°) = 1/\pi$ m.

From (3.86)–(3.88), $|\mathbf{E}|$ is given by

$$|\mathbf{E}| = \frac{\eta I(0)}{2\lambda r} h_e(\theta)$$

The feed-point current

$$I(0) = (2W/R_a)^{1/2} = [2(584)/73]^{1/2}$$
$$= 4 \text{ A}$$

Hence at point P where $r = 100$ m and $\theta = 90°$, we have

$$|\mathbf{E}| = \frac{120\pi(4)}{2(1)(100)}(1/\pi) = 2.4 \text{ V m}^{-1}$$

$$W_a = \frac{(1/\pi)^2}{8(73)}(2.4)^2 = 10^{-3} \text{ W}.$$

Example 2

Two centre-fed vertical half-wave dipoles and one centre-fed horizontal half-wave dipole are arranged in the manner shown in Figure 4.6. The antennas lie

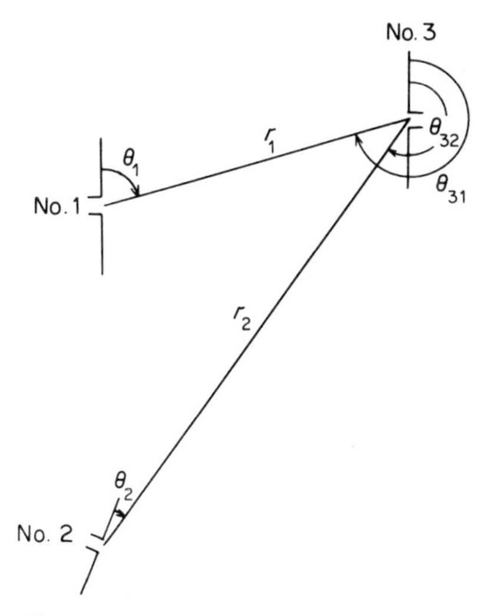

Figure 4.6 A system of two transmitting and one receiving half-wave dipoles

in the same plane, with no. 1 and no. 2 transmitting and no. 3 receiving. The feed-point currents at no. 1 and no. 2 are $I_1 \angle 0°$ and $I_2 \angle \alpha°$ respectively. If $kr_1 \gg 1$ and $kr_2 \gg 1$, derive expressions for

(a) the open-circuit voltage induced at the terminals of no. 3,
(b) the power available at no. 3 when it is connected to a matched load.

Solution

The electric field at no. 3 due to no. 1 is

$$\mathbf{E}_1 = \frac{\eta \mathrm{j} k I_1 \angle 0}{4\pi r_1} \exp(-\mathrm{j}kr_1) h_\mathrm{e}(\theta_1) \hat{\boldsymbol{\theta}}_1$$

where

$$h_\mathrm{e}(\theta_1) = \frac{2\cos(\frac{1}{2}\pi\cos\theta_1)}{k\sin\theta_1}$$

The electric field at no. 3 due to no. 2 is

$$\mathbf{E}_2 = \frac{\eta \mathrm{j} k I_2 \angle \alpha° \exp(-\mathrm{j}kr_2)}{4\pi r_2} h_\mathrm{e}(\theta_2) \hat{\boldsymbol{\theta}}_2$$

where

$$h_\mathrm{e}(\theta_2) = \frac{2\cos(\frac{1}{2}\pi\cos\theta_2)}{k\sin\theta_2}$$

(a) Let the phase of $\mathbf{E}_1$ be taken as the reference. Let the open-circuit voltage induced at no. 3 due to no. 1 be $V_{oc}^{(1)}$ and that due to no. 2 be $V_{oc}^{(2)}$. Then

$$V_{oc}^{(1)} = -h_e(\theta_{31})\frac{kI_1}{4\pi r_1}\eta h_e(\theta_1)$$

$$V_{oc}^{(2)} = -h_e(\theta_{32})\frac{kI_2}{4\pi r_2}\eta h_e(\theta_2)\underline{/k(r_1 - r_2) + \alpha^\circ}$$

The total open-circuit voltage induced at no. 3 is

$$V_{oc} = V_{oc}^{(1)} + V_{oc}^{(2)}$$

Note that there is a phase difference of $k(r_1 - r_2) + \alpha^\circ$ between $V_{oc}^{(1)}$ and $V_{oc}^{(2)}$.

(b) The power available at no. 3 is

$$W_a = |V_{oc}|^2/8R_a$$

where R_a is the real part of the antenna impedance of no. 3 and

$$|V_{oc}| = -h_e(\theta_{31})\frac{kI_1}{4\pi r_1}\eta h_e(\theta_1) - h_e(\theta_{32})\frac{kI_2}{4\pi r_2}\eta h_e(\theta_2)\cos[k(r_1 - r_2) + \alpha^\circ]$$

4.7 PROBLEMS

1. A centre-fed half-wave dipole in free space is situated along the z-axis from $z = -\lambda/4$ to $z = \lambda/4$. The antenna is in the electromagnetic field of an incident plane wave with the wave normal making an angle 30° with the antenna axis. The electric vector is in the plane containing the antenna axis and the wave normal direction. If the operating frequency is 300 MHz, and the available power at the antenna terminals is 100 μW, what is the magnitude of the incident electric field? Assume the antenna impedance to be equal to $Z_a = 73 + j43.5$ ohm.

2. A small dipole of length $dl = 0.01$ m operating in free space is in the field of an incident plane wave whose electric field $E = 8\sqrt{(15\pi)}$ V m^{-1}. The operating frequency is $300/\pi$ Hz. Assume that the current distribution along the small dipole is a constant. (a) What is the incident power per unit area? (b) If the input terminals of the antenna are connected to a load which is adjusted to receive maximum power, and the axis of the antenna is oriented parallel to the incident electric field, what is the value of this maximum power? (c) What is the directivity and the maximum value of the effective area of this antenna?

3. An electrically short antenna in free space is situated along the z-axis from $z = -L/2$, to $z = L/2$, with the input terminals situated at its centre. When used as a transmitting antenna, the current distribution along the antenna can be assumed to be triangular. Evaluate the effective area and the directive gain for this antenna and hence verify equation (4.35).

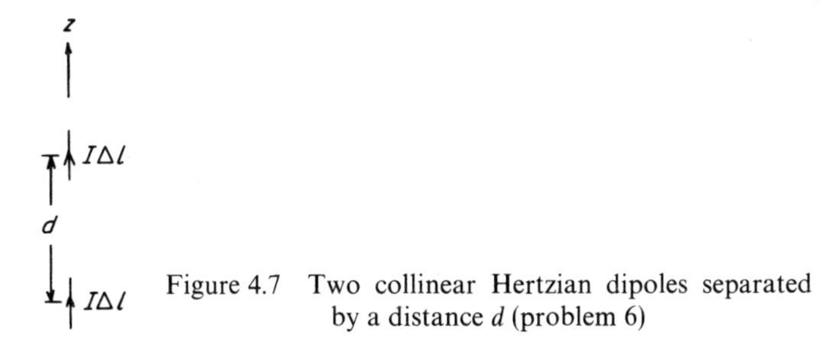

Figure 4.7 Two collinear Hertzian dipoles separated by a distance d (problem 6)

4. A transmitting antenna and a receiving antenna are separated by a distance of 10^3 m. If the transmitting antenna radiates a power of 100 W, calculate the available power at the receiving antenna under the following conditions:

   (a) Directivity of transmitting antenna is 1.64 and effective area of receiving antenna is $0.25\ m^2$.
   (b) Effective areas of both antennas are $0.50\ m^2$ at an operating wavelength of 0.05 m.
   (c) Directivities of both antennas are 1.5 and the operating wavelength is 0.10 m.

5. If the three antennas described in worked example 2 in section 4.6 are $\frac{3}{2}$ wavelength dipoles, determine the phase angle α for I_2 such that the available power at antenna no. 3 is (a) a maximum; (b) a minimum. Assume the angles θ_1 and θ_2 are in the first quadrant while the angles θ_{32} and θ_{31} are in the third quadrant.

6. Figure 4.7 shows an array of two z-directed Hertzian dipoles situated in free space, separated by a distance d and oriented in the same direction.

   (a) Find the radiation intensity of the array.
   (b) Find the power radiated by the array.
   (c) When used as a receiving antenna, find the maximum value of the effective area in units of λ^2 for the case $d = \lambda/2$.

7. A half-wave long antenna in free space is situated along the z-axis from $z = -\lambda/4$ to $z = \lambda/4$. The wavelength is 10 m. The current distribution along the antenna is a travelling wave of the form $I(z) = I_0 \exp(-jkz)$. For what direction with respect to the axis of the antenna is the effective area a maximum and what is the value of this maximum?

8. Verify that equation (4.35) holds for the travelling wave antenna of problem 7.

9. Figure 4.8 shows two half-wave dipoles, with centres at A_1 and A_2 respectively. The dipoles lie in the same plane, and are in the far zones of each other. Dipole 1 is transmitting and dipole 2 is receiving. Let A_2 be

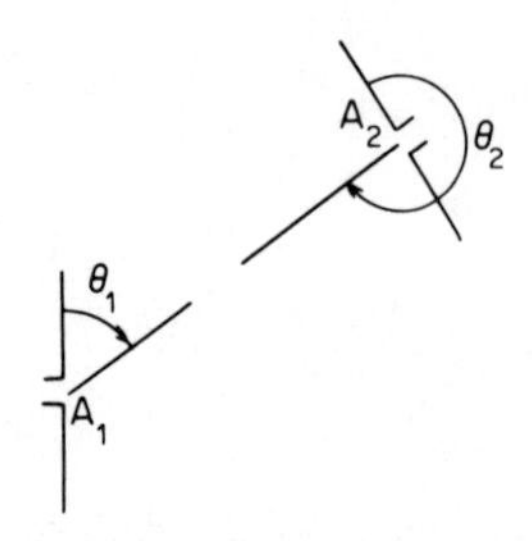

Figure 4.8 Geometry for problem 9

specified by $(r, \theta_1, 0)$ with respect to the coordinate system of dipole 1 and A_1 be specified by $(r, \theta_2, 0)$ with respect to the coordinate system of dipole 2. Let W_t be the power radiated by dipole 1 and W_r be the power available at dipole 2 when it is connected to a matched load.

(a) If $\theta_1 = 60°$, $\theta_2 = 240°$ and $r = 80\lambda$, find W_r/W_t.
(b) If $\theta_1 = 60°$ and $r = 80\lambda$, determine θ_2 such that W_r/W_t is (i) maximum, (ii) minimum.
(c) What are the maximum and minimum values of W_r/W_t in (b)?

10. In the absolute method of measuring directivity, suppose the two antennas are not identical. Let their directivities (referred to an isotropic source) be D_1 and D_2. In addition to measuring W_r and W_t as described in section 4.5.2, the directivity of each antenna is also measured with respect to a third reference antenna, whose directivity need not be known. Let these be denoted by D'_1 and D'_2. Show that

$$D_1 = \frac{4\pi r}{\lambda}\sqrt{\frac{W_r D'_1}{W_t D'_2}}$$

$$D_2 = \frac{4\pi r}{\lambda}\sqrt{\frac{W_r D'_1}{W_t D'_2}}$$

BIBLIOGRAPHY

Kraus, J. (1950). *Antennas*, McGraw-Hill, New York, Chapter 15.
Seshadri, S. R. (1971). *Fundamentals of Transmission Lines and Electromagnetic Fields*, Addison-Wesley, Reading, MA, Chapter 10.

CHAPTER 5
Mutual and Self-impedances

5.1 INTRODUCTION

When we discussed the receiving antenna in Chapter 4, the voltage induced at the terminals of the receiving antenna was due to the electromagnetic wave radiated by a distant transmitter. We now consider the situation in which two antennas are close to each other, as shown in Figure 5.1. Suppose antenna no. 1 is driven with a voltage V_1, and the resultant current at the feed point or base is I_1. [Strictly speaking, we should use the symbol $I_1(0)$ for the feed-point current. This argument is suppressed in this section for convenience.] If the open-circuit voltage induced at the terminals of antenna no. 2 is V_{21}, then the mutual impedance, Z_{21}, is defined as

$$Z_{21} = V_{21}/I_1 \tag{5.1}$$

Similarly,

$$Z_{12} = V_{12}/I_2 \tag{5.2}$$

where V_{12} is the open-circuit voltage across the terminals of no. 1 due to a current I_2 at the base of no. 2. By the reciprocity theorem, we have

$$Z_{12} = Z_{21} \tag{5.3}$$

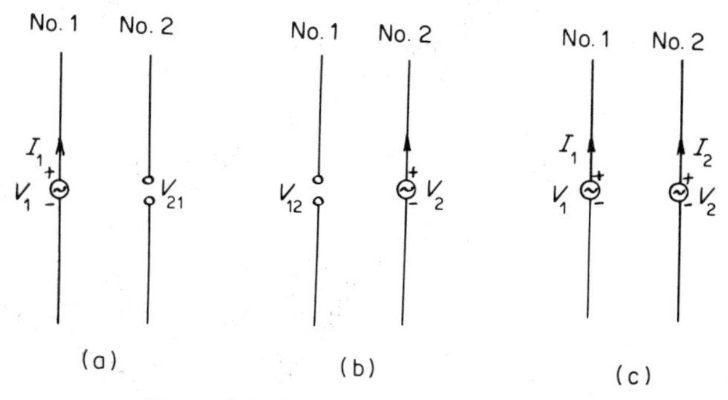

Figure 5.1 Mutual effects of antennas

If both antennas are being driven, as shown in Figure 5.1(c), then we have

$$V_1 = I_1 Z_{11} + I_2 Z_{12} \tag{5.4}$$

$$V_2 = I_1 Z_{21} + I_2 Z_{22} \tag{5.5}$$

where Z_{11} is the impedance measured at the terminals of no. 1 with the terminals on no. 2 open. Similarly, Z_{22} is the impedance measured at the terminals of no. 2 with the terminals on no. 1 open. Strictly speaking, Z_{11} is not identical with the self-impedance of antenna no. 1, Z_{a1}, which is its input impedance with all other antennas removed. However, for most practical purposes, the difference is neglected and Z_{11} is taken to be the same as Z_{a1}. Similarly, Z_{22} is taken to be the same as Z_{a2}.

Equations (5.4) and (5.5) can also be written as

$$V_1/I_1 = Z_{11} + Z_{12} I_2/I_1 \tag{5.6}$$

$$V_2/I_2 = Z_{21} I_1/I_2 + Z_{22} \tag{5.7}$$

The input impedances, V_1/I_1 and V_2/I_2, are the impedances that any impedance-transforming networks must be designed to feed. Equations (5.6) and (5.7) show that they are dependent upon the current ratios and the mutual impedances, the values of which must be known.

In general, for N antennas, we have

$$\begin{aligned} V_1 &= I_1 Z_{11} + I_2 Z_{12} + \cdots + I_N Z_{1N} \\ V_2 &= I_1 Z_{21} + I_2 Z_{22} + \cdots + I_N Z_{2N} \\ &\vdots \\ V_N &= I_1 Z_{N1} + I_2 Z_{N2} + \cdots + I_N Z_{NN} \end{aligned} \tag{5.8}$$

5.2 CALCULATION OF MUTUAL IMPEDANCE

To calculate the mutual impedance Z_{21} in (5.1), we must find the open-circuit voltage V_{21} at the terminals of antenna no. 2 due to a base current $I_1(0)$ in antenna no. 1. We cannot make use of the results of Chapter 4 because they are developed on the assumption that the receiving antenna is in the far zone of the transmitting antenna. We can, however, readily obtain an expression for Z_{21} by an application of the reciprocity theorem.

Consider antenna no. 2 to be in the electromagnetic field of no. 1, which is in the close vicinity of no. 2. Let us consider an element dz' on antenna no. 2 which is at a distance z' from the central terminals. Let the component of the incident electric field parallel to the element dz' be E_{z21}. The field induced inside dz' will be

$$E_{zi} = -E_{z21}$$

since the total field must add up to zero for a perfect conductor. The induced electromotive force in dz' is therefore $-E_{z21}\,dz'$. If the central terminals are short-circuited, this electromotive force will cause a current to flow across the

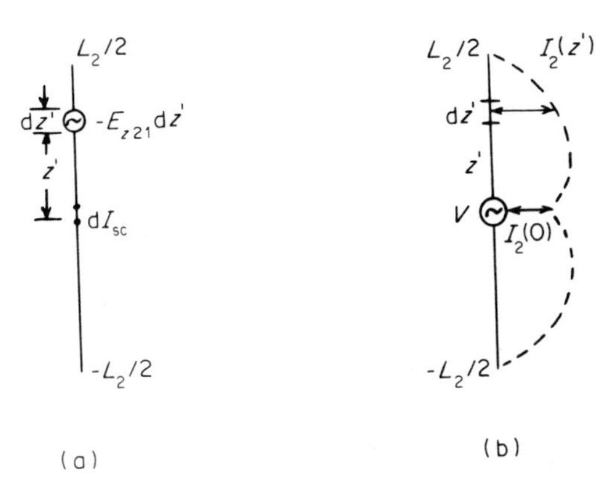

Figure 5.2 Reciprocity theorem applied to the calculation of the induced open-circuit voltage: (a) antenna in the field of an incident wave; (b) antenna being driven as a transmitter

terminals, which we denote by dI_{sc}. The situation is illustrated in Figure 5.2(a).

Let us now imagine another situation in which antenna no. 2 is used as a transmitter. A voltage V is applied to the central terminals, and a current $I_2(z')$ results, as shown in Figure 5.2(b). At the element dz' its value is $I_2(z')$. By the reciprocity theorem, we have

$$\frac{-E_{z21}\,dz'}{dI_{sc}} = \frac{V}{I_2(z')}$$

or

$$dI_{sc} = -\frac{E_{z21}\,dz'}{V} I_2(z') \tag{5.9}$$

The total short-circuit current I_{sc} is due to the contributions of all the electromotive forces induced along the entire antenna. It is given by

$$I_{sc} = -\frac{1}{V}\int E_{z21} I_2(z')\,dz' \tag{5.10}$$

From (4.5), the open-circuit voltage V_{21} is equal to the short-circuit current I_{sc} multiplied by the antenna impedance Z_{a2}:

$$V_{21} = I_{sc} Z_{a2} = -\frac{Z_{a2}}{V}\int E_{z21} I_2(z')\,dz' \tag{5.11}$$

Since $Z_{a2} = V/I_2(0)$, we have

$$V_{21} = -\frac{1}{I_2(0)}\int E_{z21} I_2(z')\,dz' \tag{5.12}$$

Use of (5.12) in (5.1) yields

$$Z_{21} = -\frac{1}{I_1(0)I_2(0)}\int E_{z21}I_2(z')\,\mathrm{d}z' \tag{5.13}$$

Equation (5.13) expresses the mutual impedance in terms of an integral of the product of the component of the incident electric field parallel to the antenna (E_{z21}) and the current distribution along the antenna when it is being used as a transmitter ($I_2(z')$). If the antennas are thick and are close together, the currents $I_2(z')$ and $I_1(z')$ (which produces E_{z21}) have to be determined by solving a pair of coupled Hallén-type integral equations. After these are determined, the integral (5.13) can be evaluated. Numerical methods are usually required for both steps. This problems will not be discussed here and the interested reader is referred to the paper by Tai (1948).

If the antennas are thin and the spacing is not too close, so that one can assume that the presence of a nearby open-circuited dipole does not distort the current distribution of the driven dipole, $I_2(z')$ and $I_1(z')$ are well approximated by sinusoidal functions of the form

$$I(z') = I_{\mathrm{m}} \sin\left[k(\tfrac{1}{2}L - |z'|)\right] \tag{5.14}$$

We shall study this case in detail, as it yields closed-form solutions. First we need to calculate the quantity E_{z21} since the expressions for the electromagnetic fields given previously for the finite-length dipole are not applicable because they are based on the far-field approximation.

5.3 ELECTROMAGNETIC FIELDS OF A FINITE-LENGTH DIPOLE WITH SINUSOIDAL CURRENT DISTRIBUTION

Consider a centre-fed dipole of length L. We wish to calculate the electromagnetic fields at a point P which is at an arbitrary distance from the antenna. Assuming the antenna to be thin, we have

$$I(z') = \begin{cases} I_{\mathrm{m}} \sin\left[k(\tfrac{1}{2}L - z')\right] & z' > 0 \\ I_{\mathrm{m}} \sin\left[k(\tfrac{1}{2}L + z')\right] & z' < 0 \end{cases} \tag{5.15}$$

For the calculation of the mutual impedance of two antennas placed parallel to each other, the z-component of the electric field at one antenna due to the other is required. It is therefore convenient to use a cylindrical coordinate system in which one of the coordinates is z. From the geometry shown in Figure 5.3, we have the following relationships:

$$R = \sqrt{[(z - z')^2 + y^2]} \tag{5.16}$$

$$R_1 = \sqrt{[(z - H)^2 + y^2]} \tag{5.17}$$

$$R_2 = \sqrt{[(z + H)^2 + y^2]} \tag{5.18}$$

$$r = \sqrt{(z^2 + y^2)} \tag{5.19}$$

where we have set $L/2 = H$ for convenience.

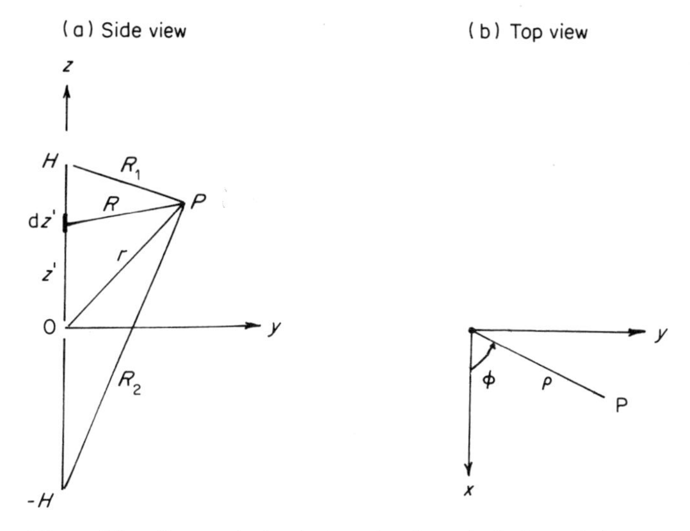

Figure 5.3 Geometrical relations in the calculation of electromagnetic fields at an arbitrary distance from a finite-length dipole

The expression for the vector potential at the point P is

$$
\begin{aligned}
A_z &= \frac{\mu I_m}{4\pi}\left(\int_0^H \frac{\sin[k(H-z')]\exp(-jkR)}{R}\,dz'\right. \\
&\qquad \left. + \int_{-H}^0 \frac{\sin[k(H+z')]\exp(-jkR)}{R}\,dz'\right) \\
&= \frac{\mu I_m}{8\pi j}\left(\exp(jkH)\int_0^H \frac{\exp[-jk(R+z')]}{R}\,dz'\right. \\
&\qquad - \exp(-jkH)\int_0^H \frac{\exp[-jk(R-z')]}{R}\,dz' \\
&\qquad + \exp(jkH)\int_{-H}^0 \frac{\exp[-jk(R-z')]}{R}\,dz' \\
&\qquad \left. - \exp(-jkH)\int_{-H}^0 \frac{\exp[-jk(R+z')]}{R}\,dz'\right)
\end{aligned}
\tag{5.20}
$$

In cylindrical coordinates, the magnetic field strength at P is given by

$$
\mathbf{H} = H_\phi \hat{\boldsymbol{\phi}}
\tag{5.21}
$$

where

$$
H_\phi = -\frac{1}{\mu}\frac{\partial A_z}{\partial \rho}
\tag{5.22}
$$

Since the problem has symmetry about the z-axis, the point P can be chosen to be in the y–z plane, without loss of generality. In this case,

$$H_\phi = -H_x = -\frac{1}{\mu}\frac{\partial A_z}{\partial y} \tag{5.23}$$

Substitution of (5.20) into (5.23) results in the expression

$$H_\phi = -\frac{I_\mathrm{m}}{8\pi\mathrm{j}}[\exp(\mathrm{j}kH)Q_1 - \exp(-\mathrm{j}kH)Q_2 + \exp(\mathrm{j}kH)Q_3 - \exp(-\mathrm{j}kH)Q_4] \tag{5.24}$$

where

$$Q_1 = \frac{\partial}{\partial y}\int_0^H \frac{\exp[-\mathrm{j}k(R+z')]}{R}\,\mathrm{d}z' \tag{5.25}$$

$$Q_2 = \frac{\partial}{\partial y}\int_0^H \frac{\exp[-\mathrm{j}k(R-z')]}{R}\,\mathrm{d}z' \tag{5.26}$$

$$Q_3 = \frac{\partial}{\partial y}\int_{-H}^0 \frac{\exp[-\mathrm{j}k(R-z')]}{R}\,\mathrm{d}z' \tag{5.27}$$

$$Q_4 = \frac{\partial}{\partial y}\int_{-H}^0 \frac{\exp[-\mathrm{j}k(R+z')]}{R}\,\mathrm{d}z' \tag{5.28}$$

Let us first consider the quantity Q_1. Making the change of variable

$$u = R + z' - z \tag{5.29}$$

we have, using (5.16),

$$\mathrm{d}u/\mathrm{d}z' = u/R \tag{5.30}$$

Hence

$$Q_1 = \exp(-\mathrm{j}kz)\frac{\partial}{\partial y}\int_{u_1}^{u_2} \frac{\exp(-\mathrm{j}ku)}{u}\,\mathrm{d}u \tag{5.31}$$

where

$$u_2 = R_1 + H - z \tag{5.32}$$

$$u_1 = r - z \tag{5.33}$$

Using the rule for the differentiation of a definite integral (e.g. Boas, 1966), (5.31) becomes

$$Q_1 = \exp(-\mathrm{j}kz)\left(\frac{\exp(-\mathrm{j}ku_2)}{u_2}\frac{\partial u_2}{\partial y} - \frac{\exp(-\mathrm{j}ku_1)}{u_1}\frac{\partial u_1}{\partial y}\right) \tag{5.34}$$

Note that

$$\frac{\partial u_2}{\partial y} = \frac{y}{R_1} \tag{5.35}$$

$$\frac{\partial u_1}{\partial y} = \frac{y}{r} \tag{5.36}$$

Hence

$$Q_1 = \exp(-jkz)\left(\frac{\exp[-jk(R_1 + H - z)]}{R_1 + H - z}\frac{y}{R_1} - \frac{\exp[-jk(r - z)]}{r - z}\frac{y}{r}\right)$$

$$= y\left(\frac{(R_1 - H + z)\exp[-jk(R_1 + H)]}{R_1[R_1^2 - (H - z)^2]} - \frac{(r + z)\exp(-jkr)}{r(r^2 - z^2)}\right) \tag{5.37}$$

Since

$$R_1^2 - (H - z)^2 = r^2 - z^2 = y^2 \tag{5.38}$$

(5.37) can be written as

$$Q_1 = \frac{1}{y}\left[\left(1 - \frac{H - z}{R_1}\right)\exp[-jk(R_1 + H)] - \left(1 + \frac{z}{r}\right)\exp(-jkr)\right] \tag{5.39}$$

The quantities Q_2, Q_3, and Q_4 can be evaluated in a similar manner. They are given by the following formulae:

$$Q_2 = \frac{1}{y}\left[\left(1 + \frac{H - z}{R_1}\right)\exp[-jk(R_1 - H)] - \left(1 - \frac{z}{r}\right)\exp(-jkr)\right] \tag{5.40}$$

$$Q_3 = \frac{1}{y}\left[\left(1 - \frac{H + z}{R_2}\right)\exp[-jk(R_1 - H)] - \left(1 - \frac{z}{r}\right)\exp(-jkr)\right] \tag{5.41}$$

$$Q_4 = \frac{1}{y}\left[\left(1 + \frac{H + z}{R_2}\right)\exp[-jk(R_2 - H)] - \left(1 + \frac{z}{r}\right)\exp(-jkr)\right] \tag{5.42}$$

Substituting (5.39)–(5.42) into (5.24) and simplifying, we obtain

$$H_\phi = -\frac{I_m}{4\pi jy}[\exp(-jkR_1) + \exp(-jkR_2) - 2\cos(kH)\exp(-jkr)] \tag{5.43}$$

The electric field is given by

$$\mathbf{E} = \frac{1}{j\omega\varepsilon}\nabla \times \mathbf{H} = E_z\hat{\mathbf{z}} + E_\rho\hat{\boldsymbol{\rho}} \tag{5.44}$$

where

$$E_z = \frac{1}{j\omega\varepsilon\rho}\frac{\partial}{\partial\rho}(\rho H_\phi) \tag{5.45}$$

$$E_\rho = -\frac{1}{j\omega\varepsilon\rho}\frac{\partial}{\partial z}(\rho H_\phi) \tag{5.46}$$

In the y–z plane, $\rho = y$ and use of (5.43) in (5.45) and (5.46) yields

$$E_z = -j30I_m\left(\frac{\exp(-jkR_1)}{R_1} + \frac{\exp(-jkR_2)}{R_2} - 2\cos(kH)\frac{\exp(-jkr)}{r}\right) \tag{5.47}$$

$$E_y = \frac{j30I_m}{y}\left(\frac{z-H}{R_1}\exp(-jkR_1) + \frac{z+H}{R_2}\exp(-jkR_2) - 2z\cos(kH)\frac{\exp(-jkr)}{r}\right) \tag{5.48}$$

Equations (5.43), (5.47), and (5.48) give the electromagnetic fields at any arbitrary distance from a finite-length dipole antenna with a sinusoidal current distribution. We obtain these simple, closed-form results because the integrand of (5.20) is a perfect differential, a situation which arises when the current distribution is an exponential function of the distance along the antenna or a linear combination of exponential functions such as a sinusoidal function.

For the particular case of a half-wave dipole, $H = \frac{1}{4}\lambda$ and we have

$$H_\phi = -\frac{I_m}{4\pi jy}[\exp(-jkR_1) + \exp(-jkR_2)] \tag{5.49}$$

$$E_z = -j30I_m\left(\frac{\exp(-jkR_1)}{R_1} + \frac{\exp(-jkR_2)}{R_2}\right) \tag{5.50}$$

$$E_y = j30\frac{I_m}{y}\left(\frac{z-\frac{1}{4}\lambda}{R_1}\exp(-jkR_1) + \frac{z+\frac{1}{4}\lambda}{R_2}\exp(-jkR_2)\right) \tag{5.51}$$

The expressions for the far fields given in section 3.9 can be obtained from (5.43), (5.47), and (5.48) by making the appropriate assumptions. This is left as an exercise for the reader (problem 7 in section 5.9).

5.4 MUTUAL IMPEDANCE OF PARALLEL ANTENNAS SIDE BY SIDE

Let us apply the results of sections 5.2 and 5.3 to find the mutual impedance of two parallel dipoles of equal length sitting side by side. The dipoles are assumed

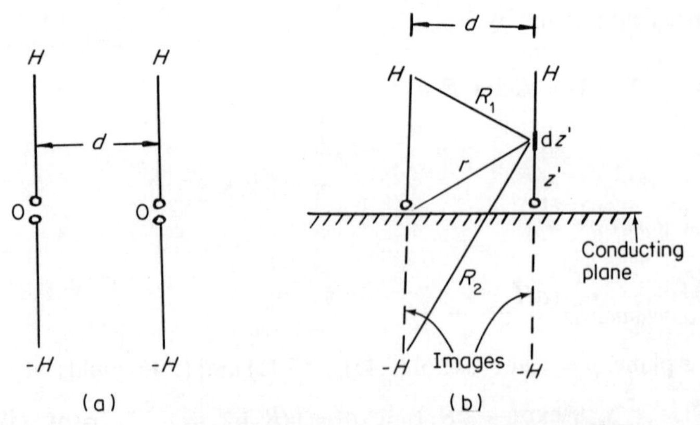

Figure 5.4 (a) Two side-by-side dipoles of equal length; (b) equivalent problem of two side-by-side monopoles

to be infinitesimally thin. Let the length be $L = 2H$ and the separation be d. An equivalent problem is that of two monopoles of length H placed perpendicular to a perfectly conducting plane, as shown in Figure 5.4. The mutual impedance between the two monopoles is equal to half the value of the mutual impedance of the two dipoles. This follows from the fact that the electric field at monopole no. 2 due to monopole no. 1 is the same as that produced by the dipole while the integration in (5.13) extends over the length H for the case of the monopole instead of $2H$ as in the case of the dipole.

Using (5.15) and (5.47) in (5.13), we obtain the following expression for the mutual impedance between the two monopoles:

$$Z_{21} = \frac{30\mathrm{j}I_{1\mathrm{m}}I_{2\mathrm{m}}}{I_1(0)I_2(0)} \times \int_0^H \left(\frac{\exp(-\mathrm{j}kR_1)}{R_1} + \frac{\exp(-\mathrm{j}kR_2)}{R_2} - \frac{2\cos(kH)\exp(-\mathrm{j}kr)}{r} \right) \times \sin[k(H - z')]\,\mathrm{d}z' \tag{5.52}$$

where

$$R_1 = \sqrt{[(z - H)^2 + d^2]} \tag{5.53}$$

$$R_2 = \sqrt{[(z + H)^2 + d^2]} \tag{5.54}$$

$$r = \sqrt{(z^2 + d^2)} \tag{5.55}$$

Since

$$\sin[k(H - z)] = \frac{1}{2\mathrm{j}}\{\exp[\mathrm{j}k(H - z)] - \exp[-\mathrm{j}k(H - z)]\} \tag{5.56}$$

(5.52) contains six integrals of similar form. We shall indicate the type of manipulations required for hand calculation. One integral has the form

$$\mathscr{I}_1 = \int_0^H \exp[\mathrm{j}k(H - z)] \frac{\exp(-\mathrm{j}kR_1)}{R_1}\,\mathrm{d}z' \tag{5.57}$$

Letting

$$u = k(R_1 - H + z') \tag{5.58}$$

we have

$$\frac{\mathrm{d}u}{\mathrm{d}z'} = k\left(\frac{\mathrm{d}R_1}{\mathrm{d}z'} + 1 \right) = \frac{u}{R_1} \tag{5.59}$$

Hence

$$\mathscr{I}_1 = \int_{u_1}^{u_0} \frac{\exp(-\mathrm{j}u)}{u}\,\mathrm{d}u \tag{5.60}$$

where

$$u_1 = k[\sqrt{(d^2 + H^2)} - H] \tag{5.61}$$

$$u_0 = kd \tag{5.62}$$

Since $\exp(-ju) = \cos u - j \sin u$, (5.60) can be written as

$$\begin{aligned}
\mathscr{I}_1 &= \int_{u_1}^{u_0} \frac{\cos u}{u} du - j \int_{u_1}^{u_0} \frac{\sin u}{u} du \\
&= \int_{u_1}^{\infty} \frac{\cos u}{u} du - \int_{u_0}^{\infty} \frac{\cos u}{u} du - j \int_{0}^{u_0} \frac{\sin u}{u} du + j \int_{0}^{u_1} \frac{\sin u}{u} du \\
&= \mathrm{Ci}(u_0) - \mathrm{Ci}(u_1) - j[\mathrm{Si}(u_0) - \mathrm{Si}(u_1)]
\end{aligned} \tag{5.63}$$

where Ci(x) and Si(x) are the cosine integral and the sine integral respectively. They are defined by

$$\mathrm{Ci}(x) = \int_{\infty}^{x} \frac{\cos u}{u} du \tag{5.64}$$

$$\mathrm{Si}(x) = \int_{0}^{x} \frac{\sin u}{u} du \tag{5.65}$$

These functions are tabulated in mathematical tables (e.g. Abramowitz and Stegun, 1964).

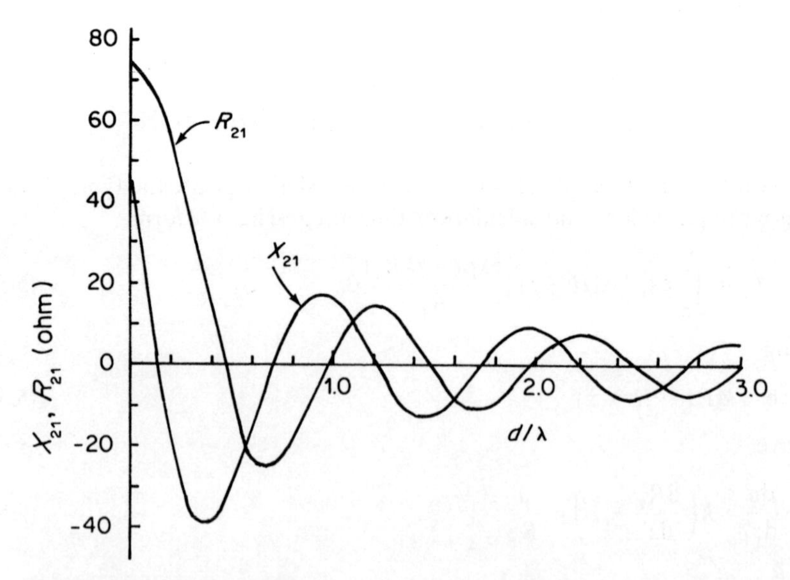

Figure 5.5 Mutual resistance (R_{21}) and mutual reactance (X_{21}) between two side-by-side half-wave dipoles of infinitesimal thickness as a function of spacing in units of wavelength (Source: Weeks (1968), *Antenna Engineering.* Reproduced by permission of McGraw-Hill)

The other five integrals in (5.52) can likewise be expressed as combinations of the cosine and sine integrals. The result is

$$Z_{21} = R_{21} + jX_{21} \tag{5.66}$$

where

$$R_{21} = \frac{30}{\sin^2(kH)}\{\sin(kH)\cos(kH)[\mathrm{Si}(u_2) - \mathrm{Si}(v_2) - 2\mathrm{Si}(v_1) + 2\mathrm{Si}(u_1)] - \tfrac{1}{2}\cos(2kH)[2\mathrm{Ci}(u_1) - 2\mathrm{Ci}(u_0) + 2\mathrm{Ci}(v_1) - \mathrm{Ci}(u_2) - \mathrm{Ci}(v_2)] - [\mathrm{Ci}(u_1) - 2\mathrm{Ci}(u_0) + \mathrm{Ci}(v_1)]\} \tag{5.67}$$

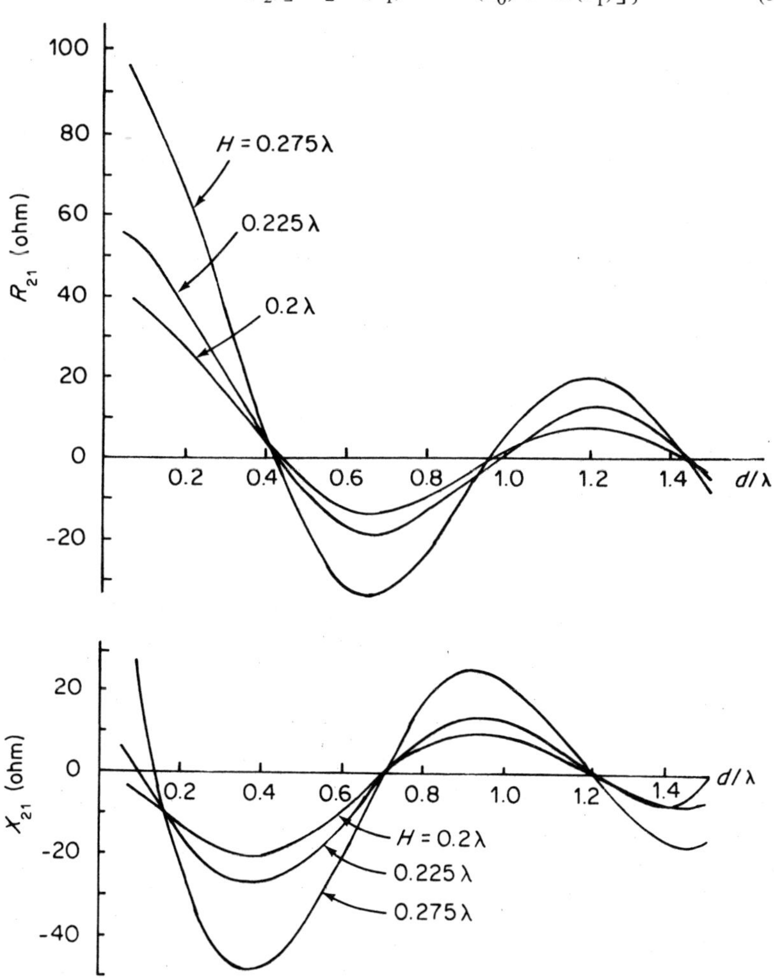

Figure 5.6 Mutual resistance (R_{21}) and mutual reactance (X_{21}) as a function of spacing for side-by-side dipoles of various half-lengths H. The dipoles are assumed to be infinitesimally thin (Source: Weeks (1968), *Antenna Engineering*. Reproduced by permission of McGraw-Hill)

$$X_{21} = \frac{-30}{\sin^2(kH)}\{\sin(kH)\cos(kH)[2\text{Ci}(v_1) - 2\text{Ci}(u_1) + \text{Ci}(v_2) - \text{Ci}(u_2)]$$
$$- \tfrac{1}{2}\cos(2kH)[2\text{Si}(u_1) - 2\text{Si}(u_0) + 2\text{Si}(v_1) - 2\text{Si}(u_2)$$
$$- \text{Si}(v_2)] - [\text{Si}(u_1) - 2\text{Si}(u_0) + \text{Si}(v_1)]\} \quad (5.68)$$

The arguments u_0 and u_1 are given by (5.62) and (5.61) respectively while u_2, v_1, and v_2 are given by

$$u_2 = k\{\sqrt{[d^2 + (2H)^2]} + 2H\} \quad (5.69)$$

$$v_1 = k[\sqrt{(d^2 + H^2)} + H] \quad (5.70)$$

$$v_2 = k\{\sqrt{[d^2 + (2H)^2]} - 2H\} \quad (5.71)$$

Some specific values are plotted in Figures 5.5 and 5.6.

5.5 MUTUAL IMPEDANCE OF OTHER CONFIGURATIONS

5.5.1 Collinear Dipoles

The configuration of collinear dipoles frequently arises in practice. For the case of two antennas shown in Figure 5.7, the mutual impedance is, on using (5.13), given by

$$Z_{21} = -\frac{1}{I_1(0)I_2(0)}\left(\int_{h'}^{H_2+h'} E_{z21}I_2(z')\,\mathrm{d}z' + \int_{H_2+h'}^{2H_2+h'} E_{z21}I_2(z')\,\mathrm{d}z'\right) \quad (5.72)$$

where

$$h' = H_1 + s \quad (5.73)$$

$$I_2(z') = \begin{cases} I_{2\text{m}}\sin[k(z' - h')] & h' < z' < H_2 + h' \\ I_{2\text{m}}\sin[k(2H_2 + h' - z')] & h' + H_2 < z' < 2H_2 + h' \end{cases} \quad (5.74)$$

$$E_{z21} = -\text{j}30I_{1\text{m}}\left(\frac{\exp(-\text{j}kR_1)}{R_1} + \frac{\exp(-\text{j}kR_2)}{R_2} - 2\cos(kH_1)\frac{\exp(-\text{j}kr)}{r}\right) \quad (5.75)$$

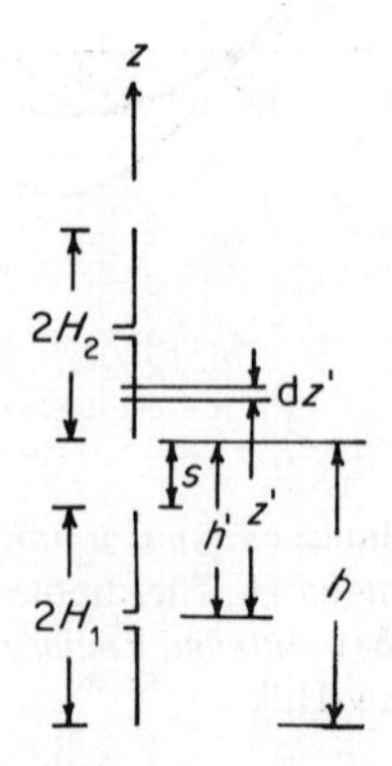

Figure 5.7 Geometry of two collinear dipoles

In (5.75),

$$R_1 = z' - H_1 \tag{5.76}$$

$$R_2 = z' + 2H_1 \tag{5.77}$$

$$r = z' \tag{5.78}$$

For the case when the lengths of both antennas are equal to odd multiples of half-wavelength, the cosine term in (5.75) is equal to zero. The integrals in (5.72) can be evaluated in a manner similar to that in section 5.4. The result is (Carter, 1932)

$$\begin{aligned} R_{21} = &-\frac{15\cos(kh)}{\sin^2(kH)}\Bigg[-2\mathrm{Ci}(2kh) + \mathrm{Ci}(2k(h-2H)) \\ &+ \mathrm{Ci}(2k(h+2H)) - \ln\left(\frac{h^2-4H^2}{4H^2}\right)\Bigg] \\ &+ \frac{15\sin(kh)}{\sin^2(kH)}[2\mathrm{Si}(2kh) - \mathrm{Si}(2k(h-2H)) - \mathrm{Si}(2k(h+2H))] \end{aligned} \tag{5.79}$$

$$\begin{aligned} X_{21} = &-\frac{15\cos(kh)}{\sin^2(kH)}[2\mathrm{Si}(2kh) - \mathrm{Si}(2k(h-2H)) - \mathrm{Si}(2k(h+2H))] \\ &+ \frac{15\sin(kh)}{\sin^2(kH)}\Bigg[2\mathrm{Ci}(2kh) - \mathrm{Ci}(2k(h-2H)) \\ &- \mathrm{Ci}(2k(h+2H)) - \ln\left(\frac{h^2-4H^2}{4H^2}\right)\Bigg] \end{aligned} \tag{5.80}$$

where

$$h = 2H + s \tag{5.81}$$

For half-wave dipoles, the values of R_{21} and X_{21} as functions of the spacing s are shown in Figure 5.8. Compared to Figure 5.5, it is seen that the mutual coupling for the collinear case is smaller than the side-by-side case.

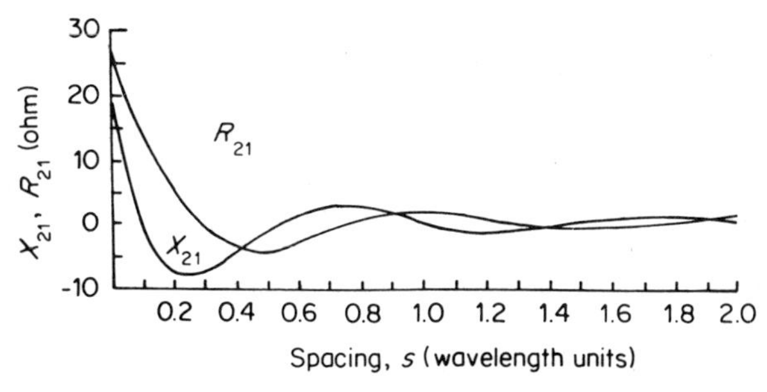

Figure 5.8 Mutual resistance (R_{21}) and mutual reactance (X_{21}) between two collinear half-wave dipoles of infinitesimal thickness as a function of spacing (Source: Weeks (1968), *Antenna Engineering*. Reproduced by permission of McGraw-Hill)

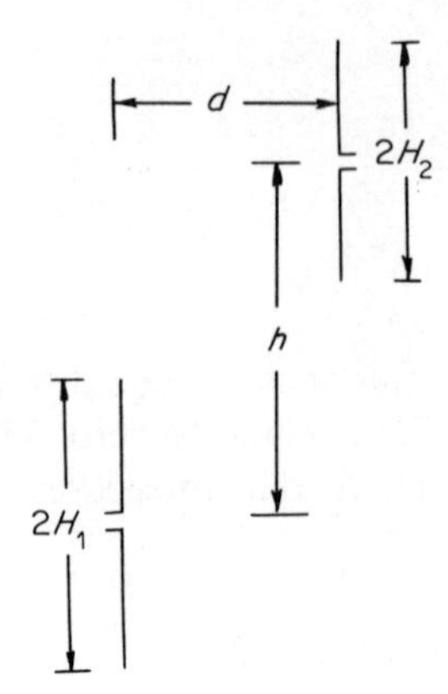

Figure 5.9 Geometry of two parallel dipoles of arbitrary lengths in echelon

5.5.2 Other Orientations

There are an infinite number of ways in which two dipoles can be put adjacent to each other, resulting in an infinite number of values for the mutual impedances. In the most general case, the antennas may be of unequal lengths, their axes non-parallel and lie in different planes. Regardless of their orientations, however, the mutual impedance can be obtained by evaluating the expression given by (5.13). Although the integral can be reduced to tabular functions only for very few special cases, it can always be computed numerically. Indeed, the numerical approach may sometimes be more convenient even if the answer is expressible in closed form, since the closed-form solution inevitably consists of a large number of terms involving the Ci and Si functions.

Some particular cases where results can be found in the literature should be mentioned here. The configuration of parallel antennas of unequal lengths was treated by Cox (1947). That of two antennas in echelon where each antenna is an odd number of half-wavelengths long was studied by Carter (1932). Finally, results for the more general case of two parallel antennas of arbitrary length in echelon (Figure 5.9) were given by King (1957).

5.6 SELF-IMPEDANCE OF ANTENNAS

In section 3.11, the self-impedance of an antenna was introduced. An expression for it was given by (3.104), which was based on the solution of Hallén's integral equation for the current distribution. In this section, we present another method, known as the induced e.m.f. method, of calculating the self-impedance. An excellent presentation, at a fairly advanced level, of the various methods of calculating antenna self-impedances is given in Elliott (1981).

Let $V(0)$ and $I(0)$ be the voltage and current at the terminals of an antenna and $I(z')$ the current distribution along the antenna. By definition, the self-impedance of the antenna is

$$Z_a = V(0)/I(0) \tag{5.82}$$

Let $E_z(z')$ be the parallel component of the electric field on the surface of the antenna caused by its own current. The induced e.m.f. at an element of length $\mathrm{d}z'$ a distance z' from the terminals is then $-E_z(z')\,\mathrm{d}z'$.

Imagine now another situation, in which the same field distribution $E_z(z')$ is maintained on the surface of the antenna, but the source creating this field distribution is located outside the antenna. In this case, if we measure the open-circuit voltage at the terminals with an ideal voltmeter, it will read $V(0)$. On the other hand, if the short-circuit current I_{sc} is measured with an ideal ammeter, the argument of section 5.2 shows that

$$I_{sc} = -\frac{1}{V(0)}\int E_z(z')I(z')\,\mathrm{d}z' \tag{5.83}$$

From (4.5), I_{sc} and $V(0)$ are related by

$$V(0) = I_{sc}Z_a \tag{5.84}$$

where Z_a is the antenna impedance. It follows from (5.82) and (5.84) that $I(0) = I_{sc}$. Equation (5.83) can therefore be written

$$V(0) = -\frac{1}{I(0)}\int E_z(z')I(z')\,\mathrm{d}z' \tag{5.85}$$

Hence

$$Z_a = -\frac{1}{I^2(0)}\int E_z(z')I(z')\,\mathrm{d}z' \tag{5.86}$$

Equation (5.86) can also be obtained by the following reasoning. By definition, the mutual impedance Z_{21} between two antennas is a measure of the open-circuit voltage at the terminals of antenna no. 2 due to a current $I_1(0)$ impressed at the terminals of antenna no. 1. If we consider two identical antennas and allow the distance d between them to approach zero, the voltage V_{21} becomes the voltage at the terminals of antenna no. 1 due to the impressed current $I_1(0)$. Hence the self-impedance of an antenna of half-length H should be equal to the mutual impedance Z_{21} between two antennas for which $H_1 = H_2 = H$ and $d = 0$. This again leads to (5.86).

In (5.86), $E_z(z')$ is the parallel component of the electric field on the surface of the antenna caused by its own current. Figure 5.10 shows a section of a cylindrical antenna of radius a. We shall again assume a thin dipole, i.e. $1/[2\ln(L/a)] \gtrsim 60$, so that the current along its length can be taken to be sinusoidal. The current is uniformly distributed around the circumference of the cylinder, the major portion of it flowing in a very small thickness of conductor adjacent to the outer surface. To simplify the computation, it will be assumed that E_z produced by the distributed current around the cylinder is essentially the same as that which can be calculated by considering the current to be concentrated on the axis of the cylinder. It then follows that the self-impedance of a monopole antenna of height H and radius a, having an assumed sinusoidal

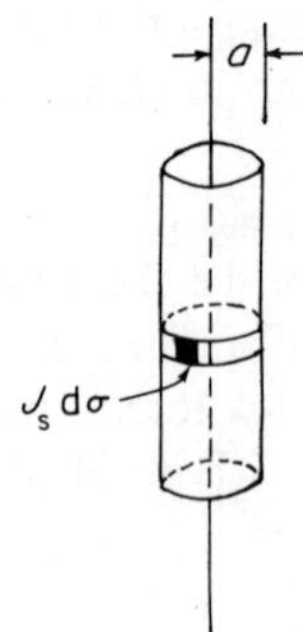

Figure 5.10 A section of a cylindrical antenna of radius a. The current flows within a thin spherical shell adjacent to the outer surface

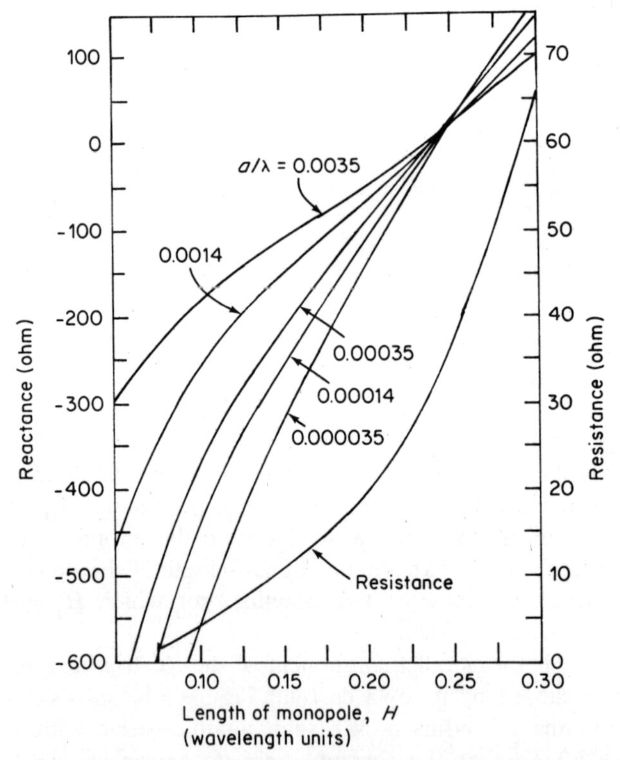

Figure 5.11 The resistance and reactance of monopole antennas as functions of length for several values of the conductor radius a. Note that there is only one curve for the resistance, showing that it is insensitive to the value of a within the framework of the assumption of sinusoidal current distribution (Source: Edward C. Jordan/Keith G. Balmain, *Electromagnetic Waves and Radiating Systems*, 2nd ed., © 1968, p. 548. Reprinted by permission of Prentice-Hill, Inc., Englewood Cliffs, N. J.)

current distribution, is equal to the mutual impedance between two filamentary antennas ($a \to 0$) of the same height and with a spacing equal to a. Thus

$$Z_a = R_a + jX_a \tag{5.87}$$

where R_a and X_a are given by (5.67) and (5.68) respectively except that the arguments of the Ci and Si functions are now given by

$$u_0 = ka \tag{5.88}$$

$$u_1 = k[\sqrt{(H^2 + a^2)} - H] \tag{5.89}$$

$$u_2 = k\{\sqrt{[(2H)^2 + a^2]} + 2H\} \tag{5.90}$$

$$v_1 = k\{\sqrt{[(H)^2 + a^2]} + H\} \tag{5.91}$$

$$v_2 = k\{\sqrt{[(2H)^2 + a^2]} - 2H\} \tag{5.92}$$

Figure 5.11 shows the resistance and reactance values computed by this method for monopole antennas of various thicknesses. The corresponding value for a dipole of length $L = 2H$ is just double that of the monopole antenna of length H.

In Figure 5.11, the single curve for the resistance means that the real part of (5.87) is independent of a/λ for the range of lengths shown. This is a consequence of the assumption of sinusoidal current distribution, and will not be true for a more complicated distribution. The reactance, on the other hand, is sensitive to a/λ. Note that the slopes of the reactance curves are smaller for larger values of a/λ, implying that fatter dipoles are less sensitive to frequency than thinner ones. This has already been alluded to in section 3.11.

5.7 CONCLUDING REMARKS

In this chapter, we have presented a simple introduction to the problem of calculating the self- and mutual impedances of antennas. After the basic concepts are introduced, the detailed calculation is carried out for simple antenna configurations involving sinusoidal current distributions. It should be mentioned that modern computing methods have been developed which are capable of finding the coupling between very complicated structures. It is, however, beyond the scope of this introductory text to describe these methods in detail. The interested reader is referred to Harrington (1968) and Thiele (1973).

5.8 WORKED EXAMPLES

Example 1

Figure 5.12 shows a centre-fed vertical dipole of length $2H$ and a horizontal Hertzian dipole of length dl separated by a distance $d = H/2$. The antennas are in the y–z plane, with the centre of the vertical dipole at the origin and the centre of the Hertzian dipole at $(0, d, H/2)$. Find the mutual impedance between the two antennas.

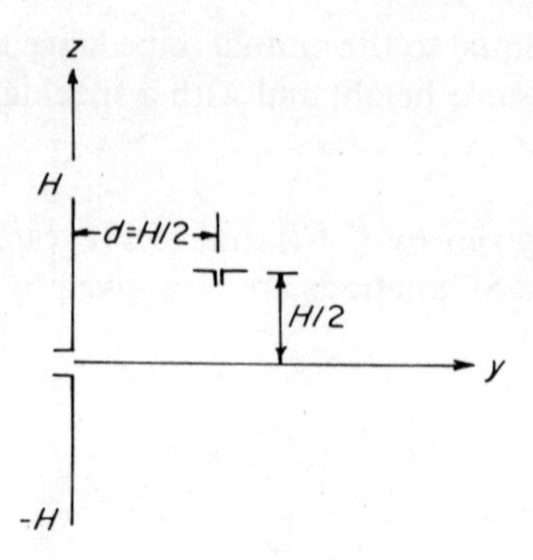

Figure 5.12 Geometry for worked example 1

Solution

Denoting the vertical dipole as no. 1 and the horizontal dipole as no. 2, the mutual impedance is given by

$$Z_{12} = -\frac{1}{I_1(0)I_2}\int_{-dl/2}^{dl/2} E_{\parallel} I_2 \, dy' = -\frac{E_{\parallel}\, dl}{I_1(0)}$$

where $E_{\parallel}$ is the parallel component of the electric field at no. 2 due to no. 1:

$$E_{\parallel} = E_y\bigg|_{\substack{z=H/2\\ y=H/2}} = \mathrm{j}30 I_{1\mathrm{m}}\left(\frac{z-H}{y}\frac{\exp(-\mathrm{j}kR_1)}{R_1} + \frac{z+H}{y}\frac{\exp(-\mathrm{j}kR_2)}{R_2} - \frac{2z\cos(kH)}{y}\frac{\exp(-\mathrm{j}kr)}{r}\right)\bigg|_{\substack{z=H/2\\ y=H/2}}$$

For $z = H/2$, $y = H/2$, we have, on using (5.53)–(5.55),

$$R_1 = H/\sqrt{2} \qquad R_2 = \sqrt{10}H/2 \qquad r = H/\sqrt{2}$$

Hence

$$E_{\parallel} = \mathrm{j}30 I_{1\mathrm{m}}\left[-\frac{\sqrt{2}}{H}\exp\left(\frac{-\mathrm{j}kH}{\sqrt{2}}\right) - \frac{2\sqrt{2}\cos(kH)}{H}\exp\left(\frac{-\mathrm{j}kH}{\sqrt{2}}\right) + \frac{6}{\sqrt{10}H}\exp\left(\frac{-\mathrm{j}k\sqrt{10}H}{2}\right)\right]$$

Since $I_{1\mathrm{m}} = I_1(0)/\sin(kH)$, we have

$$Z_{12} = -\frac{dl\,\mathrm{j}30}{H\sin(kH)}\left[-2\exp\left(\frac{-\mathrm{j}kH}{\sqrt{2}}\right) - 2\sqrt{2}\cos(kH)\exp\left(\frac{-\mathrm{j}kH}{\sqrt{2}}\right) + \frac{6}{\sqrt{10}}\exp\left(\frac{-\mathrm{j}k\sqrt{10}H}{2}\right)\right]$$

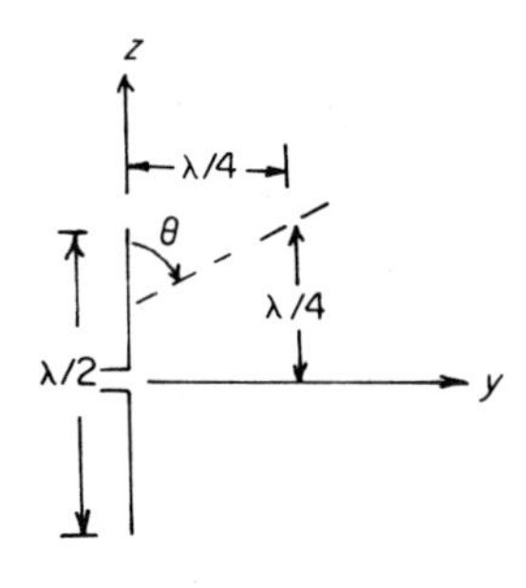

Figure 5.13 Geometry for worked example 2

Example 2

Figure 5.13 shows a centre-fed vertical $\lambda/2$ dipole and a Hertizan dipole of length $\mathrm{d}l$ inclined at an angle θ with respect to the z-axis. The antennas are in the y–z plane, with the centre of the $\lambda/2$ dipole at the origin and the centre of the Hertzian dipole at $(0, \lambda/4, \lambda/4)$.

(a) When the $\lambda/2$ dipole is transmitting with a sinusoidal current distribution $I_1 = I_{1\mathrm{m}} \cos(kz)$, what is the open-circuit voltage V_{oc} at the terminals of the Hertzian dipole?
(b) Determine the value of θ such that the mutual impedance is a pure resistance.
(c) Find the value of the resistance in (b) for $\mathrm{d}l/\lambda = \sqrt{5}/240$.

Solution

(a) At the position of the Hertzian dipole (no. 2), the parallel component of the electric field due to the half-wave dipole (no. 1) is

$$E_{\parallel} = \mathbf{E} \cdot \hat{\mathbf{z}}' = E_y \sin\theta + E_z \cos\theta$$

where E_y and E_z are evaluated at $y = \lambda/4$, $z = \lambda/4$. Using (5.47), (5.48), and (5.33)–(5.55), we have

$$E_z = -\mathrm{j}30 I_{1\mathrm{m}} \left(\frac{\exp(-\mathrm{j}kR_1)}{R_1} + \frac{\exp(-\mathrm{j}kR_2)}{R_2} \right)$$

$$E_y = \mathrm{j}60 I_{\mathrm{m}} \exp(-\mathrm{j}kR_2)/R_2$$

where $R_1 = \lambda/4$, $R_2 = \sqrt{5}\lambda/4$, $r = \sqrt{2}\lambda/4$.

The open-circuit voltage at the terminals of the Hertzian dipole is

$$V_{\mathrm{oc}} = -\frac{1}{I_2(0)} \int_{-\mathrm{d}l/2}^{\mathrm{d}l/2} E_{\parallel} I_2 \, \mathrm{d}z' = -E_{\parallel} \, \mathrm{d}l$$

$$= \mathrm{j}30 I_{1\mathrm{m}} \mathrm{d}l \left(\cos\theta \frac{\exp(-\mathrm{j}kR_1)}{R_1} + (\cos\theta - 2\sin\theta) \frac{\exp(-\mathrm{j}kR_2)}{R_2} \right)$$

(b) As $I_1(0) = I_{1\mathrm{m}}$ for a half-wave dipole

$$Z_{12} = V_{\mathrm{oc}}/I_1(0) = V_{\mathrm{oc}}/I_{1\mathrm{m}}$$

$$= 30\mathrm{d}l\left(\mathrm{j}\cos\theta\frac{\exp(-\mathrm{j}kR_1)}{R_1} + \mathrm{j}(\cos\theta - 2\sin\theta)\frac{\exp(-\mathrm{j}kR_2)}{R_2}\right)$$

$$= 30\,\mathrm{d}l\left(\frac{[\mathrm{j}\cos(kR_1) + \sin(kR_1)]\cos\theta}{R_1}\right.$$

$$\left. + \frac{[\mathrm{j}\cos(kR_2) + \sin(kR_1)](\cos\theta - 2\sin\theta)}{R_2}\right)$$

For Z_{12} to be a pure resistance, we must have

$$\frac{\cos\theta\cos(kR_1)}{R_1} + \frac{\cos(kR_2)}{R_2}(\cos\theta - 2\sin\theta) = 0.$$

The first term is equal to zero since $R_1 = \lambda/4$. Hence Z_{12} will be purely resistive if $\tan\theta = \frac{1}{2}$ or $\theta = 26°34'$.

(c) For $\tan\theta = \frac{1}{2}$ and $\mathrm{d}l/\lambda = \sqrt{5}/240$.

$$Z_{21} = 30\,\mathrm{d}l[\cos\theta\sin(kR_1)/R_1] = 30\,\mathrm{d}l\left(\frac{1}{R_1}\frac{2}{\sqrt{5}}\right) = 30\frac{\mathrm{d}l}{\lambda}\left(\frac{8}{\sqrt{5}}\right) = 1\text{ ohm}$$

5.9 PROBLEMS

1. Figure 5.14 shows two Hertzian dipoles along the z-axis, separated by a distance d and oriented in the same direction. Obtain an expression for the mutual impedance in terms of $\mathrm{d}l/\lambda$ for the case $d = \lambda/2$.

2. Two Hertzian dipoles, each of length $\mathrm{d}l$, are arranged in the manner shown in Figure 5.15. The dipoles lie in the same plane.
   (a) Obtain an expression for the mutual impedance of the two antennas.
   (b) For the case $\theta_1 = 90°$, obtain an equation that determines the distance d such that the mutual impedance is (i) purely real and (ii) purely imaginary.
   (c) If $kd \ll 1$ and $\theta_1 = 90°$, is the mutual impedance dominantly real or dominantly imaginary?

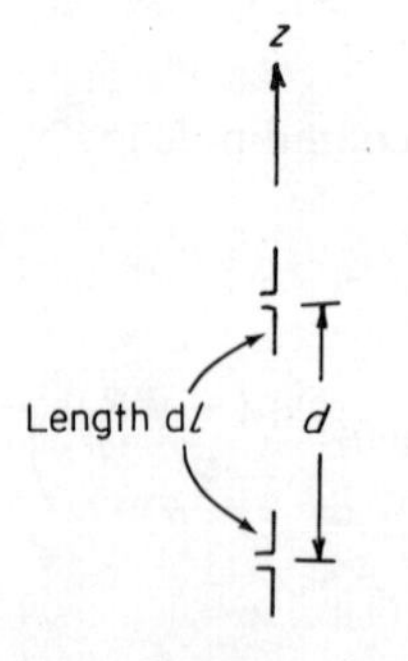

Figure 5.14 Geometry for problem 1

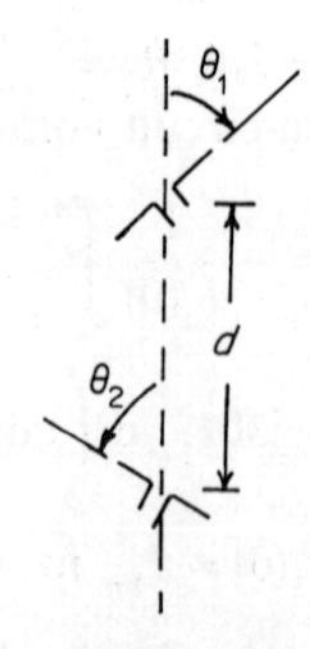

Figure 5.15 Geometry for problem 2

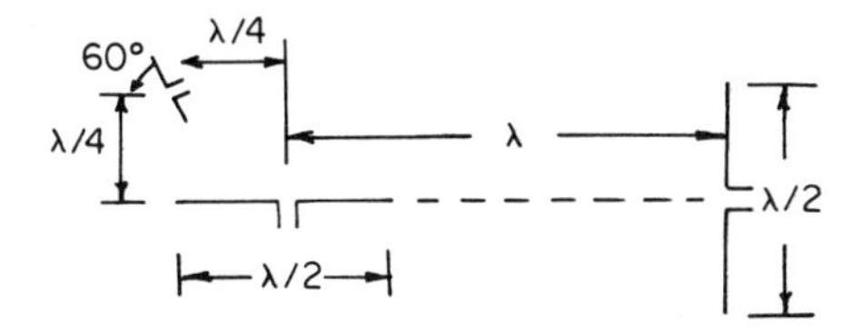

Figure 5.16 Geometry for problem 3

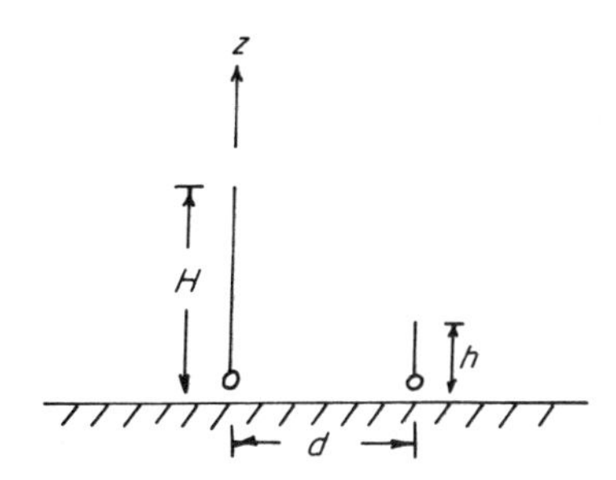

Fig. 5.17 Geometry for problem 4

3. Two half-wave dipoles and a Hertzian dipole of length $\mathrm{d}l = \lambda/1000$ are arranged in the manner shown in Figure 5.16. The antennas are centre-fed and lie in the same plane. The currents at the feed points are the same in both magnitude and phase. Find the driving-point impedance at the terminals of the horizontal dipole.

4. Find the mutual impedance between a vertical monopole of height H over a perfect ground and a very short vertical monopole over ground (Figure 5.17). Determine a set of values for d and H (in units of λ) such that the mutual impedance is a pure resistance. Assume a triangular current distribution on the very short monopole.

5. (a) A half-wave dipole and two very short dipoles of length $\mathrm{d}l = \lambda/100$ are arranged in the manner shown in Figure 5.18(a). The antennas are centre-fed and lie in the same plane. The half-wave dipole is oriented horizontally while the very short dipoles are oriented vertically. The currents at the feed points are the same both in magnitude and phase. Find the driving-point impedance at the terminals of the horizontal half-wave dipole.
   (b) A fourth antenna in the form of a vertical half-wave dipole is added as shown in Figure 5.18(b). Its centre is at the same height as the centre of the horizontal half-wave dipole and it carries the same feed-point current as the other antennas. Find the driving-point impedance at the terminals of the horizontal dipole.

6. A z-directed centre-fed Hertzian dipole of length $\mathrm{d}l$ situated at the origin carries a current $\mathrm{I}\sin(\omega t)$. When a second z-directed Hertzian dipole of

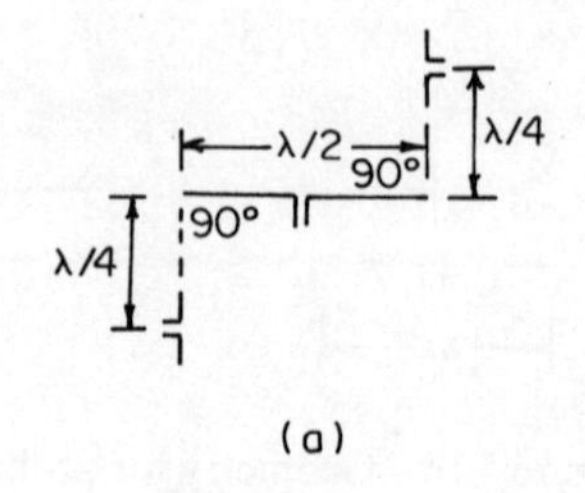

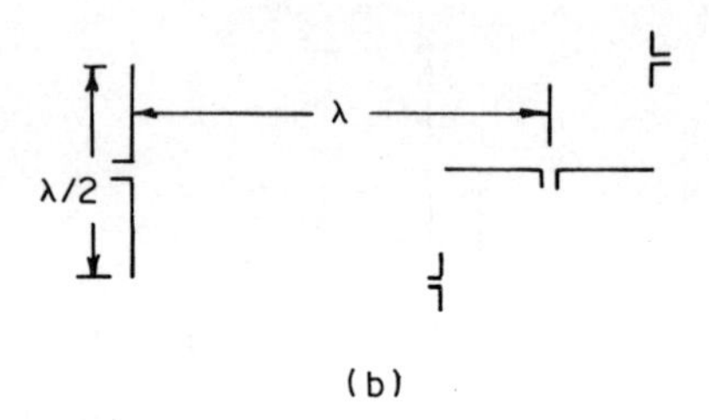

Figure 5.18 Geometry for problem 5

length dl is placed with its centre at $(0, 0, d)$, the r.m.s. voltage at the terminals, measured with an ideal voltmeter, is V_1 volts. When this second Hertzian dipole is moved to a new position, centred at $(0, d, 0)$, it is found that, if the axis remains in the z-direction, the reading of the voltmeter is unchanged. Find the distance d in units of wavelength.

7. Show that, under the far-field approximation, (5.43) reduces to (3.87) of Chapter 3.
8. Verify (5.40).
9. Verify (5.67).

BIBLIOGRAPHY

Elliott, R. S. (1981). *Antenna Theory and Design*, Prentice-Hall, Englewood Cliffs, NJ, Chapter 7.

Jordan, E. C. and Balmain, K. G. (1968). *Electromagnetic Waves and Radiating Systems*, Prentice-Hall, Englewood Cliffs, NJ, Chapter 14.

CHAPTER 6
Some Practical Considerations of the Dipole Antenna

The dipole antenna is the simplest and one of the most extensively used antennas. The basic theories concerning this antenna were given in Chapters 3–5. The present chapter is devoted to a discussion of some of the practical aspects of this antenna. The concept of resonant length is introduced in section 6.1. The related topics of bandwidth and quality factor are discussed in section 6.2. This is followed by the methods of feeding the dipole in section 6.3. Finally, in section 6.4, a discussion on ground effects is given. A good reference for the materials of this chapter is the *ARRL Antenna Book*.

6.1 LENGTH OF A RESONANT DIPOLE ANTENNA

The resonant length of a dipole antenna at a given frequency is defined as the length at which the impedance of the antenna is purely resistive. This is analogous to a tuned circuit; at the resonant frequency the input impedance of the circuit is a resistance. Unlike the case of an open-circuited transmission line, the resonant lengths of a dipole antenna do not occur at exact multiples of a half-wavelength. For example, a dipole of exactly a half-wavelength long has an input impedance of $Z_{in} = R_{in} + jX_{in} = 73 + j42.5$ ohm if it is infinitesimally thin. Reference to Figure 5.11 shows that, by reducing its length slightly, the half-wave dipole can be made to resonate, with $X_{in} = 0$. The amount of shortening required to achieve this depends on the diameter of the antenna. The full curve of Figure 6.1 shows the factor by which the length of a half-wave dipole in free space should be multiplied to obtain the physical length of a resonant half-wave dipole as a function of the ratio of half-wavelength to conductor diameter. Note that, as the length is reduced to obtain resonance, the input resistance also decreases. This is shown in the broken curve of Figure 6.1.

6.2 BANDWIDTH AND ANTENNA *Q*

The *IEEE Standard Dictionary of Electrical and Electronics Terms* defined the bandwidth of an antenna as 'the range of frequencies within which the performance of the antenna, with respect to some characteristics, conforms to a specified

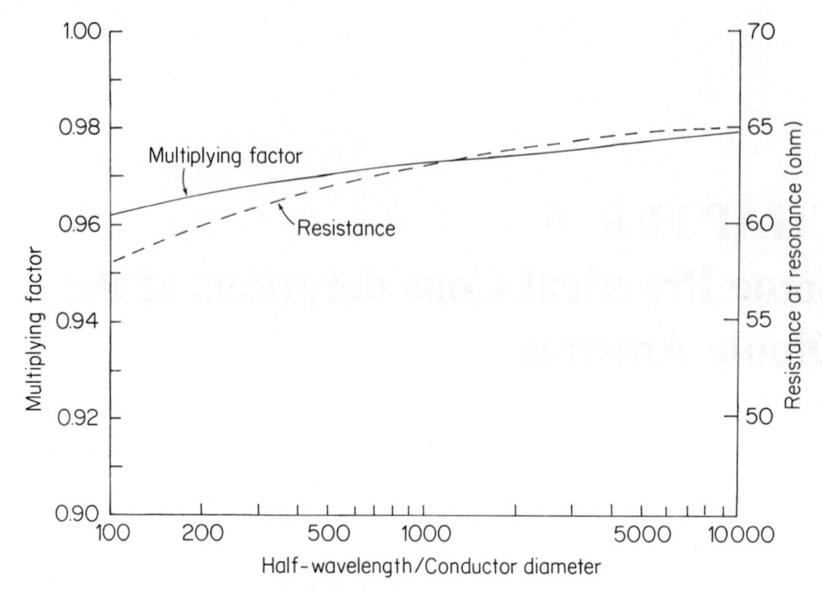

Figure 6.1 The full curve shows the multiplying factor by which the length of a half-wave in free space should be multiplied to obtain the physical length of a resonant half-wave dipole having the ratio of half-wavelength to conductor diameter shown on the horizontal axis. The broken curve shows the input resistance at resonance (Source: *ARRL Antenna Book* (1974). Reproduced by permission of the American Radio Relay League Inc.)

standard'. The characteristics referred to can be pattern or impedance. For the latter, which is usually more sensitive to frequency, a useful standard is that the voltage standing wave ratio (VSWR) be less than 2. If the antenna impedance is Z_a and the characteristic impedance of the transmission line feeding the antenna is Z_0, then the voltage reflection coefficient ρ is

$$\rho = (Z_a - Z_0)/(Z_a + Z_0) \tag{6.1}$$

The VSWR is related to ρ by

$$\mathrm{VSWR} = S = (1 + |\rho|)/(1 - |\rho|) \tag{6.2}$$

From transmission-line theory, the percentage of reflected power is equal to $|\rho|^2 \times 100$ and the percentage of transmitted power is equal to $(1 - |\rho|^2) \times 100$. A VSWR of 2 therefore corresponds to a reflected power of 11.1%.

Consider a dipole antenna of resonant length connected to a transmission line that has a characteristic impedance equal to the input impedance of the antenna at the resonant frequency. The resistance change in the region around resonance is relatively small compared to the change in reactance. Consequently, the principal cause of the change in VSWR is the reactive component of the antenna impedance when the frequency is varied. An antenna that has a relatively slow rate of reactance change therefore will have a wider bandwidth. Reference to

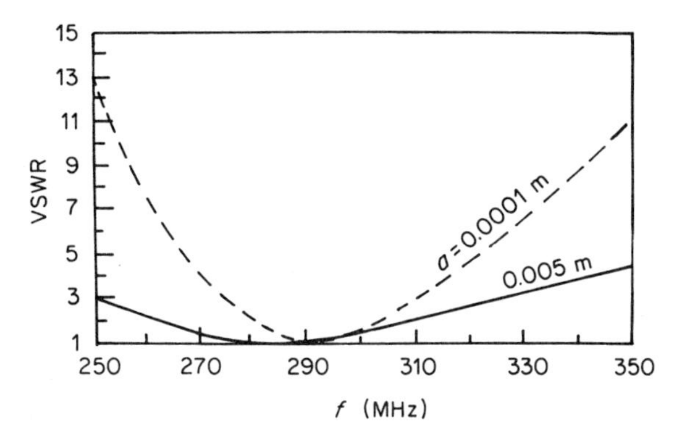

Figure 6.2 The voltage standing wave ratio (VSWR) as a function of frequency on a 72 ohm transmission line feeding two dipoles of 0.5 m length having two values of conductor radius a (Source: Stutzman and Thiele (1981), *Antenna Theory and Design*. Reproduced by permission of John Wiley and Sons Inc.)

Figure 5.11 shows that an antenna with a larger diameter-to-length ratio will have a wider bandwidth compared to an antenna with a smaller diameter-to-length ratio.

As an example, consider a dipole 0.5 m long, fed at the centre by a transmission line of characteristic impedance $Z_0 = 72$ ohm. The calculated VSWRs as a function of frequency are shown in Figure 6.2 for two values of the antenna diameter. For the dipole with radius $a = 0.005$ m, the resonant frequency is 285 MHz and the bandwidth is about 16%. For the thinner dipole with radius $a = 0.0001$ m, the resonant frequency is 294 MHz and the bandwidth is about 8%.

As in a resonant circuit, the antenna impedance picks up a reactive component as the frequency deviates from the resonant frequency. The quality factor Q of an antenna is defined in a manner similar to the Q of a series-resonant circuit having lumped circuit elements. It can be found by calculating or measuring its input resistance R and reactance X at some frequency close to the resonant frequency. For frequency changes of less than 5% from the exact resonant frequency, Q is given with sufficient accuracy by the formula

$$Q = \frac{X}{2nR} \tag{6.3}$$

where n is the percentage difference, expressed as a decimal, between the resonant frequency and the frequency at which X and R are measured or computed. As in the case of a resonant circuit, Q is a measure of the antenna's selectivity. For a half-wave dipole, the approximate value of Q varies from about 14 for $L/a = 2500$ to about 8 for $L/a = 50\,000$.

6.3 METHODS OF FEEDING THE DIPOLE ANTENNA

6.3.1 General Principles

In feeding a dipole antenna, two factors should be considered. First, it is desirable that the antenna should present a matched load to the transmission line. The advantages of an impedance match are : (1) there is no power reflected; (2) in the absence of standing waves, the loss due to dissipation is minimized; (3) the input impedance of the line is independent of frequency and its length; and (4) the problem of dielectric breakdown and excessive heating associated with high VSWR is avoided. The requirement on the VSWR that can be tolerated is, of course, dependent on the particular application. As a general rule, the system can be considered matched if the VSWR does not exceed 1.5.

The other consideration is that when the antenna is connected to the transmission line, the resultant currents on the line should be balanced; i.e. each conductor should carry an equal and opposite current. If the spacing between conductors is much smaller than a wavelength, the radiation from a line carrying a balanced current is negligible. However, if the connection results in an unbalanced current on the line, line radiation will be significant. This radiation is undesirable since it upsets the radiation characteristics of the antenna.

6.3.2 Matching Methods

Direct Feeding and Quarter-Wave Transformer

An antenna to be connected to a transmission line should be set to its resonant length so that its impedance is resistive. In practice, one usually starts out by constructing the antenna longer than required. It is then connected to a signal generator via a transmission line and its impedance at the designed frequency is measured. This can be done either directly with an admittance meter or indirectly by measuring the VSWR on the line and the shift of the position of voltage minimum relative to the case when the line is shorted. For an antenna longer than the resonant length, its impedance Z_a will contain a reactive component. The ends of the antenna are then cut until a nearly resistive value for Z_a is obtained.

Sometimes, it is possible to connect the antenna directly to a transmission line. For example, a resonant half-wave dipole has an impedance of approximately 70 ohm. If this is connected to a 75 ohm parallel-wire line, no matching device is necessary. If a line with a different characteristics impedance is to be used, for example a 300 ohm parallel-wire line, an impedance match can be obtained by using a quarter-wave transformer. The characteristic impedance of the quarter-wave transformer required is the geometric mean of the antenna impedance Z_a and the transmission-line characteristic impedance Z_0 :

$$Z_{\lambda/4} = \sqrt{(Z_a Z_0)} \tag{6.4}$$

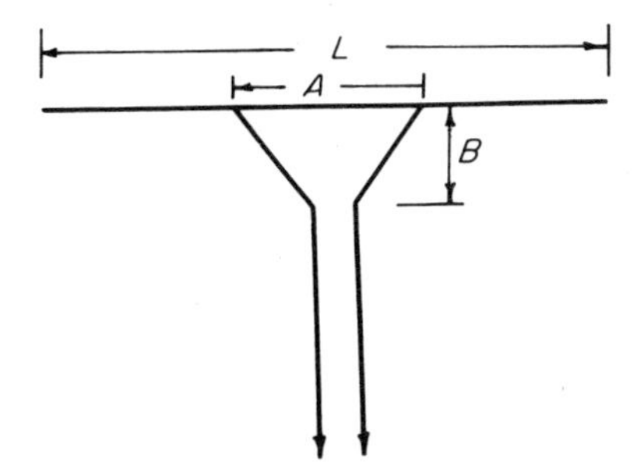

Figure 6.3 The delta match system

The required characteristic impedance can be obtained by choosing the appropriate values for the spacing and the conductor diameter.

Note that, since transmission lines are nearly lossless at high frequencies, both $Z_{\lambda/4}$ and Z_0 are real numbers and a quarter-wave transformer can be used to obtain a match only if Z_a is real. This will be the case if the antenna is of resonant length.

Delta Match

There are ways to obtain a match without using a matching network such as a quarter-wave transformer. These are based on the experimental fact that the impedance presented between any two points symmetrically placed with respect to the centre of a half-wave dipole with no gap at the centre will depend on the distance between the points. The greater the separation, the higher the value of the impedance. Feeding systems utilizing this principle are known as shunt feeds, and include the delta match, folded dipoles, and the T and gamma matches. In the delta match system, shown in Figure 6.3, a fanned-out section of the line (known as the delta) is used to couple the antenna and the transmission line. The ends of the delta are attached at points equidistant from the centre of the antenna. When so connected, the terminating impedance for the line will depend on the dimensions A and B. Experimentally, it is found to be essentially resistive if the antenna length is resonant. As an example, for coupling a simple half-wave dipole to a 600 ohm line, the distance A between the ends of the delta is about 0.12λ while the distance B is about 0.15λ.

Folded Dipoles

Another way of matching a low-impedance antenna to a high-impedance transmission line is to split the antenna into two or more parallel conductors, with the transmission line attached to the centre of only one of them. The resultant structures are known as folded dipoles. A two-conductor folded dipole (Figure 6.4a) has already been introduced in section 3.13. A three-conductor folded dipole is shown in Figure 6.4(b). Since the conductor spacing is small,

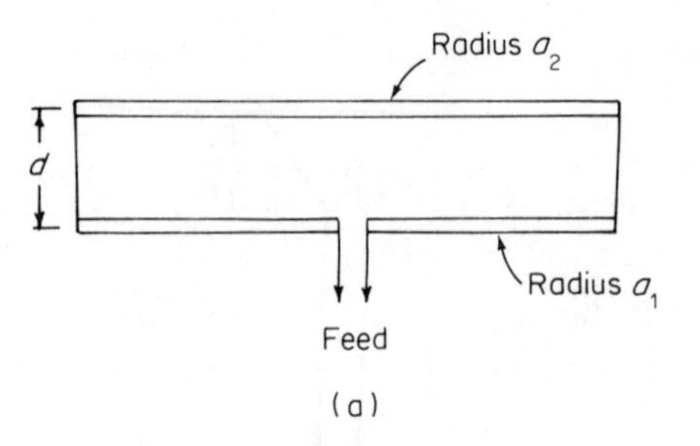

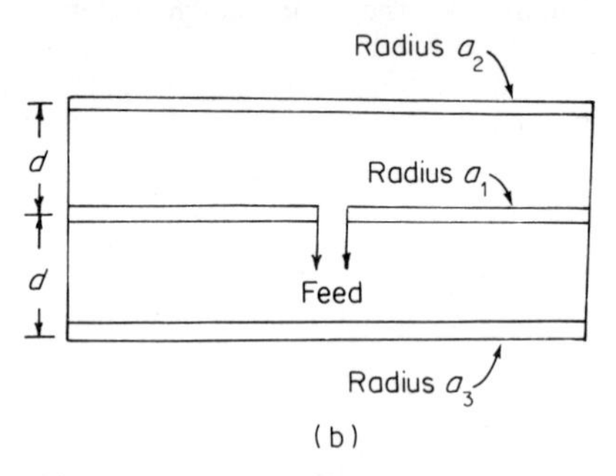

Figure 6.4 (a) Two-conductor folded dipole; (b) three-conductor folded dipole

the radiating properties of a folded dipole are equivalent to those of an ordinary dipole. However, the current flowing into the input terminals of the antenna from the line is the current in one conductor only. Since the power from the line is delivered at this value of current, it follows that splitting the antenna into two or more conductors has the effect of raising the input impedance of the antenna.

For a given number of conductors, the division of current among them is governed by both the conductor diameters and the spacing between them. In particular, for a two-conductor half-wave folded dipole with radii a_1, a_2 and spacing d, where a_1 corresponds to the fed conductor, analysis shows that the input impedance is given by (Harrison and King, 1961)

$$Z_{in} = (1 + f)^2 Z_D \tag{6.5}$$

where Z_D is the impedance of an ordinary half-wave dipole with radius a_1 and f is the current division factor. If a_1 and a_2 are much less than d, as is usually the case, f is approximately given by

$$f \simeq \frac{\ln(d/a_1)}{\ln(d/a_2)} \tag{6.6}$$

For the special but important case when the two conductors have equal radii, the current division factor f is unity. The input impedance is 4 times that of an ordinary dipole, which is close enough to 300 ohm to afford a good match to a 300 ohm parallel-wire line.

The bandwidth of a folded dipole is broader than that of a simple dipole. For the two-conductor case, analysis shows that it behaves like a dipole with an equivalent radius a_e given by (Wolff, 1966)

$$\ln a_e \simeq \ln a_1 + \frac{1}{(1+u)^2}(u^2 \ln u + 2u \ln v) \tag{6.7}$$

where

$$u = a_2/a_1 \qquad v = d/a_1 \tag{6.8}$$

Finally, the reader is referred to the *ARRL Antenna Book* for information on the impedance step-up ratio as a function of conductor diameters and spacing for a three-conductor folded dipole.

T and Gamma Matches

A matching arrangement which in part resembles the delta match and in part resembles the folded dipole is the T match shown in Figure 6.5(a). The conductor

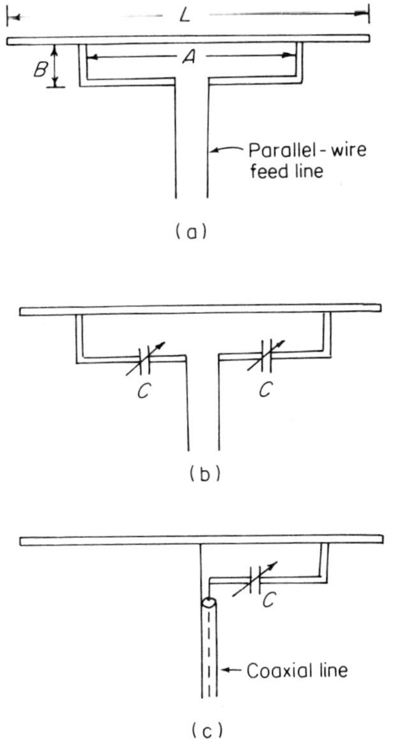

Figure 6.5 (a) The *T* match system; (b) the *T* match system with the series capacitors for tuning out residual reactance; (c) the gamma match system

diameters, the spacing B, and the distance A determine the input impedance that the line sees. The impedance is found to have a reactive component, which can be tuned out either by adjusting the length of the antenna or by inserting a capacitance of the proper value in series at the input terminals, as shown in Figure 6.5(b).

The arrangement shown in Figure 6.5(c) is known as the gamma match. It is an unbalanced version of the T match and is suitable for use with coaxial lines.

6.3.3 Balancing Devices

Consider a dipole fed at the centre by a parallel-wire line. Owing to symmetry, the current on the antenna is symmetrical with respect to the centre and the currents on the two conductors of the transmission line are equal and opposite. A balanced situation is said to prevail. At VHF and UHF frequencies, however, it is common to use coaxial lines instead of parallel lines mainly because of the former's lower losses and the shielding provided by the outer conductor. When a dipole is fed at the centre by a coaxial line, as shown in Figure 6.6, a net current (I_3) is induced on the outside of the coaxial cable. This current, unlike the internal currents I_1 and I_2, which are shielded from the external world by the thickness of the outer conductor, will radiate. To see how it arises, note that the voltages at the antenna terminals are equal in amplitude with respect to ground but opposite in phase. Each of these voltages causes a current of opposite polarity to flow on the outside of the coaxial cable. Since one antenna terminal is directly connected to the outer conductor and the other is only weakly coupled to it, the voltage at the directly connected terminal produces a much larger current. Consequently, there is little cancellation and a net current results. An unbalanced situation is said to prevail. The unbalanced current flowing on the outside of the coaxial cable gives rise to undesirable radiation. It can be minimized by a number of devices known as 'baluns', which is the abbreviation for 'balanced to unbalanced'. One such arrangement is shown in Figure 6.7. To understand how it works, note first that the currents induced on the outer conductor could be made equal in amplitude if a direct connection is made between the outside of the line and the antenna terminal that is connected to the inner conductor. Clearly, if this is done right at the antenna terminals, the line and the antenna would be

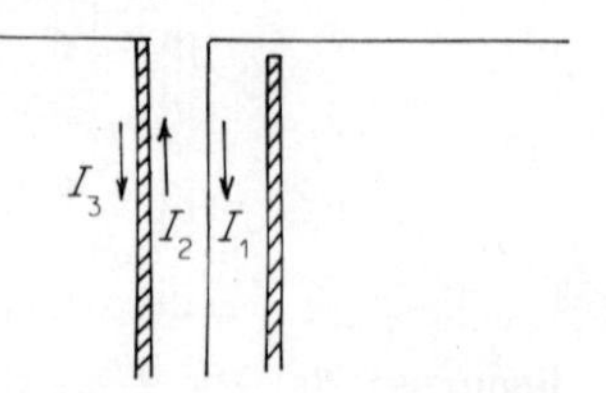

Figure 6.6 The currents in the coaxial line when connected to a dipole antenna

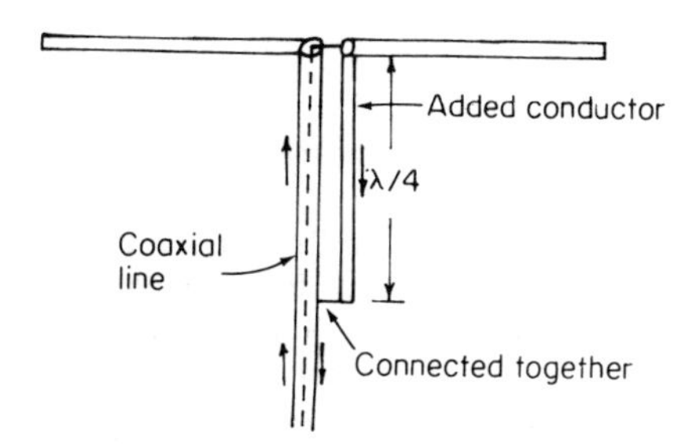

Figure 6.7 A one-to-one balun

short-circuited. On the other hand, if the connection is made through an additional conductor parallel to the line and a quarter of a wavelength long, the balancing section would appear as an open circuit to the antenna and would have no effect on its operation. However, any unbalance current flowing on the outside of the coaxial cable has a counterpart in an equal and opposite current flowing on the added conductor. Beyond the point where the added conductor is connected to the outer conductor, the two currents cancel, with the result that no current flows on the outside of the remainder of the transmission line. Note that the length of the added conductor has no bearing on its operation in balancing out the undesired current. It is sometimes made shorter than a quarter of a wavelength in order to provide the shunt inductance required in certain types of matching systems.

Figure 6.8 shows another balun arrangement. Here, equal and opposite voltages, balanced to ground, are taken from the inner conductors of the main transmission line and a phasing section which is a half-wavelength long. This scheme provides a 4 : 1 step-down in impedance from the balanced to the unbalanced side, since the voltages at the former are in series while the voltages at the latter are in parallel. It is therefore useful for coupling between a 75 ohm coaxial cable and a balanced 300 ohm line or a centre-fed half-wave folded dipole.

The effect of unbalanced coupling that results from connecting a coaxial line to a balanced antenna may also be reduced by choking off the current from flowing on the outside of the feed line. A direct approach to this objective is shown in Figure 6.9, where the line itself is formed into a coil at the antenna feed point. However, owing to the distributed capacitance among the turns, the effectiveness of this arrangement decreases at the higher frequencies.

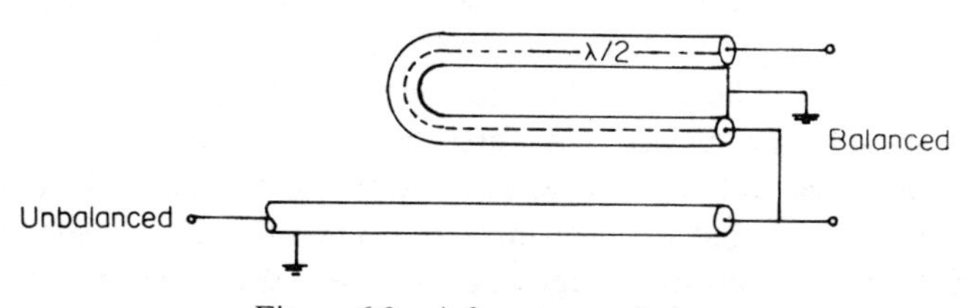

Figure 6.8 A four-to-one balun

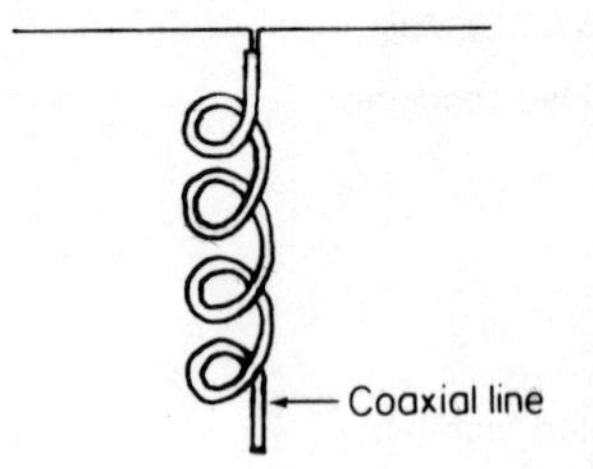

Figure 6.9 An RF choke formed by coiling the feed line (Source: *ARRL Antenna Book* (1974). Reproduced by permission of the American Radio Relay League Inc.)

6.4 GROUND EFFECTS

6.4.1 Introductory Remarks

Antennas in practice are erected near a ground plane. The ground plane can take many forms. Some are purposely introduced (such as radial wires around a monopole, the roof of a car) while others are part of the natural environment (such as the real Earth). The effect of the ground plane is not important only if the antenna is many wavelengths away from it. Generally, this means that the operation of high frequency (roughly UHF and above) antennas is little affected but the effect on low-frequency (roughly VHF and below) antennas is significant.

For analytical purposes, when the dimensions of the ground plane are large compared to the wavelength, it can be approximated as infinite and planar. While the conductivity of a metallic ground can usually be assumed to be infinite, that of the real Earth is more complicated. Whether the Earth is considered a perfect conductor or a lossy dielectric depends on the ratio of displacement current to conduction current, i.e. $\omega\varepsilon/\sigma$. If this ratio is less than 0.1, we can approximate the Earth as a perfect conductor. For typical values of $\varepsilon(\simeq 10\varepsilon_0)$ and σ ($\simeq 5.8 \times 10^7$ ohm^{-1} m^{-1}), the frequency at which $\omega\varepsilon = 0.1\sigma$ is about 1 MHz. Consequently, for frequencies below 1 MHz, we can model the Earth as a perfect conductor. For frequencies above 1 MHz, the effect of finite conductivity should be taken into account. These two cases will be discussed in sections 6.4.2 and 6.4.3 respectively.

6.4.2 Perfectly Conducting Ground

Consider a dipole antenna above a perfectly conducting ground that is assumed to be planar and infinite in extent. As discussed in section 3.13, the ground can be simulated by an appropriate image. The position and orientation of the image is dependent upon the height of the antenna above the ground and its orientation with respect to the surface of the ground. In general, the effect is such that, in the vertical plane radiation pattern, the original free-space field strength is multiplied by a factor that varies from 0 to 2. The value 2 corresponds to complete reinforcement from the image and the value 0 corresponds to complete cancellation.

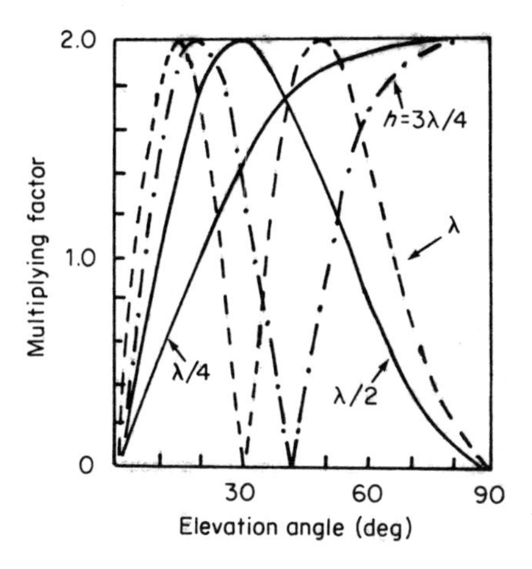

Figure 6.10 Multiplying factor for a horizontal dipole at four values of the height h above a perfect ground as a function of elevation angle (Source: *Radio Amateur's Bandbook* (1974). Reproduced by permission of the American Radio Relay League Inc.)

For the case of a horizontal dipole, the image is a dipole carrying a negative current. Figure 6.10 shows the variation of the multiplying factor (derived in worked example 1 in section 6.5) as a function of elevation angle for several heights above ground. As the height is increased, the angle at which complete reinforcement takes place is lowered. It occurs at 15° for a height of one wavelength. Similar curves can be obtained for a vertical or inclined dipole.

The perfect ground also affects the driving-point impedance of the antenna. If the applied voltage at the antenna terminals is V_1, then

$$V_1 = I_1 Z_{11} + I_2 Z_m \tag{6.9}$$

where I_1 is the antenna current, I_2 is the image current, Z_{11} is the self-impedance of the antenna, and Z_m is the mutual impedance of the antenna and its image. The driving-point impedance Z_1 is

$$Z_1 = \frac{V_1}{I_1} = Z_{11} + \frac{I_2}{I_1} Z_{\mathrm{m}} \tag{6.10}$$

For a horizontal dipole, $I_2 = -I_1$ and

$$Z_1 = Z_{11} - Z_{\mathrm{m}} \tag{6.11}$$

The variation of the input resistance $R_1 = \mathrm{Re}\, Z_1$ and input reactance $X_1 = \mathrm{Im}\, Z_1$ as functions of the height above ground for a horizontal $\frac{1}{2}\lambda$ dipole of infinitesimal thickness is shown in Figure 6.11. The impedance approaches the free-space value as the height becomes large but differs considerably from it at low heights.

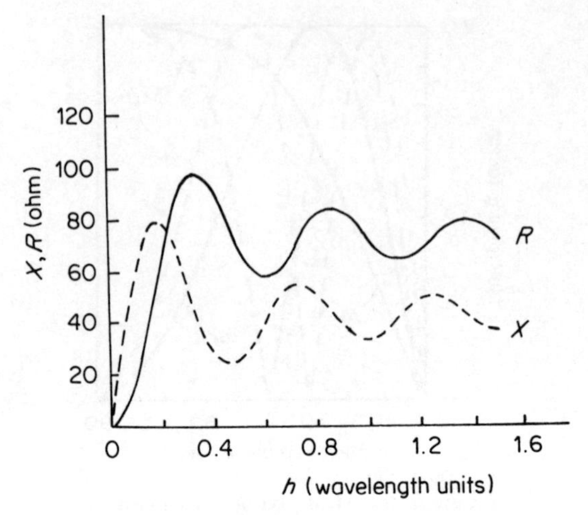

Figure 6.11 Variation of the input resistance R and input reactance X as a function of height in wavelength above a perfect ground for a horizontal half-wave dipole of infinitesimal thickness (Source: Carter (1932), *Proc. IRE*, 20, 1004–1041. © 1932, IEEE)

6.4.3 Imperfect Ground

If the ground plane has finite conductivity, the problem is considerably more complicated. A simplification is achieved if it is assumed that the antenna is high above ground so that the waves striking ground are substantially plane over any limited area. Under such circumstances, the effect of the ground can also be accounted for by the method of images. However, the amplitude and phase of the current in the image antenna are dependent on the reflection coefficient of the ground plane. This coefficient is a function not only of the parameters of the medium but also of the angle of incidence and the polarization of the incoming wave. The principal effect is that the imperfect ground absorbs most of the energy radiated at low elevation angles, with the result that the curves for the multiplication factor based on the assumption of a perfect ground become inaccurate for angles less than about 15°. Above 15°, however, they are accurate enough for most practical purposes. As for the input impedance, the amplitude of oscillation about the free-space value is slightly less, with a slight shift in the actual heights above ground at which the maximum and minimum impedances occur. For details, the reader is referred to Jordan and Balmain (1968).

6.5 WORKED EXAMPLES

Example 1

Consider a horizontal Hertzian dipole placed at a height h above a perfectly conducting ground. Derive an expression for the far-zone electric field at a point $P(r, \theta, \phi)$ and discuss how the patterns in the principal planes are modified by the ground.

Solution

Let the y–z plane be the ground plane. Assuming it to be perfectly conducting, its effect on a horizontal dipole can be accounted for by an image dipole carrying an opposite current, as shown in Figure 6.12. For a point $P(r, \theta, \phi)$ in the far zone, the path difference between the dipole and its image is $2h \cos \chi = 2h \sin \theta \cos \phi$. Taking the origin to be the point of phase reference, the electric field at P is

$$\mathbf{E}(\mathbf{r}) = \frac{I dl \exp(-jkr)}{4\pi r} k\eta \sin \theta [\exp(jkh \cos \chi) - \exp(-jkh \cos \chi)]$$

$$= \frac{jI dl \exp(-jkr)}{4\pi r} k\eta \sin \theta \, 2 \sin(kh \cos \chi)$$

The magnitude of $\mathbf{E}$ is

$$|\mathbf{E}(\mathbf{r})| = \frac{I dl}{4\pi r} k\eta \sin \theta \, f(\cos \chi)$$

where

$$f(\cos \chi) = 2 \sin(kh \cos \chi)$$

The actual field strength is therefore obtained by multiplying the free-space field strength and the factor $f(\cos \chi)$. This statement applies for the general case of a finite-length dipole.

Let Ψ be the angle of elevation above the horizontal. The multiplying factor in the x–z and x–y planes is

$$f = 2 \sin\left(\frac{2\pi h}{\lambda} \sin \Psi\right).$$

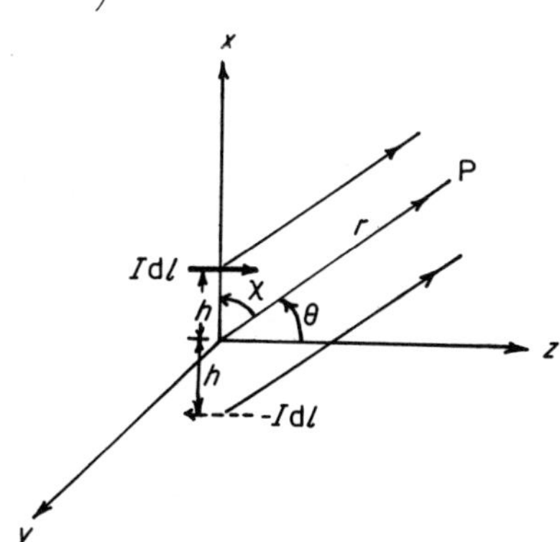

Figure 6.12 Geometry for the calculation of the far field above a perfectly conducting ground (y–z plane)

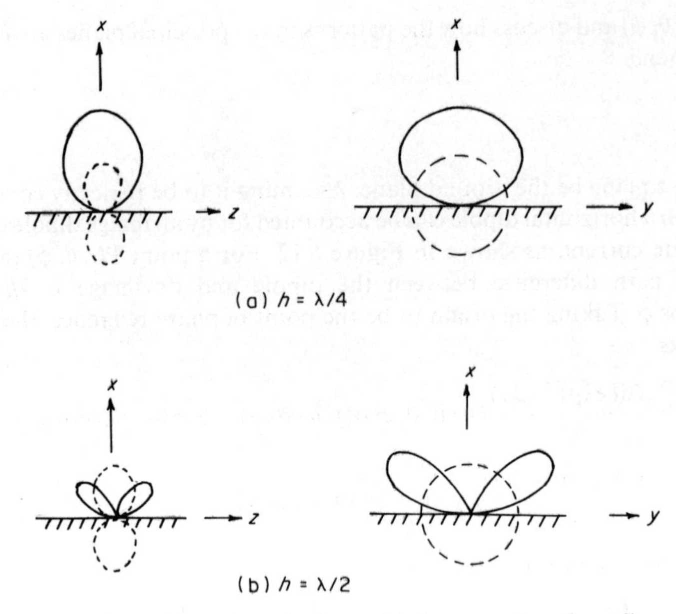

Figure 6.13 Effect of a perfectly conducting ground on the radiation patterns of a horizontal Hertzian dipole at heights of (a) $h = \lambda/4$ and (b) $h = \lambda/2$ above ground. The broken lines correspond to the free-space patterns (Source: *ARRL Antenna Book* (1974). Reproduced by permission of the American Radio Relay League Inc.)

Its effects on the patterns in these planes are illustrated in Figure 6.13 for $h = \lambda/2$ and $h = \lambda/4$. In the horizontal (y–z) plane, $\chi = 90°$, $f(\cos\chi) = 0$ and $|\mathbf{E}| = 0$.

Example 2

Repeat example 1 for a vertical Hertzian dipole.

Solution

As shown in Figure 6.14, the image of a vertical dipole is a vertical dipole carrying a current of the same phase. Hence

$$\mathbf{E}(\mathbf{r}) = \frac{Idl}{4\pi r}\exp(-jkr)k\eta\sin\theta'[\exp(jkh\cos\chi) + \exp(-jkh\cos\chi)]$$

$$= \frac{Idl}{4\pi r}\exp(-jkr)k\eta\cos\theta f(\cos\chi)$$

where the multiplying factor

$$f(\cos\chi) = 2\cos(kh\cos\chi) = 2\cos(kh\sin\theta\cos\phi)$$

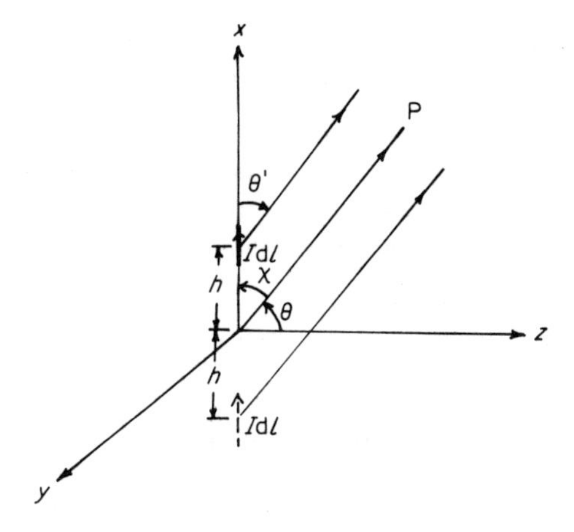

Figure 6.14 Geometry for the calculation of the far field due to a vertical Hertzian dipole above a perfect ground (y–z plane)

In the horizontal plane, the multiplying factor is equal to 2. Except for the doubling of strength, the pattern has the same shape as the free-space pattern. In the x–y and x–z elevation planes, the multiplying factor is

$$f = 2\cos\left(\frac{2\pi h}{\lambda}\sin\Psi\right)$$

where Ψ is the angle of elevation above the horizontal.

6.6 PROBLEMS

1. A half-wave dipole designed to resonante at 30 MHz has a radius $a = 0.2$ cm. It is to be fed by a parallel-wire transmission line with a characteristic impedance of 300 ohm via a quarter-wave transformer. Find the physical length of the dipole and the characteristic impedance of the quarter-wave transformer to effect a match.

2. In the two-conductor folded dipole shown in Figure 6.4(a), the radii are $a_1 = 0.4$ cm and $a_2 = 0.8$ cm. The length is cut to correspond to the resonant length at the frequency 30 MHz. The antenna is fed directly by a 300 ohm parallel-wire transmission line.
   (a) Find the VSWR if the spacing $d = 8a_2$.
   (b) Find the VSWR if d is changed to (i) $4a_2$, (ii) $16a_2$.

3. A horizontal half-wave dipole of infinitesimal thickness is placed at a height h above a perfect ground. It is to radiate 1 mw at the frequency 300 MHz. Find the feed-point current if (a) $h = 0.3\lambda$, (b) $h = 0.6\lambda$, (c) $h \to \infty$.

4. Construct a diagram similar to Figure 6.10 for a vertical dipole above a perfect ground.

5. Referring to Figure 6.14, sketch the effect of the perfect ground on the radiation patterns of the vertical Hertzian dipole in the x–y and y–z planes for (a) $h = \lambda/4$ and (b) $h = \lambda/2$.

6. Consider a Hertzian dipole inclined at an angle α with respect to the horizontal at a height h above a perfect ground. Derive an expression for the far-zone electric field at a point $P(r, \theta, \phi)$.

7. Construct a diagram similar to Figure 6.11 for a vertical half-wave dipole of infinitesimal thickness above a perfect ground.

BIBLIOGRAPHY

ARRL Antenna Book (1974). The American Radio Relay League, Newington, CT, Chapters 1 and 3.

Stutzman, W. L. and Thiele, G. A. (1981). *Antenna Theory and Design*, Wiley, New York, Chapter 5.

CHAPTER 7
Linear Arrays

7.1 INTRODUCTION

For purposes of point-to-point communication such as in radar and in radio astronomy, it is desirable for the antenna to have a narrow beam in one direction and little radiation in other directions. Such a pattern cannot be obtained by using a single dipole. As mentioned in section 3.8, although the half-power beamwidth decreases from 90° to 47° as the length of the dipole increases from infinitesimal to one wavelength, it does not follow that the longer the dipole, the narrower the beamwidth. Instead, when the length exceeds 1.2 wavelengths, the pattern becomes multilobed, an undesirable feature in the aforementioned applications. However, by means of an array of antennas, it is possible to obtain a pattern that is highly directive in one direction. Moreover, the beamwidth can in principle be made as narrow as one wishes simply by increasing the number of elements in the array.

An antenna array can also be designed to approximate a given radiation pattern. For example, if one wishes to beam radiation to two separate population centres situated in a desert, it is desirable for the transmitting antenna to have a pattern that is uniform in the directions subtended by the two centres and has little radiation elsewhere. Such a prescribed pattern can be synthesized by means of an array of antennas. Again, the approximation can in theory be improved indefinitely by increasing the number of elements.

The elements that make up an array are usually of the same type and are similarly oriented. A linear array is one in which the terminals of the individual antennas lie along a straight line and is the configuration most frequently used in practice. In this chapter, we shall be concerned with the theory of linear arrays.

7.2 FAR FIELD OF A TWO-DIMENSIONAL CURRENT SHEET

A linear array of antennas can be viewed as a two-dimensional aperture on which a certain distribution of current flows. This distribution is governed by the antenna elements that make up the array and by the manner in which these elements are arranged to form the array. Although the current distribution is a discrete one, we can regard a discrete distribution as a special case of a

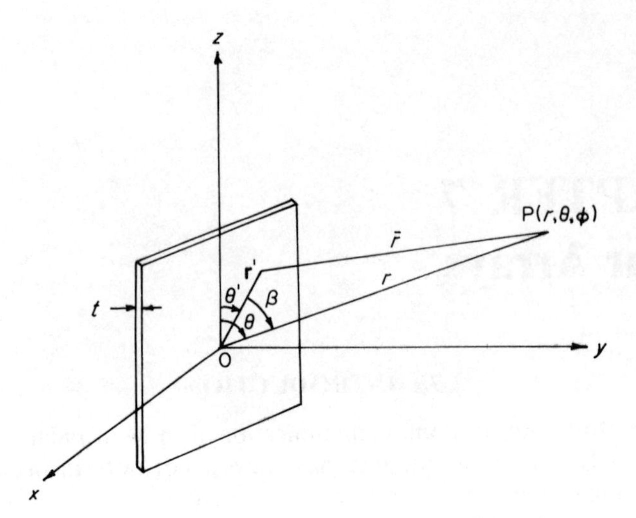

Figure 7.1 Geometry for evaluating the far field of a two-dimensional current sheet

continuous distribution. Accordingly, we begin by calculating the far field of a two-dimensional current sheet. The results obtained will be applied to linear arrays in section 7.3.

Consider a two-dimensional aperture of thickness t lying in the x–z plane carrying a current density $\mathbf{J}(x', y', z') = J_z(x', y', z')\hat{\mathbf{z}}$, as shown in Figure 7.1. The vector potential at a point $P(r, \theta, \phi)$ is

$$\mathbf{A} = A_z \hat{\mathbf{z}}$$

where

$$A_z = \frac{\mu}{4\pi} \int\int\int_0^t \frac{\exp(-jk\bar{r})}{\bar{r}} J_z(x', y', z')\,dy'\,dx'\,dz' \tag{7.1}$$

Assuming $t \ll \lambda$, we can write

$$A_z = \frac{\mu}{4\pi} \int\int \frac{\exp(-jk\bar{r})}{\bar{r}} \left(\int_0^t J_z(x', y', z')\,dy' \right) dx'\,dz' \tag{7.2}$$

since $\bar{r}$ is essentially constant for $0 < y < t$ if point P is more than a few wavelengths from the $y = 0$ plane. Let us define

$$K_z(x', z') = \int_0^t J_z(x', y', z')\,dy' \tag{7.3}$$

$K_z(x', z')$ can be regarded as a surface current density on the $y = 0$ plane.

From Figure 7.1, we have, for P in the far zone,

$$\bar{r} \simeq r - r' \cos\beta \tag{7.4}$$

$$\mathbf{r}\cdot\mathbf{r}' = rr' \cos\beta = xx' + zz' \tag{7.5}$$

But

$$x = r \sin\theta \cos\phi \tag{7.6}$$

$$z = r \cos\theta \tag{7.7}$$

Hence

$$r' \cos\beta = x' \sin\theta \cos\phi + z' \cos\theta \tag{7.8}$$

and

$$\bar{r} = r - x' \sin\theta \cos\phi - z' \cos\theta \tag{7.9}$$

Substitution of (7.3) and (7.9) into (7.2) yields

$$A_z = \frac{\mu \exp(-jkr)}{4\pi r} \iint K_z(x', z') \exp[jk(x' \sin\theta \cos\phi + z' \cos\theta)] \, dx' \, dz' \tag{7.10}$$

In (7.10), the quantity $r' \cos\beta$ has been neglected in the denominator since it is always much less than r. We now let

$$\psi_1 = \sin\theta \cos\phi \tag{7.11}$$

$$\psi_2 = \cos\theta \tag{7.12}$$

Then (7.10) can be written as

$$A_z = \frac{\mu \lambda^2 \exp(-jkr)}{4\pi r} P(\psi_1, \psi_2) \tag{7.13}$$

where

$$P(\psi_1, \psi_2) = \iint K_z(x', z') \exp\left[j2\pi \left(\frac{x'}{\lambda}\psi_1 + \frac{z'}{\lambda}\psi_2 \right) \right] d\left(\frac{x'}{\lambda}\right) d\left(\frac{z'}{\lambda}\right) \tag{7.14}$$

The integrations in (7.14) are over the dimensions of the aperture. Since $K_z(x', z')$ is automatically zero outside the aperture, we can also extend the limits to $\pm\infty$ in each integral. Thus the function $P(\psi_1, \psi_2)$ can be regarded as the two-dimensional Fourier transform of the source distribution $K_z(x', z')$.

The magnetic field is given by

$$\mu \mathbf{H} = \nabla \times \mathbf{A} = \frac{\hat{\mathbf{r}}}{r \sin\theta}\left(\frac{\partial}{\partial\theta}(\sin\theta\, A_\phi) - \frac{\partial}{\partial\phi} A_\theta \right) + \frac{\hat{\boldsymbol{\theta}}}{r}\left(\frac{1}{\sin\theta}\frac{\partial}{\partial\phi} A_r - \frac{\partial}{\partial r}(r A_\phi) \right) + \frac{\hat{\boldsymbol{\theta}}}{r}\left(\frac{\partial}{\partial r}(r A_\theta) - \frac{\partial}{\partial\theta} A_r \right) \tag{7.15}$$

Substituting (7.13) into (7.15), we obtain

$$H_r = \frac{1}{\mu r}\frac{\partial A_z}{\partial\phi} = \frac{\lambda^2 \exp(-jkr)}{4\pi r^2}\frac{\partial P}{\partial\phi} \tag{7.16}$$

$$H_\theta = \frac{\cos\theta}{\mu r \sin\theta}\frac{\partial A_z}{\partial\phi} = \frac{\lambda^2 \exp(-jkr)}{4\pi r^2}\cot\theta\frac{\partial P(\psi_1, \psi_2)}{\partial\phi} \tag{7.17}$$

$$H_\phi = \frac{1}{\mu r}\left(\frac{\partial}{\partial r}(-rA_z \sin\theta) - \frac{\partial}{\partial\theta}(A_z \cos\theta)\right)$$

$$= \frac{jk\lambda^2}{4\pi r}\exp(-jkr)\sin\theta\, P(\psi_1, \psi_2)$$

$$-\frac{\lambda^2}{4\pi r^2}\exp(-jkr)\frac{\partial}{\partial\theta}[\cos\theta\, P(\psi_1, \psi_2)] \tag{7.18}$$

Since the components H_r and H_θ as well as the second term in (7.18) vary as $1/r^2$, they are negligible in the far zone compared with the first term in (7.18), which varies as $1/r$. Hence we have

$$\mathbf{H} = H_\phi \hat{\boldsymbol{\phi}}$$

where

$$H_\phi = \frac{j\lambda \exp(-jkr)}{2r}\sin\theta\, P(\psi_1, \psi_2) \tag{7.19}$$

From

$$\mathbf{E} = \frac{1}{j\omega\varepsilon}(\nabla \times \mathbf{H})$$

we find that, with $\mathbf{H}$ given by (7.19), $\mathbf{E} = E_\theta \hat{\boldsymbol{\theta}}$ where $E_\theta = \eta H_\phi$.

The power pattern is given by the real part of the radial component of the Poynting vector:

$$\mathrm{Re}\, S_r = \mathrm{Re}(\tfrac{1}{2}E_\theta H_\phi^*) = \frac{\mu}{8}\left(\frac{\lambda}{r}\right)^2 \sin^2\theta |P(\psi_1, \psi_2)|^2 \tag{7.20}$$

An aperture is called degenerate if the function $K_z(x', z')$ can be written as a product of a function of x and a function of z:

$$K_z(x', z') = K_{z1}(x')K_{z2}(z') \tag{7.21}$$

Then

$$P(\psi_1, \psi_2) = P(\psi_1)P(\psi_2) \tag{7.22}$$

where

$$P(\psi_1) = \int K_{z1}(x')\exp[j2\pi(x'/\lambda)\psi_1]\,d(x'/\lambda) \tag{7.23}$$

$$P(\psi_2) = \int K_{z2}(z')\exp[j2\pi(z'/\lambda)\psi_2]\,d(z'/\lambda) \tag{7.24}$$

E-plane and H-plane Patterns

Since it is awkward to plot the complete pattern of an antenna on a single sheet of paper, it is often convenient to choose two mutually perpendicular planes and to study the variations in the function $P(\psi_1, \psi_2)$ as the point of observation moves in a circular path in one of these planes. The y–z plane $(x = 0)$ is called the E-plane because the electric field $\mathbf{E} = E_\theta \hat{\boldsymbol{\theta}}$ lies in this plane while the magnetic field $\mathbf{H} = H_\phi \hat{\boldsymbol{\phi}}$ is perpendicular to it. The pattern traced out in the $x = 0$ plane is called the principal E-plane pattern. Since $\phi = 90°$

$$\psi_1 = 0 \tag{7.25}$$

$$\psi_2 = \cos\theta \tag{7.26}$$

and

$$P(\psi_2) = (1/\lambda) \int I_z(z'/\lambda) \exp[j2\pi(z'/\lambda)\psi_2]\, d(z'/\lambda) \tag{7.27}$$

where

$$I_z(z'/\lambda) = \int K_z(x', z')\, dx' \tag{7.28}$$

The x–y plane $(z = 0)$ is called the H-plane because the magnetic field $\mathbf{H} = H_\phi \hat{\boldsymbol{\phi}}$ lies in this plane. The pattern traced out in the $z = 0$ plane is called the principal H-plane pattern. Since $\theta = 90°$,

$$\psi_2 = 0 \tag{7.29}$$

$$\psi_1 = \cos\phi \tag{7.30}$$

and

$$P(\psi_1) = (1/\lambda) \int K_z'(x'/\lambda) \exp[j2\pi(x'/\lambda)\psi_1]\, d(x'/\lambda) \tag{7.31}$$

where

$$K_z'(x'/\lambda) = \int K_z(x', z')\, dz' \tag{7.32}$$

7.3 ANTENNA ARRAY AS A TWO-DIMENSIONAL APERTURE: PRINCIPLE OF PATTERN MULTIPLICATION

7.3.1 Array of Elements Side by Side

Consider an array of N dipoles arranged side by side as shown in Figure 7.2. The dipoles lie in the x–z plane, with their axes oriented along the z-axis. Let the positions of the elements along the axis be $x_0', x_1', \ldots, x_{N-1}'$ and the cor-

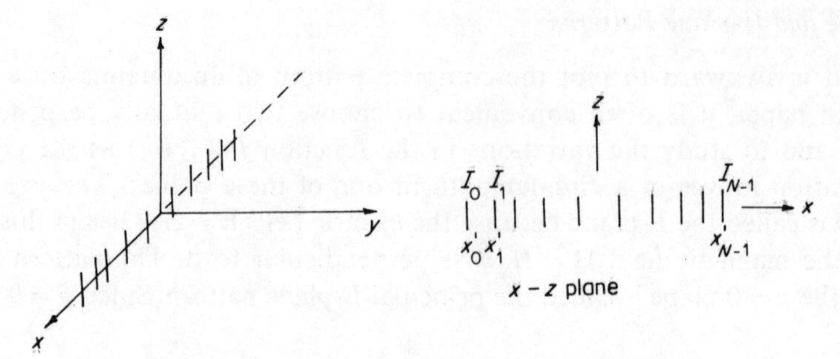

Figure 7.2 An array of N side-by-side dipoles

responding currents be $I_0, I_1, \ldots, I_{N-1}$. Such an array can be considered as an aperture with a discrete line-current distribution across its surface. The function $K_z(x', z')$ takes the form

$$K_z(x', z') = \sum_{n=0}^{N-1} I_n(z')\delta(x' - x_n') \tag{7.33}$$

Note that

$$\int K_z(x', z')\, dx' = \sum_{n=0}^{N-1} I_n(z')$$

and is the total current flowing in the aperture in the z-direction.

Substituting (7.33) in (7.14), we obtain, after performing the integration with respect to x', the following expression for the factor $P(\psi_1, \psi_2)$:

$$P(\psi_1, \psi_2) = \frac{1}{\lambda} \sum_{n=1}^{N-1} P_n(\psi_2) \exp(jkx_n'\psi_1) \tag{7.34}$$

where

$$P_n(\psi_2) = \int I_n(z') \exp(jkz'\psi_2)\, d(z'/\lambda) \tag{7.35}$$

The function $P_n(\psi_2)$ describes the pattern of the nth element when it is all by itself.

We now assume that the elements that make up the array are of the same type. For example, they may be all half-wave or full-wave dipoles. The current distribution along z on each dipole will then be of the same form, although the amplitude and phase of the current can differ from one dipole to another. Under these circumstances we can write

$$I_n(z') = I_n g(z') \tag{7.36}$$

where I_n is a complex number representing the terminal current on the nth element. The function $g(z')$ describes the variation of current along z' and is the same for all elements. Thus we have

$$P(\psi_1, \psi_2) = \frac{P_0'(\psi_2)}{\lambda} f(\psi_1) \tag{7.37}$$

$$\mathbf{H} = \frac{\mathrm{j}\exp(-\mathrm{j}kr)}{2r} \sin\theta\, P_0'(\psi_2) f(\psi_1) \hat{\boldsymbol{\phi}} \tag{7.38}$$

$$\mathbf{E} = \eta |\mathbf{H}| \hat{\boldsymbol{\theta}} \tag{7.39}$$

where

$$P_0'(\psi_2) = \int g(z') \exp(\mathrm{j}kz'\psi_2)\, \mathrm{d}(z'/\lambda) \tag{7.40}$$

$$f(\psi_1) = \sum_{n=0}^{N-1} I_n \exp(\mathrm{j}kx_n'\psi_1) \tag{7.41}$$

Equations (7.38)–(7.41) show that the pattern of an array is equal to the product of the individual pattern $\sin\theta\, P_0'(\psi_2)$ and a factor $f(\psi_1)$ which depends on group effects. This result is known as the principle of pattern multiplication. The factor $f(\psi_1)$ is called the array factor or the space factor of the array. It depends on the number of elements, the inter-elemental spacing, and the amplitudes and phases of the excitation currents. Since the individual pattern is a known quantity, the study of the array factor yields all the information about the properties of an array. Note that the array factor can also be considered as the pattern of a hypothetical array consisting of isotropic radiators.

As we see in (7.41), the array factor is a function of the variable $\psi_1 = \sin\theta\cos\phi$. Although ψ_1 depends on both θ and ϕ, it has a simple geometrical interpretation. Referring to Figure 7.3, if the angle between the array axis and the line OP χ, then

$$\cos\chi = \hat{\mathbf{r}} \cdot \hat{\mathbf{x}} = \sin\theta\cos\phi = \psi_1 \tag{7.42}$$

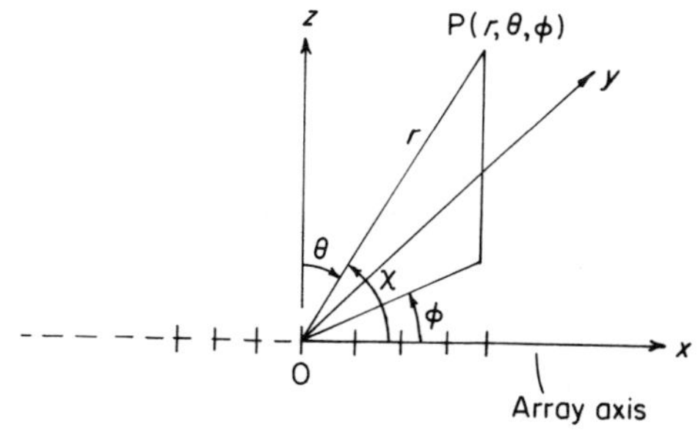

Figure 7.3 Geometry showing $\psi_1 = \cos\chi$

7.3.2 Array of Collinear Elements

In section 7.3.1, the principle of pattern multiplication was obtained for an array of elements arranged side by side. This principle also applies to an array of collinear elements. Such an arrangement is shown in Figure 7.4. This array can be considered as an aperture in which there is a current density of the form

$$J_z(x', y', z') = \delta(x')\delta(y')[I_o g_o(z') + I_1 g_1(z') + \cdots + I_{N-1} g_{N-1}(z')] \tag{7.43}$$

If all the elements are identical but are simply displaced a distance from each other along the z-axis, then the individual distributions are related by the following:

$$\begin{aligned} g_0(z') &= g(z') \\ g_1(z') &= g(z' - z'_1) \\ g_2(z') &= g(z' - z'_2) \\ \vdots \quad & \quad \vdots \\ g_{N-1}(z') &= g(z' - z'_{N-1}) \end{aligned} \tag{7.44}$$

Hence

$$J_z(x', y', z') = \delta(x')\delta(y') \sum_{n=0}^{N-1} I_n g(z' - z'_n) \tag{7.45}$$

$$K_z(x', z') = \int J_z(x', y', z')\, dy' = \delta(x') \sum_{n=0}^{N-1} I_n g(z' - z'_n) \tag{7.46}$$

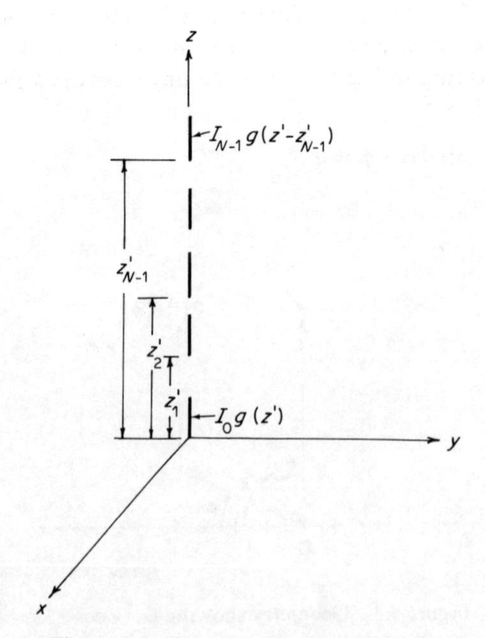

Figure 7.4 An array of collinear dipoles

From (7.14),

$$P(\psi_1, \psi_2) = \int\int \delta(x') \sum_{n=0}^{N-1} I_n g(z' - z'_n) \exp[\mathrm{j}(kx'\psi_1 + kz'\psi_2)] \, \mathrm{d}(x'/\lambda) \, \mathrm{d}(z'/\lambda)$$

$$= \frac{1}{\lambda} \int \sum_{n=0}^{N-1} I_n g(z' - z'_n) \exp(\mathrm{j}kz'\psi_2) \, \mathrm{d}(z'/\lambda). \tag{7.47}$$

On letting $z_0 = z' - z'_n$, (7.47) becomes

$$P(\psi_2) = \frac{1}{\lambda} \sum_{n=0}^{N-1} I_n \exp(\mathrm{j}kz'_n\psi_2) \int g(z_0) \exp(\mathrm{j}kz_0\psi_2) \, \mathrm{d}(z_0/\lambda) \tag{7.48}$$

Noting that $\psi_2 = \cos\theta = \cos\chi$, where χ is the angle between the array axis and the line OP, (7.48) can be written as

$$P(\psi_2) = P(\cos\theta) = \frac{P'_0(\cos\theta)}{\lambda} f(\chi) \tag{7.49}$$

where

$$P'_0(\cos\theta) = \int g(z') \exp(\mathrm{j}kz' \cos\theta) \, \mathrm{d}(z'/\lambda) \tag{7.50}$$

and

$$f(\chi) = \sum_{n=0}^{N-1} I_n \exp(\mathrm{j}kz'_n \cos\chi) \tag{7.51}$$

Thus the principle of pattern multiplication applies also to collinear arrays.

7.3.3 Planar Array

We now show one more example in which the pattern multiplication principle applies. Consider a planar array of $M \times N$ elements in the x–z plane, as shown in Figure 7.5. This array can be considered as an aperture in which there is a current density of the form

$$J_z(x', y', z') = \delta(y) \sum_{m=0}^{M-1} \sum_{n=0}^{N-1} I_{nm} \delta(x' - x'_m) g(z' - z'_n) \tag{7.52}$$

For such a current distribution, the function $P(\psi_1, \psi_2)$ is

$$P(\psi_1, \psi_2) = \frac{1}{\lambda} \int \sum_{m=0}^{M-1} \sum_{n=0}^{N-1} I_{nm} g(z' - z'_n) \exp(\mathrm{j}kx'_m\psi_1) \exp(\mathrm{j}kz'\psi_2) \, \mathrm{d}(z'/\lambda) \tag{7.53}$$

On letting $z_0 = z' - z'_{n'}$ (7.53) becomes

$$P(\psi_1, \psi_2)$$

$$= \frac{1}{\lambda} \sum_{m=0}^{M-1} \sum_{n=0}^{N-1} I_{nm} \exp(\mathrm{j}kx'_m\psi_1) \exp(\mathrm{j}kz'_n\psi_2) \int g(z'_0) \exp(\mathrm{j}kz_0\psi_2) \, \mathrm{d}(z'_0/\lambda) \tag{7.54}$$

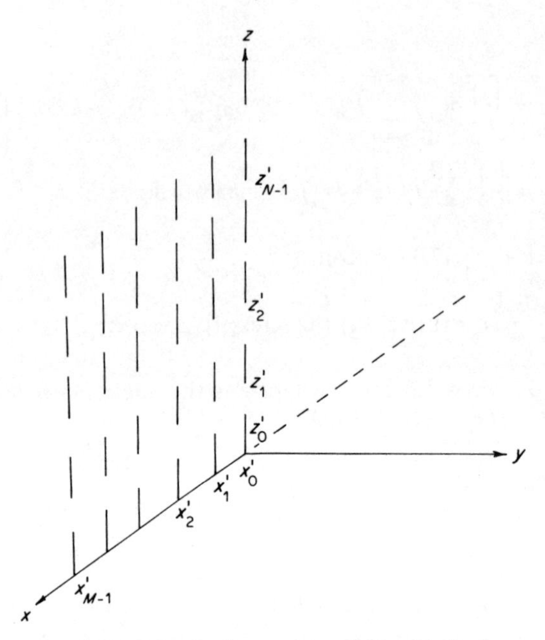

Figure 7.5 A planar array of $M \times N$ dipoles

Equation (7.54) can be written as the product of an individual factor and a group factor:

$$P(\psi_1, \psi_2) = (1/\lambda)P'_0(\psi_2)f(\psi_1, \psi_2) \tag{7.55}$$

where $P'_0(\psi_2)$ is given by (7.50) and

$$f(\psi_1, \psi_2) = \sum_{m=0}^{M-1} \sum_{n=0}^{N-1} I_{nm} \exp(jkx'_m\psi_1)\exp(jkz'_n\psi_2) \tag{7.56}$$

For an array in which the currents in all the elements are equal, I_{nm} can be set equal to unity and

$$f(\psi_1, \psi_2) = \sum_{m=0}^{M-1} \exp(jkx'_m\psi_1) \sum_{n=0}^{N-1} \exp(jkx'_n\psi_2) \tag{7.57}$$

The first sum is the array factor of M elements of identical currents along the x-axis and the second sum is the array factor of N elements of identical currents along the z-axis. Note that $\psi_1 = \cos\chi$ and $\psi_2 = \cos\theta$ where χ and θ are the angles that line OP makes with the horizontal array axis and the vertical array axis respectively.

7.3.4 General Situation in which the Pattern Multiplication Principle is Applicable

In the above sections, we have considered three situations in which the principle of pattern multiplication applies. In each case, the array element is taken to be a dipole because it is the only antenna that we have considered in detail so far. However, the pattern multiplication principle can be shown to be applicable for any array consisting of identical and similarly oriented elements (i.e. any two elements can be made congruent by a simple translation). The element can be a dipole, a loop, a helix, a slot or a horn. It can also be a mixed collection of dipoles and horns, loops and slots, etc. For a proof of the principle in this general form, the reader is referred to the article by Elliott (1966) listed at the end of this chapter.

7.4 ARRAY FACTOR OF UNIFORMLY SPACED ARRAYS

7.4.1 General Properties

An array in which the spacing between adjacent elements is the same for all elements is called a uniformly spaced array. For a uniformly spaced array of N elements along the x-axis with inter-element spacing d, we have

$$x'_n = nd \tag{7.58}$$

Use of (7.58) in (7.41) yields the following form for the array factor:

$$f(\psi_1) = \sum_{n=0}^{N-1} I_n \exp(jknd\psi_1) \tag{7.59}$$

where

$$\psi_1 = \cos\chi = \sin\theta\cos\phi$$

In the H-plane, $\psi_1 = \cos\phi$. In the E-plane, $\psi_1 = 0$ and the array factor is a constant.

The series in (7.59) is a finite exponential Fourier series in the variable ψ_1. The array factor f is therefore a periodic function in ψ_1. The period is

$$T = \lambda/d \tag{7.60}$$

Consequently, as ψ_1 increases by a value λ/d, the function $f(\psi_1)$ repeats itself. A sketch of $f(\psi_1)$ is shown in Figure 7.6.

As a mathematical function, $f(\psi_1)$ is defined for all values of ψ_1. However,

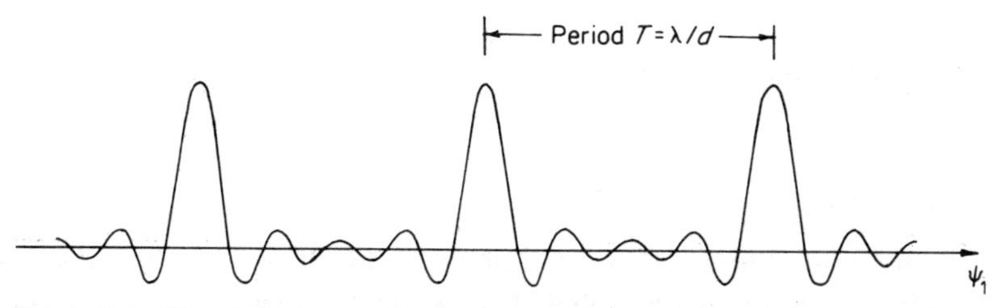

Figure 7.6 Sketch of the array factor $f(\psi_1)$ as a function of ψ_1. The function is periodic with a period $T = \lambda/d$

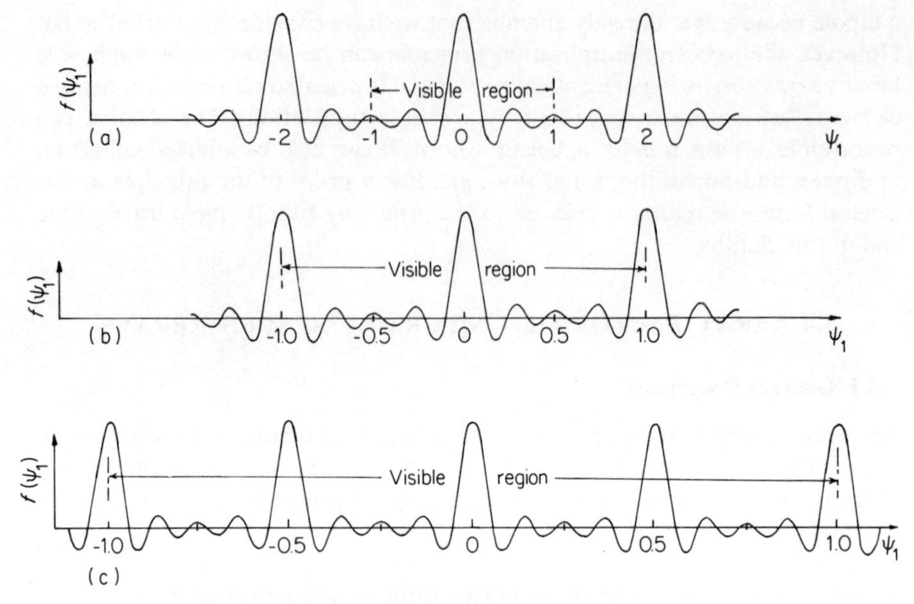

Figure 7.7 The number of main beams in the visible region as a function of d/λ

since $\psi_1 = \cos\chi$ and χ is a physical angle limited to the range $-\pi \leqslant \chi \leqslant \pi$, the range of ψ_1 corresponding to the entire physical space is $-1 \leqslant \psi_1 \leqslant 1$. Hence in the sketch of $f(\psi_1)$ versus ψ_1, only the portion between $-1 \leqslant \psi_1 \leqslant 1$ is visible. This region may include one or more periods, depending on the ratio d/λ. To illustrate this point, suppose the maximum of the array factor at $\psi_1 = 0$, corresponding to $\chi = 90°$. This maximum will repeat itself at $\psi_1 = \pm n\lambda/d$ where n is a positive integer. Consider now several values of d/λ:

(a) If $d/\lambda = \frac{1}{2}$, $\lambda/d = 2$ and in the visible region, $-1 \leqslant \psi_1 \leqslant 1$, there is only one maximum (main beam). This is illustrated in Figure 7.7(a).
(b) If $d/\lambda = 1$, $\lambda/d = 1$ and there are two main beams in the visible region. This is illustrated in Figure 7.7(b).
(c) If $d/\lambda = 2$, $\lambda/d = \frac{1}{2}$ and there are four main beams in the visible region. This is illustrated in Figure 7.7(c).

7.4.2 Progressive Phaseshift Array

Let us write

$$I_n = a_n \exp(\mathrm{j}\alpha_n) \tag{7.61}$$

where a_n and α_n are the amplitude and phase of the current on the nth element respectively. A progressive phaseshift array is one in which

$$\alpha_n = n\alpha \tag{7.62}$$

where α is a constant. For such an array,

$$f(\psi_1) = \sum_{n=0}^{N-1} a_n \exp[jn(\alpha + kd\psi_1)] \tag{7.63}$$

The maximum (main beam) of (7.63) occurs when the terms of the series are all in phase. This corresponds to the direction

$$\psi_1 = -\alpha/kd \tag{7.64}$$

If $\alpha = 0$, i.e. the elements are in phase, $\psi_1 = 0$ and the main beam occurs at $\chi = 90°$. This is called a broadside array. If $\alpha = -kd$, $\psi_1 = 1$ and the main beam occurs at $\chi = 0°$. Such an array is called an ordinary endfire array.

For a broadside array, it is clear from the discussions illustrated in Figure 7.7 that, if there is to be only one main beam, the spacing between the elements must be less than a wavelength ($d/\lambda < 1$).

The value of α can be changed electronically by means of phaseshifters. The position of the main beam changes with α according to (7.64). This technique is known as electronic steering.

In practice, the necessary phase relationships between the elements of the array may be obtained with either a series-fed or a parallel-fed arrangement. In the series-fed arrangement shown in Figure 7.8(a), the energy is transmitted from one end of the line and all the phaseshifters introduce the same amount of phaseshift. The parallel-fed array is illustrated in Figure 7.8(b). The energy is to be divided among the elements by a power splitter. Equal lengths of line

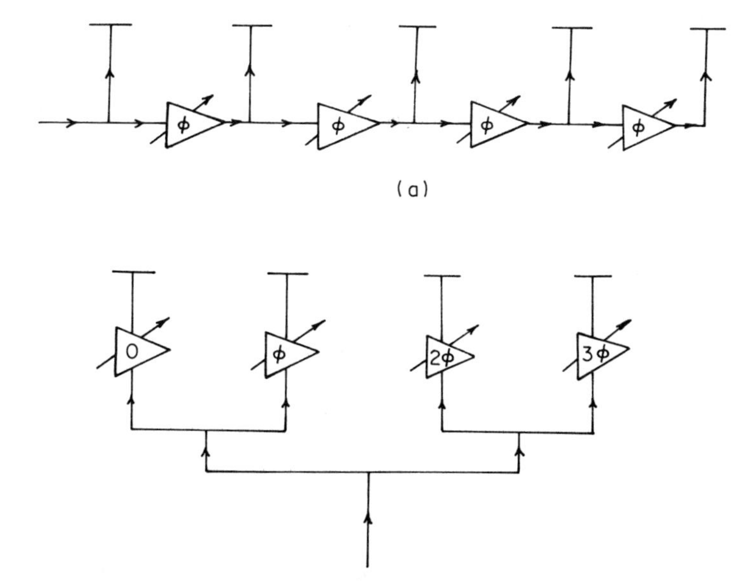

Figure 7.8 Two methods of providing the phase relationships between the elements of an array: (a) series feed; (b) parallel feed

transmit the energy to each element so that the lines themselves do not introduce any phase difference. If the phase of the first element is taken as the reference, the phases required in the succeeding elements are $\alpha, 2\alpha, 3\alpha, \ldots, (N-1)\alpha$. These are provided by the phaseshifters.

In a series-fed array containing N shifters, the signal suffers the insertion loss of a single phaseshifter N times. However, it has the advantage that, since each phaseshifter has the same value of phaseshift, only a single control signal is needed to steer the beam. In an N-element parallel-fed array similar to that of Figure 7.8(b), the insertion loss of the phaseshifter is introduced only once. However, it requires a separate control signal for each of the $(N-1)$ phaseshifters. Thus the series-fed array introduces more loss than a parallel-fed array but it is easier to program the necessary phaseshifts. The choice between the two will depend upon the particular application at hand.

As for the necessary amplitude distributions, they can be obtained by means of variable attenuators, which can be placed either before or after each phaseshifter.

7.5 UNIFORM ARRAY

7.5.1 General Properties

By definition, a uniform array is an array of equispaced elements that are fed with currents of equal magnitude and having a progressive phaseshift along the array. For convenience, let us set the current magnitude a_n to unity. Equation (7.63) then reads

$$f(u) = \sum_{n=0}^{N-1} \exp(jnu) \tag{7.65}$$

where we have defined a new variable

$$u = kd \cos \chi + \alpha \tag{7.66}$$

and α is the phase angle by which the current in any element leads that of the preceding element.

The array factor as given by (7.65) is a phasor quantity. To study how the pattern varies with the angle χ, it is sufficient to examine the magnitude of $f(u)$, i.e. $|f(u)|$, since, in measuring the electric or magnetic fields, it is the r.m.s. rather than the instantaneous value that is measured. Thus

$$\begin{aligned} |f| &= |1 + \exp(ju) + \exp(j2u) + \cdots + \exp(j(N-1)u)| \\ &= \left|\frac{1-\exp(jNu)}{1-\exp(ju)}\right| = \left|\frac{\exp(\frac{1}{2}jNu)(\exp(-\frac{1}{2}jNu) - \exp(\frac{1}{2}jNu))}{\exp(\frac{1}{2}ju)(\exp(-\frac{1}{2}ju) - \exp(\frac{1}{2}ju))}\right| \\ &= \left|\frac{\sin(\frac{1}{2}Nu)}{\sin(\frac{1}{2}u)}\right| \end{aligned} \tag{7.67}$$

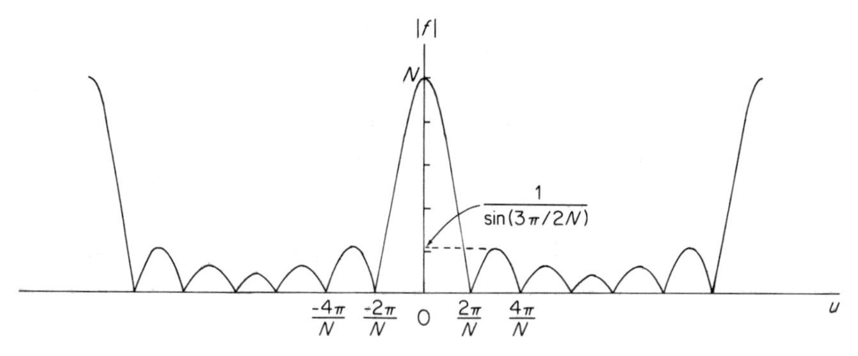

Figure 7.9 Sketch of the magnitude of the array factor of an N-element uniform array as a function of the variable $u = kd\cos\chi + \alpha$

A sketch of $|f|$ versus u is shown in Figure 7.9. The principal maximum (main beam) occurs at $u = 0$ and is equal to N. The zeros are given by

$$\tfrac{1}{2}Nu = \pm n\pi$$

or

$$u = \pm 2n\pi/N \tag{7.68}$$

where $n = 1, 2, 3, \ldots$.

The secondary maxima (side lobes) occur approximately midway between the nulls, when the numerator attains maximum values. These correspond to

$$\tfrac{1}{2}Nu = \pm(2m+1)\pi/2$$

or

$$u = \pm(2m+1)\pi/N \tag{7.69}$$

where $m = 1, 2, 3, \ldots, .$

The first side lobe occurs when

$$u = 3\pi/N \tag{7.70}$$

The amplitude of the first side lobe is

$$\frac{1}{\sin(u/2)} = \frac{1}{\sin(3\pi/2N)} \tag{7.71}$$

The ratio of the amplitudes of the main beam to the first side lobe is $N\sin(3\pi/2N)$. If N is large so that $3\pi/2N \ll 1$, then $\sin(3\pi/2N) \simeq 3\pi/2N$ and the ratio of first side lobe to main beam is $2/3\pi = 0.212$. In other words, the first side lobe is about 13.5 dB below the main beam. This ratio is independent of N as long as N is large enough for $\sin(3\pi/2N) \simeq 3\pi/2N$ to hold. Thus it is not possible to reduce the side lobe radiation relative to the main beam below 13.5 dB no matter how many elements we put into the array.

On the other hand, it is possible to decrease the width of the main beam by increasing the number of elements. The width of the main beam is equal to twice the angle between the principal maximum and the first null. In terms of the variable u, the main beam spans the range $-2\pi/N \leqslant u \leqslant 2\pi/N$, as shown in Figure 7.9. The width is thus inversely proportional to N. In theory, then, the main beam can be made as narrow as possible by increasing the number of elements.

7.5.2 Broadside Array, Ordinary Endfire Array, and Endfire Array with Increased Directivity

In order to express the width of the main beam in terms of the array parameters, it is necessary to specify the phase angle α in (7.66). In the following, we consider three types of uniform arrays.

Broadside Array

Let us recall that the main beam occurs at $u = 0$, corresponding to

$$\cos\chi = -\alpha/kd \tag{7.72}$$

For the broadside array, the main beam occurs at $\chi = 90^\circ$, which requires $\alpha = 0^\circ$. The first null occurs at $u = \pm 2\pi/N$. The correspondence between u and χ is obtained by setting $\alpha = 0$ in (7.66):

$$u = kd\cos\chi$$

or

$$\cos\chi = u/kd \tag{7.73}$$

Let the angle at which the first null occurs be $(\pi/2 + \Delta\chi)$. Then

$$\cos(\pi/2 + \Delta\chi) = \pm 2\pi/Nkd = \pm\lambda/Nd \tag{7.74}$$

Since the left-hand side is equal to $-\sin(\Delta\chi)$, we must choose the minus sign in (7.74). Thus the width of the main beam, to be abbreviated as MBW, is

$$(\text{MBW})_{\text{BS}} = 2(\Delta\chi) = 2\sin^{-1}(\lambda/Nd) \tag{7.75}$$

where subscript BS stands for broadside. If $\lambda/Nd \ll 1$, then $\sin(\Delta\chi) \simeq \Delta\chi$ and (7.75) simplifies to

$$(\text{MBW})_{\text{BS}} = 2\lambda/Nd \tag{7.76}$$

Ordinary Endfire Array

For the ordinary endfire array, we set $\alpha = -kd$ in (7.66) to obtain

$$u = kd(\cos\chi - 1) \tag{7.77}$$

From (7.72), the main beam occurs at $\cos\chi = 1$ or $\chi = 0$. Let the angle at which the first null occurs be $\Delta\chi$. Then

$$kd[\cos(\Delta\chi) - 1] = \pm 2\pi/N$$

or

$$\cos(\Delta\chi) - 1 = -\lambda/Nd$$
$$-2\sin^2(\tfrac{1}{2}\Delta\chi) = -\lambda/Nd$$

Hence

$$\Delta\chi = 2\sin^{-1}(\lambda/Nd)^{1/2}$$

The width of the main beam, being 2χ, is given by

$$(\mathrm{MBW})_{\mathrm{OE}} = 4\sin^{-1}(\lambda/2Nd)^{1/2} \tag{7.78}$$

where subscript OE stands for ordinary endfire. If $\lambda/Nd \ll 1$, then (7.78) simplifies to

$$(\mathrm{MBW})_{\mathrm{OE}} = 2\sqrt{\frac{2\lambda}{Nd}} \tag{7.79}$$

Comparing (7.79) with (7.76), we see that for an array of the same length (same Nd), the ordinary endfire array has a broader width of main beam than does the broadside array.

Endfire Array with Increased Directivity

The case just discussed, namely $\alpha = -kd$, produces a maximum field in the direction $\chi = 0$. However, it does not give the maximum directivity. It has been found (Hansen and Woodyard, 1938) that a larger directivity is obtained by increasing the value of α to

$$\alpha = -(kd + \pi/N) \tag{7.80}$$

With α given by (7.80), we have, from (7.66) and (7.65),

$$u = kd(\cos\chi - 1) - \pi/N$$

$$|f(u)| = \left|\sum_{n=0}^{N-1} \exp\{jn[kd(\cos\chi - 1 - \pi/N)]\}\right| \tag{7.81}$$

The condition (7.80) is referred to as the condition for 'increased directivity', and a uniform array with α given by (7.80) is called an endfire array with increased directivity. The proof that condition (7.80) yields a larger directivity than $\alpha = -kd$ is given in the paper by Hansen and Woodyard (1938) and will not be taken up here.

As an example, consider a 10-element uniform array with a quarter-wavelength spacing (Kraus, 1950). The value of α giving increased directivity is $-(2\pi/4 + \pi/10) = -0.6\pi$, while that corresponding to an ordinary endfire array is -0.5π. The array factor for the former is shown in Figure 7.10(a)

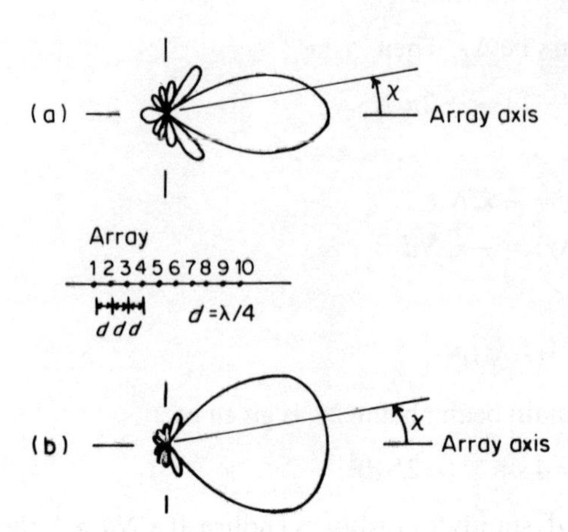

Figure 7.10 Sketches of the magnitudes of the array factors as a function of χ for: (a) a 10-element uniform array with $d = \lambda/4$ and a phasing between elements corresponding to the increased directivity condition; (b) a 10-element uniform array with $d = \lambda/4$ and a phasing between elements corresponding to an ordinary endfire array (Source: Kraus (1950), *Antennas*. Reproduced by permission of McGraw-Hill)

while that for the latter is shown in Figure 7.10(b). Both patterns are plotted to the same maximum. The increased directivity is apparent from the greater sharpness of the pattern in Figure 7.10(a). The half-power beamwidth for this pattern is 37° while that for the ordinary endfire pattern shown in Figure 7.11(b) is 68°. If the elements of the array are isotropic radiators, we have, from the definition of directivity,

$$D = \frac{4\pi |f_m|^2}{\int f(\chi)^2 \, d\Omega} \tag{7.82}$$

where f_m is the maximum value of f and Ω is solid angle. Numerical evaluation of (7.82) yields $D = 19$ for the array with increased directivity and $D = 11$ for the ordinary endfire array.

For the endfire array with increased directivity, it is readily shown that the width of the main beam is given by

$$(\text{MBW})_{\text{EID}} = 4 \sin^{-1}\left(\frac{\lambda}{4Nd}\right)^{1/2} \tag{7.83}$$

If the array is long so that $Nd \gg \lambda$, (7.83) becomes

$$(\text{MBW})_{\text{EID}} = 2\left(\frac{\lambda}{Nd}\right)^{1/2} \tag{7.84}$$

This width is 71% of the width of the ordinary endfire array.

In summary, this section has been concerned with the general properties of the array factor of uniform arrays as well as the beamwidths of three types of such arrays. In the next two sections, detailed studies will be given for two-element arrays consisting of $\frac{1}{2}\lambda$ dipoles.

7.6 TWO-ELEMENT ARRAY OF HALF-WAVELENGTH DIPOLES FED WITH EQUAL IN-PHASE CURRENTS

Consider two centre-fed $\frac{1}{2}\lambda$ dipoles sitting side by side with a spacing d as in Figure 7.11. The axes of both antennas are directed along $\hat{\mathbf{z}}$. In this section, we treat the case in which the two elements are fed with equal in-phase currents. We study the array factor, the field patterns, the driving-point impedances, and the gain in field intensity of the array over a reference antenna. The case in which the two elements are fed with equal currents of opposite phase will be treated in section 7.7.

7.6.1 Array Factor

The array factor for two equal in-phase elements of spacing d is obtained by putting $N = 2$, $a_0 = a_1 = a$ and $\alpha = 0$ in (7.63):

$$\begin{aligned} f &= a[1 + \exp(jkd\cos\chi)] \\ &= a\exp(\tfrac{1}{2}jkd\cos\chi)[\exp(-\tfrac{1}{2}jkd\cos\chi) + \exp(\tfrac{1}{2}jkd\cos\chi)] \\ &= 2a\cos(\tfrac{1}{2}kd\cos\chi)\exp(\tfrac{1}{2}jkd\cos\chi) \end{aligned} \tag{7.85}$$

As mentioned in section 7.5.1, when we measure the field strength at a point, we obtain its time-averaged value. This value is proportional to the magnitude of f and is independent of its phase. The phase factor in (7.85) will therefore not be considered further.

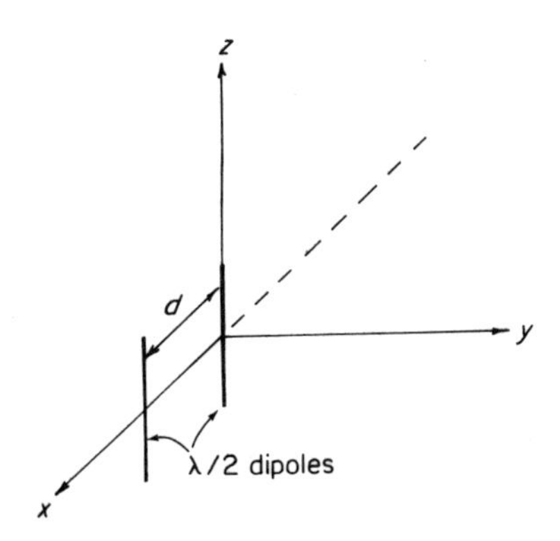

Figure 7.11 Geometry of a two-element array of side-by-side half-wave dipoles

From the expression

$$|f| = 2a \cos(\tfrac{1}{2}kd \cos \chi)$$

we find that the maxima occur at values of χ satisfying

$$\tfrac{1}{2}kd \cos \chi = n\pi \qquad n = 0, 1, 2, \ldots \tag{7.86}$$

Thus

$n = 0$:	$\chi = \frac{1}{2}\pi$	primary maximum (main beam)
$n = 1$:	$\chi = \cos^{-1}(\lambda/d)$	first subsidiary maximum
$\vdots$	$\vdots$	$\vdots$
$n = n$:	$\chi = \cos^{-1}(n\lambda/d)$	nth subsidiary maximum

The minima occur at values of χ satisfying

$$\tfrac{1}{2}kd \cos \chi = (2m + 1)\pi/2 \qquad m = 0, 1, \ldots \tag{7.87}$$

Thus

$m = 0$:	$\chi = \cos^{-1}(\lambda/2d)$	first null
$m = 1$:	$\chi = \cos^{-1}(3\lambda/2d)$	second null
$\vdots$	$\vdots$	$\vdots$
$m = m$:	$\chi = \cos^{-1}[(2m + 1)\lambda/2d]$	mth null

Since $-1 \leqslant \cos \chi \leqslant 1$, the number of maxima and minima is determined by the value λ/d.

Since $\cos \chi = \sin \theta \cos \phi$, the array factors in the x–y, y–z, and x–z planes take the following forms:

$$|f_{xy}| = 2a|\cos(\tfrac{1}{2}kd \cos \phi)| \tag{7.88}$$

$$|f_{yz}| = 2a \tag{7.89}$$

$$|f_{xz}| = 2a|\cos(\tfrac{1}{2}kd \sin \theta)| \tag{7.90}$$

Figure 7.12 illustrates the angular dependence of $|f_{xy}|$ for several values of d/λ. Note that the pattern described by $|f_{xz}|$ also has the same shape, except that θ in (7.90) is complementary to ϕ in (7.88).

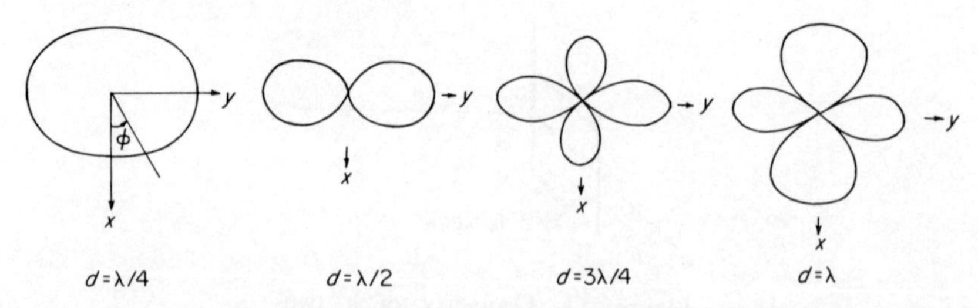

Figure 7.12 The array factors in the x–y plane of the two z-directed half-wave dipoles of Figure 7.11 fed with equal and in-phase currents

7.6.2 Field Patterns

The far field of the array is given by the product of the array factor and the element pattern, which in this case is a $\frac{1}{2}\lambda$ dipole. Denoting the element pattern by $E_{\lambda/2}(\theta)$ we have, by setting $\eta = 120\pi$ and $I(0) = a$ in (3.87) and (3.89),

$$|E_{\lambda/2}(\theta)| = \frac{60a}{r}\left|\frac{\cos(\frac{1}{2}\pi\cos\theta)}{\sin\theta}\right| \tag{7.91}$$

On applying the pattern multiplication principle, we obtain the far field of the array as

$$|E(\theta, \phi)| = \frac{120a}{r}\left|\cos(\tfrac{1}{2}kd\cos\chi)\frac{\cos(\frac{1}{2}\pi\cos\theta)}{\sin\theta}\right| \tag{7.92}$$

Equation (7.92) is valid for any point in space and for any spacing d. For the case where $d = \frac{1}{2}\lambda$, the fields in the x–y, y–z, and x–z planes are given by the following:

$$|E_{xy}(\phi)|_{d=\lambda/2} = \frac{120a}{r}|\cos(\tfrac{1}{2}\pi\cos\phi)| \tag{7.93}$$

$$|E_{yz}(\theta)|_{d=\lambda/2} = \frac{120a}{r}\left|\frac{\cos(\frac{1}{2}\pi\cos\theta)}{\sin\theta}\right| \tag{7.94}$$

$$|E_{xz}(\theta)|_{d=\lambda/2} = \frac{120a}{r}\left|\cos(\tfrac{1}{2}\pi\sin\theta)\frac{\cos(\frac{1}{2}\pi\cos\theta)}{\sin\theta}\right| \tag{7.95}$$

The field patterns given by (7.93)–(7.95) are illustrated in Figure 7.13(a)–(c) respectively. Note that there is relatively little radiation in the plane of the dipoles (x–z plane). This arises from the fact that, in this plane, the element pattern is maximum in the direction where the array factor is zero and vice versa.

If the number of elements is increased, it is evident from the discussion in section 7.4 that, if the spacing d is less than one wavelength, the pattern in the x–y plane will consist of a narrow main lobe and several side lobes. In the y–z

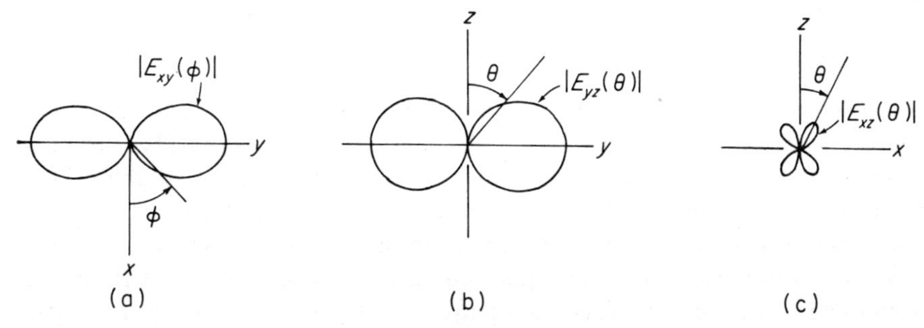

Figure 7.13 The patterns in the three coordinate planes of two z-directed half-wave dipoles spaced a half-wavelength apart and fed with equal and in-phase currents

plane, we still have the broad pattern of the $\frac{1}{2}\lambda$ dipole, while in the x–z plane there is relatively little radiation. Hence, for an array of many elements, the main lobe will be narrow in the x–y plane but broad in the y–z plane, a shape resembling that of a 'fan'. Such a pattern is referred to as a 'fan' beam. If we want the main lobe to be narrow in both the x–y and the y–z planes, namely a 'pencil' beam, it is necessary to have a two-dimensional array in the form of Figure 7.5. In this case, the array factor is given by (7.57), which consists of the product of two array factors, one corresponding to the array with its axis along the x-direction and the other corresponding to the array with its axis along the z-direction.

7.6.3 Driving-Point Impedances

Let V_1, I_1, and V_2, I_2 be the voltages and currents at the terminals of elements 1 and 2 respectively. Then

$$V_1 = I_1 Z_{11} + I_2 Z_{12} \tag{7.96}$$

$$V_2 = I_1 Z_{12} + I_2 Z_{22} \tag{7.97}$$

where Z_{11} and Z_{22} are the self-impedances of elements 1 and 2 respectively and Z_{12} is the mutual impedance between the two elements. Since the currents are equal and in phase

$$I_1 = I_2 \tag{7.98}$$

The input or driving-point impedances of elements 1 and 2 are, respectively,

$$Z_1 = V_1/I_1 = Z_{11} + Z_{12} \tag{7.99}$$

$$Z_2 = V_2/I_2 = Z_{22} + Z_{12} \tag{7.100}$$

For $\frac{1}{2}\lambda$ dipoles, $Z_{11} = Z_{22} = 73 + \text{j}42.5$ ohm and

$$Z_1 = Z_2 = 73 + \text{j}42.5 + Z_{12} \tag{7.101}$$

Since $Z_1 = Z_2$ and $I_1 = I_2$, it is necessary that the applied voltages V_1 and V_2 be equal and in phase. For the case where the spacing $d = \frac{1}{2}\lambda$, $Z_{12} = (-13 - \text{j}29)$ ohm (Figure 5.5) and

$$Z_1 = Z_2 = 60 + \text{j}13.5 \text{ ohm} \tag{7.102}$$

The reactance of 13.5 ohm can be tuned out by a series capacitor. It is also possible to resonate the elements by shortening them slightly. This modifies the resistive part somewhat and also alters the $E(\theta)$ field pattern slightly. To a first approximation, however, these effects can usually be neglected.

One method of feeding the two-element broadside array is shown in Figure 7.14. Two transmission lines of equal length l are joined at P to a third line extending to a transmitter. This will ensure $V_1 = V_2$, $I_1 = I_2$. If l is $\frac{1}{2}\lambda$ in length, the driving-point impedance of the array at P is a pure resistance of 30 ohm. Some impedance-transforming device such as a quarter-wave transformer is

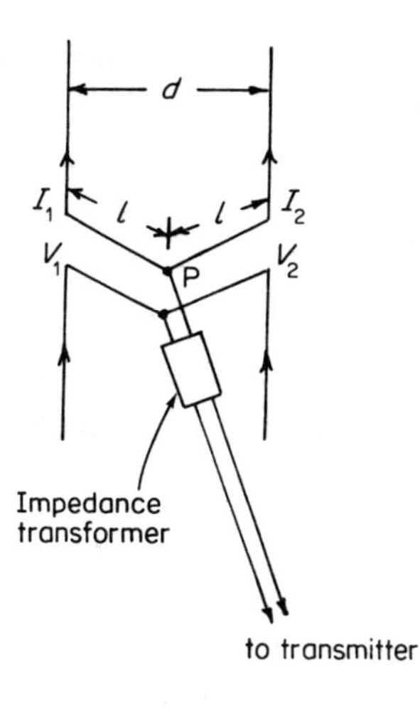

Figure 7.14 Method of supplying equal and in-phase currents to two dipoles

necessary to match this value to the characteristic impedance of the main transmission line.

7.6.4 Gain in Field Intensity

The gain in field intensity of an array over a reference antenna is given by the ratio of the field intensity from the array in a given direction to the field intensity from the reference antenna in the same direction when both are supplied with the same power W. If a hypothetical isotropic radiator is taken to be the reference antenna, the gain in field intensity is equal to the square root of the power gain defined in section 3.5. However, it is common to use a half-wave dipole as the reference antenna.

For the two-element array under consideration, the total power input W is the sum of the power in element 1, W_1, and the power in element 2, W_2. Assuming perfect conductors, we have

$$W_1 = \tfrac{1}{2}I_1^2 \operatorname{Re}(Z_1) = \tfrac{1}{2}I_1^2(R_{11} + R_{12}) \tag{7.103}$$

$$W_2 = \tfrac{1}{2}I_1^2 \operatorname{Re}(Z_2) = \tfrac{1}{2}I_2^2(R_{22} + R_{12}) \tag{7.104}$$

where I_1 and I_2 are peak currents. Since $R_{22} = R_{11}$ and $I_2 = I_1 = a$,

$$W = W_1 + W_2 = a^2(R_{11} + R_{12})$$

or

$$a = \sqrt{\frac{W}{R_{11} + R_{12}}} \tag{7.105}$$

Substitution of (7.105) into (7.92) yields the field intensity of the array:

$$|E(\theta, \phi)| = \frac{120}{r}\sqrt{\left(\frac{W}{R_{11} + R_{12}}\right)}\left|\cos(\tfrac{1}{2}kd\cos\chi)\frac{\cos(\frac{1}{2}\pi\cos\theta)}{\sin\theta}\right| \tag{7.106}$$

Let us take the reference antenna to be a $\frac{1}{2}\lambda$ dipole. For a power W supplied to this antenna, the current at its terminals is

$$a = \sqrt{(2W/73)} \tag{7.107}$$

Substituting (7.107) for a in (7.91), we obtain

$$|E_{\lambda/2}(\theta)| = \frac{60}{r}\sqrt{\left(\frac{2W}{73}\right)}\left|\frac{\cos(\frac{1}{2}\pi\cos\theta)}{\sin\theta}\right| \tag{7.108}$$

The gain in field intensity of the array with respect to a $\frac{1}{2}\lambda$ depole, denoted by $[G_{\mathrm{f}}(\theta, \phi)]_{\lambda/2}$, is

$$[G_{\mathrm{f}}(\theta, \Phi)]_{\lambda/2} = \frac{|E(\theta, \phi)|}{|E_{\lambda/2}(\theta)|} = \sqrt{\left(\frac{146}{73 + R_{12}}\right)}|\cos(\tfrac{1}{2}kd\cos\chi)| \tag{7.109}$$

where we have set the self-resistance R_{11} to 73 ohm. For the case where the spacing $d = \frac{1}{2}\lambda$, $R_{12} = -13$ ohm (Figure 5.5) and (7.109) becomes

$$[G_{\mathrm{f}}(\theta, \phi)]_{\lambda/2} = 1.56|\cos(\tfrac{1}{2}\pi\cos\chi)| \tag{7.110}$$

In the x–y and y–z planes, we have

$$[G_{\mathrm{f}}(\phi)]_{\lambda/2} = 1.56|\cos(\tfrac{1}{2}\pi\cos\phi)| \tag{7.111}$$

$$[G_{\mathrm{f}}(\theta)]_{\lambda/2} = 1.56 \tag{7.112}$$

In Figure 7.15, the array patterns in the x–y and y–z planes are compared with that of a single $\frac{1}{2}\lambda$ dipole with the same input power. Note that in the x–y plane, $G_{\mathrm{f}}(\phi)$ exceeds unity only for ϕ between 56° and 124° and attains the maximum value of 1.56 in the broadside direction ($\phi = 90°$). In the y–z plane, the shape of the array pattern is the same as the $\frac{1}{2}\lambda$ dipole, but the ratio of the radius vectors in the same direction is a constant equal to 1.56.

Returning to the case of arbitrary spacing, it is of interest to see how the maximum gain varies with spacing. From (7.109), the maximum gain occurs

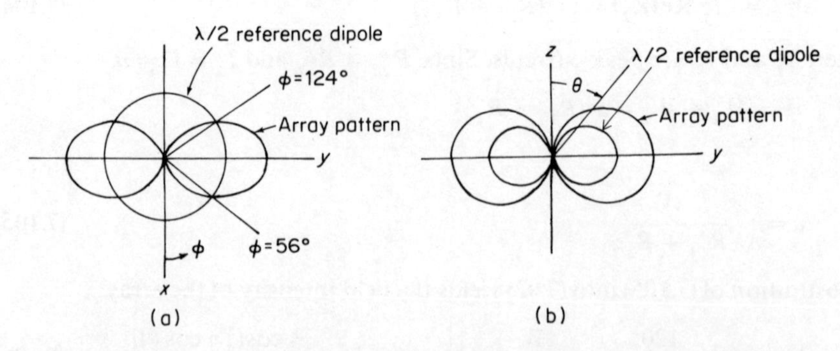

Figure 7.15 Patterns in the x–y and y–z planes of two z-directed half-wave dipoles spaced a half-wavelength apart

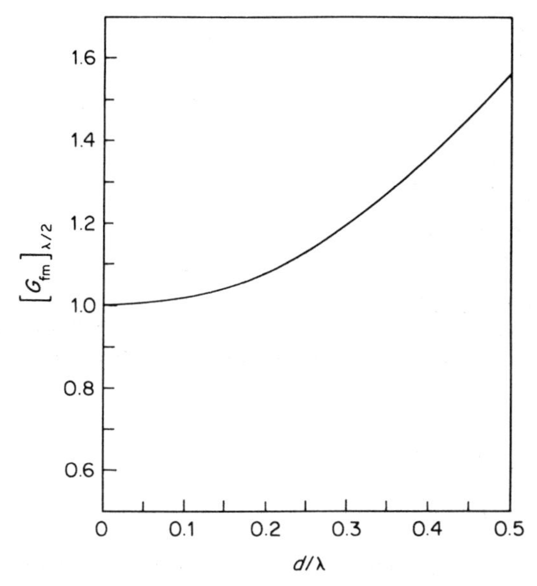

Figure 7.16 Maximum gain in field intensity of two half-wave dipoles fed with equal and in-phase currents as a function of spacing

at $\chi = 90°$ and is given by

$$G_{fm} = \sqrt{\frac{146}{73 + R_{12}}} \tag{7.113}$$

Figure 7.16 shows $[G_{fm}]_{\lambda/2}$ as a function of d/λ up to $d/\lambda = 0.5$. In this range, $[G_{fm}]_{\lambda/2}$ is a monotonically increasing function. For very small spacing, the value of G_{fm} approaches unity.

7.7 TWO-ELEMENT ARRAY OF HALF-WAVELENGTH DIPOLES FED WITH EQUAL CURRENTS OF OPPOSITE PHASE

In this section, we consider the case in which the two $\frac{1}{2}\lambda$ dipoles of Figure 7.11 are fed with currents which are equal in magnitude but opposite in phase. As in section 7.6, we study the array factor, the field patterns, the driving-point impedances, and the gain in field intensity of the array over a reference antenna.

7.7.1 Array Factor

The array factor for two elements of spacing d fed with equal but opposite-phase currents is obtained by putting $N = 2$, $a_0 = a_1 = a$ and $\alpha = -\pi$ in (7.63):

$$f = a[1 - \exp(jkd\cos\chi)] \tag{7.114}$$

$$= -2ja\sin(\tfrac{1}{2}kd\cos\chi)\exp(\tfrac{1}{2}jkd\cos\chi) \tag{7.115}$$

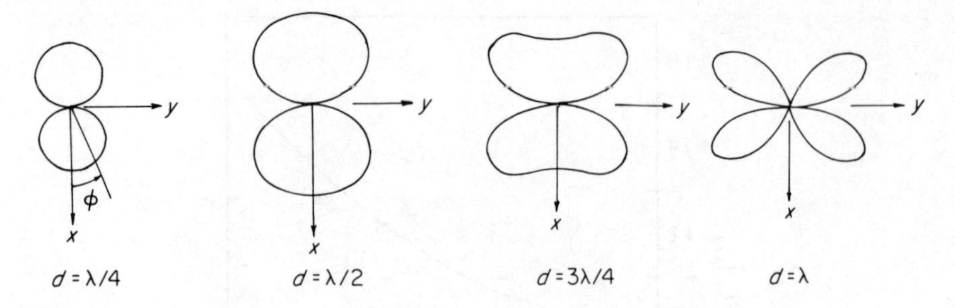

Figure 7.17 Array factors in the x–y plane of two z-directed half-wave dipoles fed with currents of equal amplitude but opposite phase

Hence

$$|f| = 2a|\sin(\tfrac{1}{2}kd\cos\chi)| \tag{7.116}$$

In the x–y, y–z, and x–z planes, the array factors take the following forms:

$$|f_{xy}| = 2a|\sin(\tfrac{1}{2}kd\cos\phi)| \tag{7.117}$$

$$|f_{yz}| = 0 \tag{7.118}$$

$$|f_{xz}| = 2a|\sin(\tfrac{1}{2}kd\sin\theta)| \tag{7.119}$$

Equation (7.118) states that the field intensity is zero in the y–z plane, a consequence of the fact that points lying in this plane are equidistant from the two out-of-phase elements. Figure 7.17 illustrates the angular dependence of $|f_{xy}|$ for several values of d/λ. The shape of f_{xz} is similar, the only difference being that θ in (7.119) is complementary to ϕ in (7.117).

7.7.2 Field Patterns

The element pattern for the half-wave dipole is given by (7.91). On applying the pattern multiplication principle, we obtain the far field of the array as

$$|E(\theta, \phi)| = \frac{120a}{r}\left|\sin(\tfrac{1}{2}kd\cos\chi)\frac{\cos(\tfrac{1}{2}\pi\cos\theta)}{\sin\theta}\right| \tag{7.120}$$

Equation (7.120) is valid for any point in space and for any spacing d. For the case $d = \frac{1}{2}\lambda$, the fields in the x–y and x–z planes are given by

$$|E_{xy}(\phi)|_{d=\lambda/2} = \frac{120a}{r}|\sin(\tfrac{1}{2}\pi\cos\phi)| \tag{7.121}$$

$$|E_{xz}(\theta)|_{d=\lambda/2} = \frac{120a}{r}\left|\sin(\tfrac{1}{2}\pi\sin\theta)\frac{\cos(\tfrac{1}{2}\pi\cos\theta)}{\sin\theta}\right| \tag{7.122}$$

For $d = \frac{1}{2}\lambda$, the field patterns in the x–y and x–z planes are illustrated in

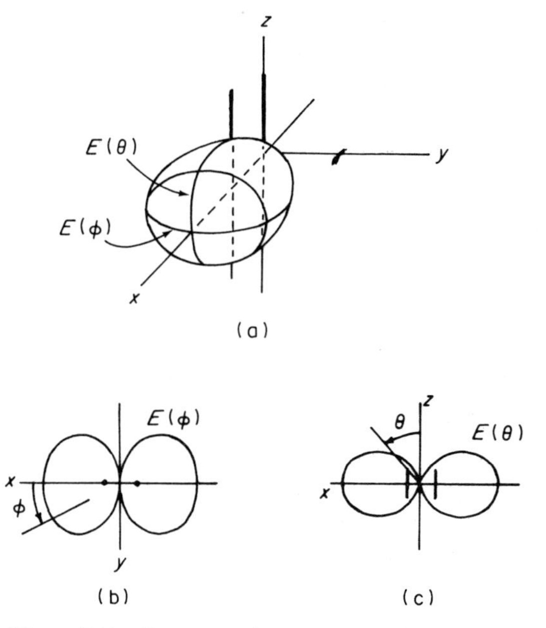

Figure 7.18 Patterns of two z-directed half-wave dipoles spaced $\lambda/2$ apart, fed with currents of equal amplitude but opposite phase

Figure 7.18(b) and (c) respectively. The relative three-dimensional pattern is suggested in Figure 7.18(a). This pattern is actually bidirectional, only one-half being shown. The maximum occurs at $\phi = 0$ and $\phi = 180°$ and the array is an ordinary endfire array with $\alpha = -kd = -\pi$.

7.7.3 Driving-Point Impedances

Since we now have $I_2 = -I_1$, the driving-point impedances of elements 1 and and 2 are readily shown from (7.96) and (7.97) to be

$$Z_1 = Z_{11} - Z_{12} = 73 + j42.5 - Z_{12} \tag{7.123}$$

$$Z_2 = Z_{22} - Z_{12} = 73 + j42.5 - Z_{12} \tag{7.124}$$

For the case where the spacing $d = \frac{1}{2}\lambda$,

$$Z_1 = Z_2 = 86 + j72 \text{ ohm} \tag{7.125}$$

One method of feeding the opposite-phase doublet is shown in Figure 7.19. A cross-over in the transmission line from the driving point P to one of the elements provides the opposite phase required. The series capacitors tune out the reactance of 72 ohm of Z_1 and Z_2. If l is $\frac{1}{2}\lambda$ in length, the driving-point

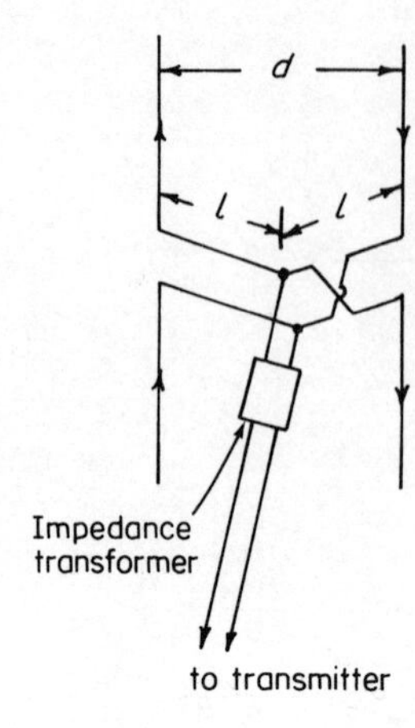

Figure 7.19 Method of supplying currents of equal amplitude but opposite phase to two dipoles

impedance of the array at P is a pure resistance of 43 ohm. Some impedance-transforming device is necessary to match this value to the characteristic impedance of the main transmission line.

7.7.4 Gain in Field Intensity

Using the same analysis as in section 7.6.4, the current magnitude in each element for a power input W to the array is given by

$$a = \sqrt{\frac{W}{R_{11} - R_{12}}} \tag{7.126}$$

In deriving (7.126), it is assumed that the antennas are perfect conductors. The gain in field intensity of the array with respect to a $\frac{1}{2}\lambda$ dipole is readily shown to be

$$[G_f(\theta, \phi)]_{\lambda/2} = \sqrt{\left(\frac{146}{73 - R_{12}}\right)}|\sin(\tfrac{1}{2}kd \cos \chi)| \tag{7.127}$$

For the case where the spacing $d = \frac{1}{2}\lambda$, (7.127) becomes

$$[G_f(\theta, \phi)]_{\lambda/2} = 1.3|\sin(\tfrac{1}{2}\pi \cos \chi)| \tag{7.128}$$

In the x–y and x–z planes, we have

$$[G_f(\phi)]_{\lambda/2} = 1.3|\sin(\tfrac{1}{2}\pi \cos \phi)| \tag{7.129}$$

$$[G_f(\theta)]_{\lambda/2} = 1.3|\sin(\tfrac{1}{2}\pi \sin \theta)| \tag{7.130}$$

7.7.5 Effect of Heat Losses on Closely Spaced Elements

Setting $\chi = 0$ in (7.127), we obtain the gain in the endfire direction for two $\frac{1}{2}\lambda$ dipoles fed with equal currents of opposite phase as

$$[G_f(0)]_{\lambda/2} = \sqrt{\left(\frac{146}{73 - R_{12}}\right)}|\sin(\pi d/\lambda)| \tag{7.131}$$

Figure 7.20 shows $[G_f(0)]_{\lambda/2}$ as a function of d/λ up to $d/\lambda = 0.5$. Note that the array produces substantial gains even when the spacing is decreased to small values. As the spacing d approaches zero, R_{12} approaches 73 ohm and the factor $146/(73 - R_{12})$ approaches infinity. At the same time, the function $|\sin(\pi d/\lambda)|$ approaches zero. The product of the two stays finite, levelling off at a value of about 1.56 for small spacings. The fact that increased gain is associated with small spacings makes this arrangement attractive, since the antenna system is more compact. However, as the spacing decreases, the radiation resistance $R_1 = \text{Re}\, Z_1 = 73 - R_{12}$ decreases while the antenna current becomes quite large, as evident from (7.126). The variation of R_1 as a function of spacing is shown in Figure 7.21.

Thus far it has been assumed that there are no heat losses in the antenna. If heat losses are taken into account the input resistance R_{1T} will consist of the radiation resistance and an equivalent loss resistance R_{1L} such that

$$R_{1T} = R_1 + R_{1L} \tag{7.132}$$

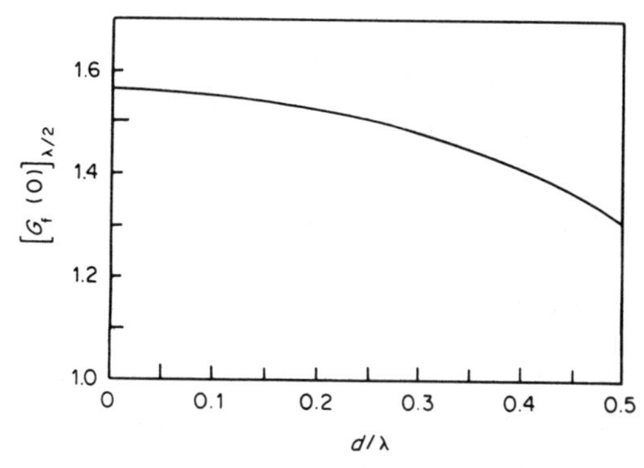

Figure 7.20 Gain in field intensity in the endfire direction as a function of spacing for two half-wave dipoles fed with currents of equal amplitude but opposite phase (Source: Kraus (1950), *Antennas*. Reproduced by permission of McGraw-Hill)

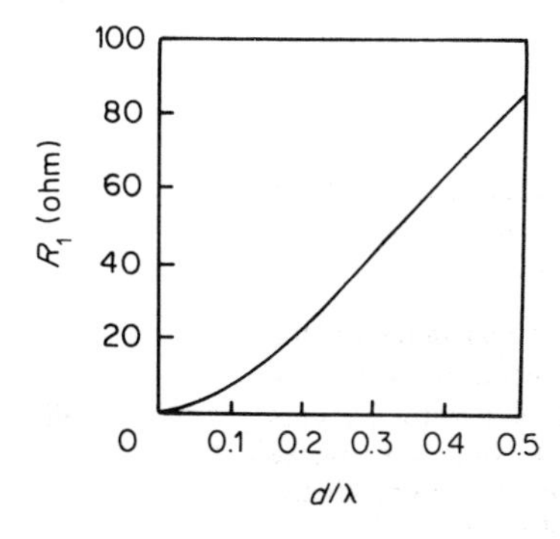

Figure 7.21 Radiation resistance as a function of spacing for two half-wave dipoles fed with currents of equal amplitude but opposite phase (Source: Kraus (1950), *Antennas*. Reproduced by permission of McGraw-Hill)

The radiation efficiency is defined as the ratio of the power radiated to the power input of the antenna. Thus

$$\text{radiation efficiency } (\%) = \frac{R_1}{R_1 + R_{1L}} \times 100 \tag{7.133}$$

If the radiation resistance is large compared to the loss resistance, the efficiency is close to 100% and the effect of loss can be neglected. For two closely spaced opposite-phase $\frac{1}{2}\lambda$ dipoles, however, R_1 is relatively small and the effect of loss resistance may become appreciable. Taking this into account, the current a in each element for a power W supplied to the array is

$$a = \sqrt{\frac{W}{73 + R_{1L} - R_{12}}} \tag{7.134}$$

Similarly, the current at the terminals of the reference $\frac{1}{2}\lambda$ dipole is $\sqrt{[2W/(73 + R_{1L})]}$. The gain in field intensity of the array with respect to a $\frac{1}{2}\lambda$ dipole then becomes

$$[G_f(\chi)]_{\lambda/2} = \sqrt{\left(\frac{2(73 + R_{1L})}{73 + R_{1L} - R_{12}}\right)} |\sin(\tfrac{1}{2}kd \cos \chi)| \tag{7.135}$$

Setting $\chi = 0°$ in (7.135), we obtain

$$[G_f(0)]_{\lambda/2} = \sqrt{\left(\frac{2(73 + R_{1L})}{73 + R_{1L} - R_{12}}\right)} |\sin(\tfrac{1}{2}kd)| \tag{7.136}$$

The effect of loss resistance on the quantity $[G_f(0)]_{\lambda/2}$ is illustrated in Figure 7.22. It is apparent that a loss resistance of only 1 ohm seriously limits the gain at spacings of less than 0.1λ, and larger loss resistances cause reductions in gain at greater spacings. In the literature, the term 'closely spaced' elements usually refers to spacings less than a quarter of a wavelength.

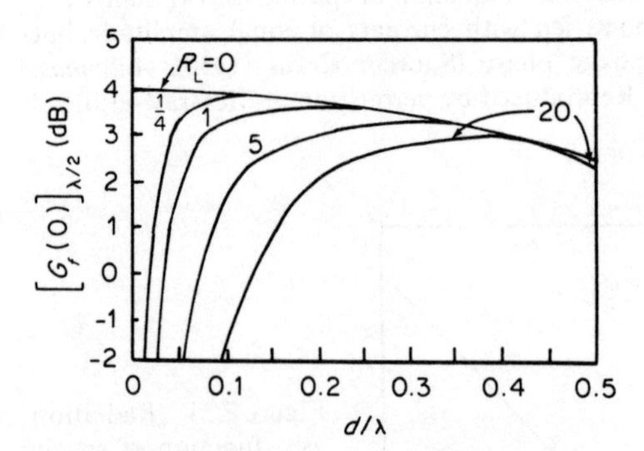

Figure 7.22 Effect of loss resistance on the gain in field intensity in the endfire direction of two half-wave dipoles fed with currents of equal amplitude but opposite phase (Source: Kraus (1950), *Antennas*. Reproduced by permission of McGraw-Hill)

In summary, we have in sections 7.6 and 7.7 studied not only the far-field patterns but also the driving-point impedances and the gain in field intensity over a reference antenna of two-element arrays consisting of $\frac{1}{2}\lambda$ dipoles fed with equal in-phase currents and with equal opposite-phase currents. It is evident that the method of analysis can similarly be used for other types of arrays. We now turn to a further examination of the properties of the array factor of N equispaced elements.

7.8 THE ARRAY FACTOR AS A POLYNOMIAL

The array factor of an equispaced array is given by (7.59):

$$f(\psi_1) = \sum_{n=0}^{N-1} I_n \exp(jknd\psi_1)$$

where

$$\psi_1 = \cos\chi$$

$$I_n = a_n \exp(j\alpha_n)$$

It is convenient to separate the phase α_n into two parts, $\alpha_n = \alpha + \alpha'_n$, where α is a constant that denotes the progressive phaseshift from element to element and α'_n denotes the deviation from progressive phaseshift of the nth element. Letting

$$u = \alpha + kd\cos\chi \tag{7.137}$$

(7.59) can be written in the form

$$f = \sum_{n=0}^{N-1} a_n \exp[j(\alpha'_n + nu)] \tag{7.138}$$

We now define

$$z = \exp(ju) \qquad A_n = a_n \exp(j\alpha'_n) \tag{7.139}$$

Then

$$f = A_0 + A_1 z + \cdots + A_{N-1} z^{N-1} \equiv P_{N-1}(z) \tag{7.140}$$

Equation (7.140) shows that the array factor of a uniformly spaced array of N elements can be expressed as an $(N-1)$th-order polynomial in the complex variable z. Thus to every linear array of N elements, we can associate an $(N-1)$th-order polynomial. Conversely, to every $(N-1)$th-order polynomial in z, we can associate a linear array of N elements. It follows from the properties of polynomials that the product of the array factor of two linear arrays is itself the array factor of a linear array. Conversely, a given array can be conceptually resolved into an arrangement of smaller arrays.

The variable $z = \exp(ju)$ where $u = \alpha + kd\cos\chi$ is a complex variable of unit magnitude. Thus as χ varies between 0 and π, z varies along the unit circle between

$$\exp[j(\alpha - kd)] \leqslant z \leqslant \exp[j(\alpha + kd)]$$

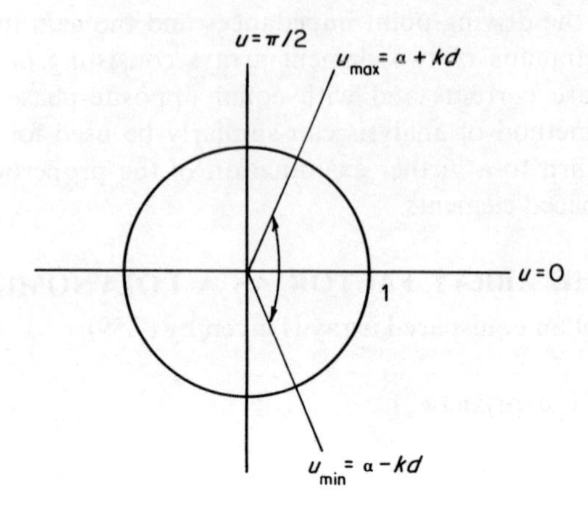

Figure 7.23 The portion of the unit circle in the $z = \exp(ju)$ plane that corresponds to the visible region

or

$$\alpha - kd \leqslant u \leqslant \alpha + kd \tag{7.141}$$

The range of z along the unit circle corresponds to the visible region and it depends on the ratio d/λ and α. This is illustrated in Figure 7.23.

Let the roots of the polynomial $P_{N-1}(z)$ be $z_1, z_2, \ldots, z_{N-1}$. Then

$$P_{N-1}(z) = (z - z_1)(z - z_2)\ldots(z - z_{N-1}) \tag{7.142}$$

The roots are in general complex numbers that can lie anywhere in the complex z-plane. If a zero falls on the unit circle within the range of variation of u, it would follow that at the corresponding angle $\chi = \cos^{-1}(u - \alpha)/kd$, there is no radiation from the array.

If the roots of the polynomial $P_{N-1}(z)$ are known, we have a graphical method of finding the array factor for any value of u or χ. It is simply given by the product of the lengths of the various segments $|z - z_i|$, i.e.

$$|f| = \prod_{i=1}^{N-1} |z - z_i| \tag{7.143}$$

Let us illustrate the above ideas with a few examples.

Uniform Broadside Array

For a uniform broadside array of N elements, $\alpha = 0$, $\alpha_i = 0$. The array factor

$$f = z^{N-1} + z^{N-2} + \cdots + z + 1 = \frac{z^N - 1}{z - 1}$$

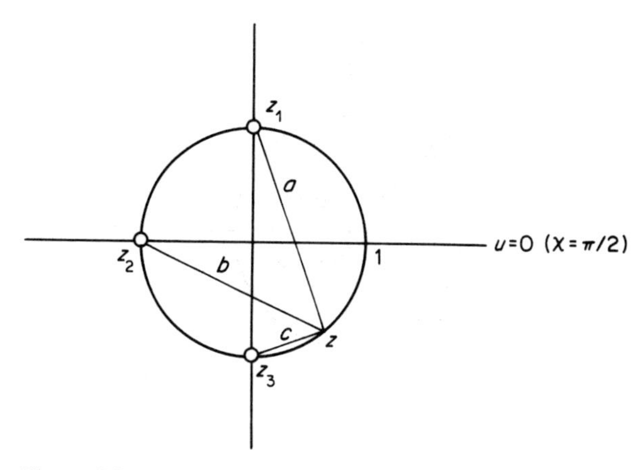

Figure 7.24 The circle diagram of a four-element uniform broadside array of half-wave spacing

$$= (z-1)^{-1} \prod_{m=1}^{N} [z - \exp(j2\pi m/N)] \tag{7.144}$$

The zeros of f are therefore equispaced on the unit circle, starting with $z_1 = \exp(j2\pi/N)$. Since $u = kd\cos\chi$, as χ varies between π and 0, u varies between $-kd$ and kd. The real axis ($u = 0$) corresponds to $\chi = \frac{1}{2}\pi$.

Uniform Broadside Array of Four Elements with $\frac{1}{2}\lambda$ Spacing

To be specific, let us consider $N = 4$ and $d = \frac{1}{2}\lambda$ in (7.144). Then

$$f = z^3 + z^2 + z + 1 = \prod_{m=1}^{4} [z - \exp(j\pi m/2)]/(z-1) \tag{7.145}$$

$$u = \pi \cos\chi \tag{7.146}$$

and the range of u is 2π radians. The circle diagram corresponding to this array is shown in Figure 7.24. Note that the magnitude of $|f|$ at any value of u is equal to the product of the lengths abc where

$$a = |z - z_1| \qquad z_1 = \exp(j\tfrac{1}{2}\pi)$$
$$b = |z - z_2| \qquad z_2 = \exp(j\pi)$$
$$c = |z - z_3| \qquad z_3 = \exp(-j\tfrac{1}{2}\pi)$$

A Design Example

Suppose we are asked to design a four-element broadside array of $\frac{1}{2}\lambda$ spacing which has a broader main-lobe width compared to the uniform array. With the aid of the circle diagram, this can readily be done. Since $u = 0$ corresponds to the direction of maximum radiation, it is clear that the two adjacent zeros will determine the width of the main lobe. If these two zeros are farther apart,

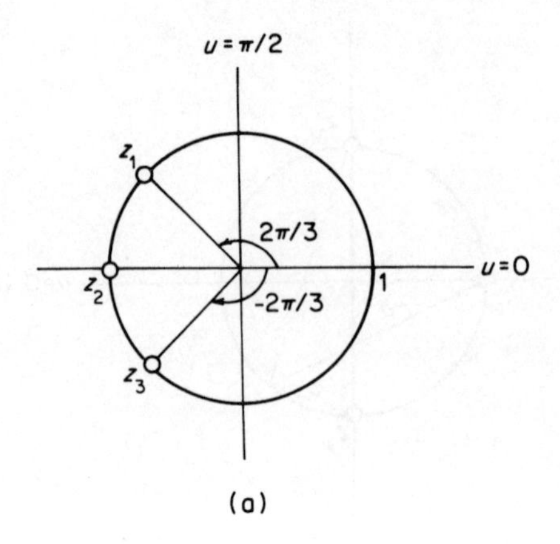

Figure 7.25 A four-element array (b) and its circle diagram (a). This array has a wider main beam and a smaller ratio of side lobe to main beam than does the uniform array of Figure 7.24

the resultant main lobe would have a broader width. For the uniform array, they are given by z_1 and z_3 in Figure 7.24, which are π radians apart. If we now move z_3 clockwise and z_1 anticlockwise, they would be farther apart. For definiteness, let the new locations be $z_1 = \exp(j2\pi/3)$ and $z_3 = \exp(-j2\pi/3)$, as indicated in Figure 7.25(a). If z_2 is allowed to remain at $\exp(j\pi)$, the array factor having this set of zeros is

$$
\begin{aligned}
f &= (z+1)[z-\exp(j2\pi/3)][z-\exp(-j2\pi/3)] \\
&= z^3 + 2z^2 + 2z + 1
\end{aligned}
\tag{7.147}
$$

The excitation currents that give rise to this array factor are shown in Figure 7.25(b). The currents are still in phase, with the magnitudes of the two central elements twice those of the end ones. Compared with the uniform array, the main lobe of the new array is larger. The ratio of side lobe to main lobe, however, is smaller. This is evident from examining the equation

$$
|f| = |z - z_1||z - z_2||z - z_3| \tag{7.148}
$$

The peak of the main lobe corresponds to $u = 0$ while the peak of the side lobe corresponds to a point approximately midway between z_1 and z_2. It is clear that, for the new array, the product in (7.148) will be larger for the main lobe and smaller for the side lobe, resulting in a smaller ratio of side lobe to main lobe. Thus widening the main lobe is accompanied by a reduction of the ratio of side lobe to main lobe. If an array with a narrower main-lobe width

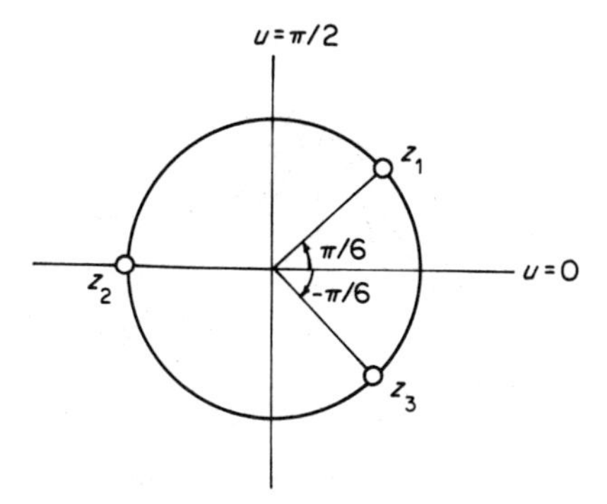

Figure 7.26 The circle diagram of a four-element array having a narrower main beam and a higher ratio of side lobe to main beam than does the uniform array of Figure 7.24

compared to the uniform array is desired, we can, starting from z_1 and z_3 corresponding to be uniform array, move z_1 clockwise and z_3 anticlockwise, as shown in Figure 7.26. Let the new locations be $z_1 = \exp(j\pi/6)$ and $z_3 = \exp(-j\pi/6)$. If z_2 still remains at $\exp(j\pi)$, the array factor is

$$\begin{aligned} f &= (z+1)[z-\exp(j\pi/6)][z-\exp(-j\pi/6)] \\ &= z^3 - 0.732z^2 - 0.732z + 1 \end{aligned} \tag{7.149}$$

It is evident that this array, while having a narrower main-lobe width, will also have a higher ratio of side lobe to main lobe.

7.9 BINOMIAL ARRAY

Let us consider the four-element array of $\frac{1}{2}\lambda$ spacing discussed in the last section. Suppose the three zeros are all located at the point -1 so that $z_1 = z_2 = z_3 = -1$. Then the main lobe will span the whole range of u from $-\pi$ to π or the entire visible range of χ from 0 to π radians. In other words, the main-lobe width is the broadest possible among four-element arrays with half-wavelength spacing and there is no side lobe. The array factor is given by

$$f = (z+1)^3 = z^3 + 3z^2 + 3z + 1 \tag{7.150}$$

In general, an equispaced array of N elements spaced $\frac{1}{2}\lambda$ apart will have no side lobes if all the zeros are located at -1, i.e. -1 is a root of multiplicity $(N-1)$. The array factor is

$$f = (z+1)^{N-1} \tag{7.151}$$

The element currents are all in phase, with a magnitude distribution following the binomial coefficients. For this reason, such an array is called a binomial

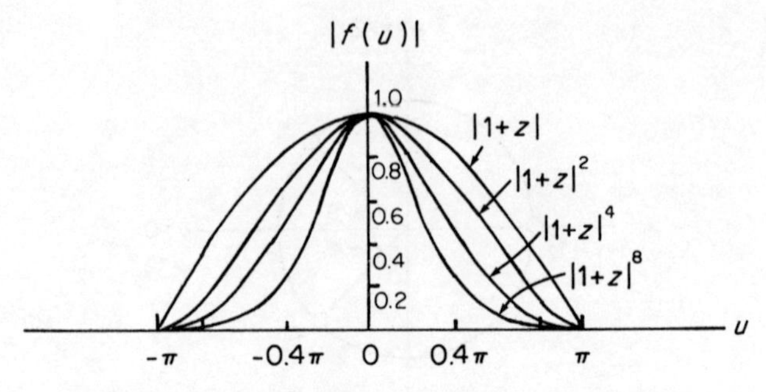

Figure 7.27 The array factors of binomial arrays as a function of the variable u

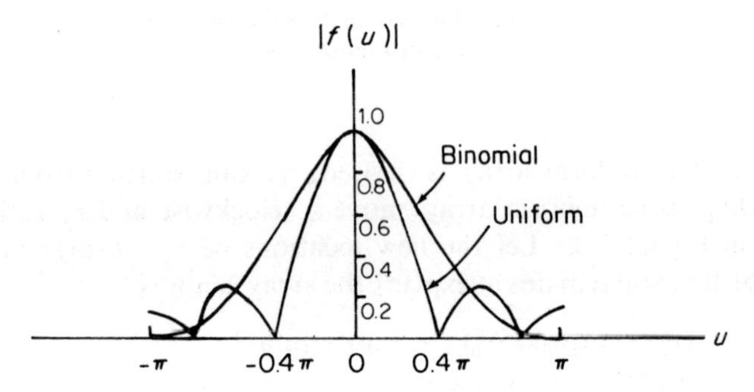

Figure 7.28 The array factors of a five-element uniform array of half-wavelength spacing and a binomial array of the same number of elements

array. The array factor as a function of u is illustrated in Figure 7.27. In Figure 7.28, the array factor of a five-element uniform array of half-wavelength spacing is compared with that of a binomial array of the same number of elements.

7.10 DOLPH–TSCHEBYSCHEFF ARRAY

7.10.1 Introduction

Let us again consider the four-element array studied in Section 7.8. The circle diagram for the case of uniform current amplitudes is shown in Figure 7.24. A narrower main lobe can be obtained by moving the two zeros z_1 and z_3 closer to the $u = 0$ axis. This, however, will raise the level of the side lobe. On the other hand, if we want to reduce the side-lobe level, we can accomplish this by moving z_1 and z_3 closer to the third zero $z_2 = -1$. In so doing, however, the main lobe becomes wider. The following question naturally suggests itself. What are the locations of zeros, and therefore the current distributions, for which the array will have (1) the narrowest main-lobe width for a given side-lobe level or (2) the lowest side-lobe level for a given main-lobe width? This question, of course, can be asked of a linear equispaced array of any number of elements. The answer is a current distribution known as the Tschebyscheff distribution and an array possessing properties (1) and (2) is said to be an 'optimum array'. It has been shown that property (1) implies (2) and vice versa.

7.10.2 Array Factor of Symmetric Arrays

The Tschebyscheff array is restricted to an array with a progressive phaseshift and symmetric current amplitudes. Thus

$$\alpha'_n = 0 \qquad \text{for all } i \tag{7.152}$$

and

$$\begin{aligned} a_0 &= a_{N-1} \\ a_1 &= a_{N-2} \\ a_2 &= a_{N-3} \\ &\text{etc.} \end{aligned} \tag{7.153}$$

For example, a seven-element array will have an amplitude and phase distributions shown in Figure 7.29. For N elements, the array factor is

$$\begin{aligned} |f| = |a_0 + a_1 \exp(\mathrm{j}u) + a_2 \exp(\mathrm{j}2u) + \cdots + a_2 \exp[\mathrm{j}(N-3)u] \\ + a_1 \exp[\mathrm{j}(N-2)u] + a_0 \exp[\mathrm{j}(N-1)u]| \end{aligned} \tag{7.154}$$

Since $|f|$ is unaltered after division by $\exp[\mathrm{j}(N-1)u/2]$, we have

$$\begin{aligned} |f| &= |a_0 \exp[-\mathrm{j}\tfrac{1}{2}(N-1)u] + a_1 \exp[-\mathrm{j}(N-3)\tfrac{1}{2}u] + \cdots \\ &\quad + a_1 \exp[\mathrm{j}(N-3)\tfrac{1}{2}u] + a_0 \exp[\mathrm{j}(N-1)\tfrac{1}{2}u]| \\ &= 2|a_0 \cos[(N-1)\tfrac{1}{2}u] + a_1 \cos[(N-3)\tfrac{1}{2}u] + \cdots| \end{aligned} \tag{7.155}$$

Each of the terms in (7.155) is of the form $\cos(\frac{1}{2}mu)$ where m is an integer. We shall show that $\cos(\frac{1}{2}mu)$ is an mth-order polynomial in the real variable $\cos(\frac{1}{2}u)$. Using this result for the time being, (7.155) shows that the array factor of a symmetric array of N elements can be expressed as an $(N-1)$th-order polynomial in the *real* variable $\cos(\frac{1}{2}u)$. The distinction with the result of section 7.8 is that the variable of the polynomial here is a real variable while, in the general case, it is a complex variable $z = \exp(\mathrm{j}u)$. If we choose the excitation coefficients a_0, a_1, etc., such that (7.155) becomes a Tschebyscheff polynomial, then the resultant array is called a Tschebyscheff array. Dolph (1946) has proved that a Tschebyscheff array has the optimum properties described

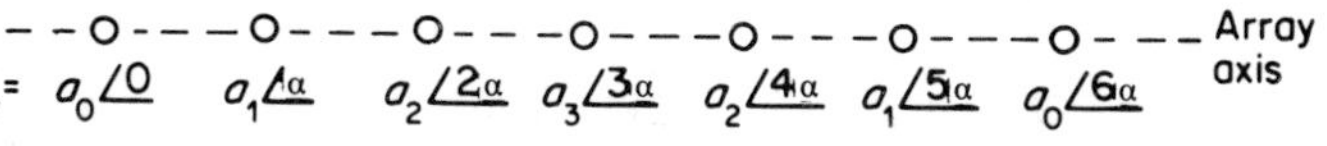

Figure 7.29 An array with a progressive phaseshift and symmetric current amplitudes

in section 7.10.1. The interested reader is referred to his paper for the mathematical details of the proof.

To show that $\cos(\frac{1}{2}mu)$ is an mth-order polynomial in $\cos(\frac{1}{2}u)$, we begin with de Moivre's theorem:

$$\exp(\mathrm{j}\tfrac{1}{2}mu) = \cos(\tfrac{1}{2}mu) + \mathrm{j}\sin(\tfrac{1}{2}mu) = [\cos(\tfrac{1}{2}u) + \mathrm{j}\sin(\tfrac{1}{2}u)]^m \tag{7.156}$$

On taking the real part of (7.156), we have

$$\begin{aligned}\cos(\tfrac{1}{2}mu) &= \mathrm{Re}[\cos(\tfrac{1}{2}u) + \mathrm{j}\sin(\tfrac{1}{2}u)]^m \\ &= \cos^m(\tfrac{1}{2}u) - \frac{m(m-1)}{2!}\cos^{m-2}(\tfrac{1}{2}u)\sin^2(\tfrac{1}{2}u) \\ &\quad + \frac{m(m-1)(m-2)(m-3)}{4!}\cos^{m-4}(\tfrac{1}{2}u)\sin^4(\tfrac{1}{2}u) + \cdots\end{aligned} \tag{7.157}$$

Since $\sin^2(\frac{1}{2}u) = 1 - \cos^2(\frac{1}{2}u)$, it follows that the right-hand side of (7.157) is an mth-order polynomial in $\cos(\frac{1}{2}u)$.

7.10.3 Properties of Tschebyscheff Polynomials

The Tschebyscheff polynomial of order N of a real variable x is defined by

$$T_N(x) = \cos(N\cos^{-1}x) \tag{7.158}$$

To show that (7.158) is a polynomial, let

$$x = \cos w \tag{7.159}$$

Then

$$T_N(x) = \cos(Nw) \tag{7.160}$$

$$T_{N+1}(x) = \cos[(N+1)w] = \cos(Nw)\cos w - \sin(Nw)\sin w \tag{7.161}$$

$$T_{N-1}(x) = \cos[(N-1)w] = \cos(Nw)\cos w + \sin(Nw)\sin w \tag{7.162}$$

Adding (7.161) and (7.162), there results

$$T_{N+1}(x) + T_{N-1}(x) = 2x\,T_N(x) \tag{7.163}$$

Equation (7.163) expresses $T_{N+1}(x)$ in terms of $T_N(x)$ and $T_{N-1}(x)$ and is known as the recurrence relation. Since $T_0(x) = 1$ and $T_1(x) = x$, we obtain, on using (7.163), the polynomial form of (7.158). The first six members are given below:

N	$T_N(x)$
0	1
1	x
2	$2x^2 - 1$
3	$4x^3 - 3x$
4	$8x^4 - 8x^2 + 1$
5	$16x^5 - 20x^3 + 5x$

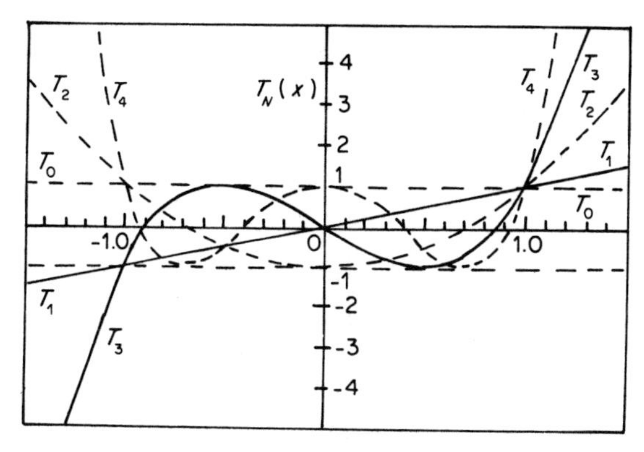

Figure 7.30 The first five members of Tschebyscheff polynomials

The function $T_N(x)$ as defined by (7.158) is for $|x| \leqslant 1$. On the other hand, the polynomial form can be used to define $T_N(x)$ for all real values of x. For $|x| \geqslant 1$, it is also possible to obtain an alternative expression for $T_N(x)$ as follows.

From (7.159), we have

$$x = \tfrac{1}{2}(e^{iw} + e^{-iw}) \tag{7.164}$$

On letting

$$w = iy \tag{7.165}$$

(7.164) becomes

$$x = \cosh y$$

or

$$y = \cosh^{-1} x \tag{7.166}$$

Thus

$$\begin{aligned} T_N(x) &= \cos(Nw) = \cos(Niy) = \cos(iN \cosh^{-1} x) \\ &= \cos[i(N \cosh^{-1} x)] = \cosh(N \cosh^{-1} x) \end{aligned} \tag{7.167}$$

In summary, the values of $T_N(x)$ can be obtained either from the polynomial form or from (7.158) for $|x| \leqslant 1$ and (7.167) for $|x| \geqslant 1$.

The first five members of $T_N(x)$ are shown in Figure 7.30. The following properties are noted:

(a) $T_N(x)$ is an even function for N even and an odd function for N odd.
(b) All zeros of $T_N(x)$ occur between $|x| \leqslant 1$.
(c) $|T_N(x)| \leqslant 1$ for $|x| \leqslant 1$
$\quad > 1$ for $|x| > 1$

7.10.4 Design Procedure

In section 7.10.2, it was shown that the array factor of a symmetric array with progressive phasing between its N elements is an $(N-1)$th-order polynomial

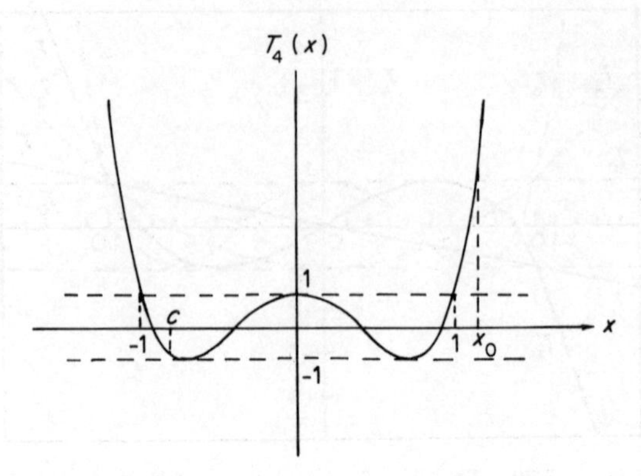

Figure 7.31 The correspondence between the array factor of a five-element Tschebyscheff array and the Tschebyscheff polynomial of order 4

in the real variable $\cos(\frac{1}{2}u)$. By making this polynomial a Tschebyscheff polynomial, the array factor will trace out the function $T_{N-1}(x)$. However, if we simply set $x = \cos(\frac{1}{2}u)$, the range of x is restricted to the interval between -1 and 1 as x, and hence u, varies. In this interval, the function $T_{N-1}(x)$ will oscillate between the values -1 and $+1$. The result is an array factor of several lobes of the same level, and there will be no main lobe. In order to obtain a main lobe, we set

$$x = x_0 \cos(\tfrac{1}{2}u) \tag{7.168}$$

where $x_0 > 1$. As illustrated in Figure 7.31, for the case $(N-1) = 4$, as x is allowed to vary from some point c up to x_0 and then back to its starting point, the function $T_{N-1}(x)$ traces out a pattern consisting of several small side lobes and one main lobe. The secondary lobes will all be of equal amplitude (unity) and will be down from the main lobe by the ratio $1/b$, where b is the value of $T_{N-1}(x_0)$. Thus the parameter x_0 is determined by the ratio of side lobe to main lobe. It can be calculated by using (7.167). Putting $N-1 = m$, we have

$$b = \cosh(m \cosh^{-1} x_0) \tag{7.169}$$

Thus

$$\cosh^{-1} x_0 = \frac{1}{m} \cosh^{-1} b$$

or

$$x_0 = \cosh(\rho/m) \tag{7.170}$$

where

$$\rho = \cosh^{-1} b \tag{7.171}$$

The most straightforward method of finding the current distribution which gives rise to the Tschebyscheff pattern of a given ratio of side lobe to main lobe is to find the position of the nulls on the circle diagram representing the array polynomial. Let the nulls be located at $u = u_1$, $u = u_2, \ldots, u = u_{N-1}$. Then the array factor is given by

$$f = \prod_{i=1}^{N-1} [z - \exp(\mathrm{j}u_i)] \tag{7.172}$$

Expanding (7.172) into a polynomial, the current excitations can be obtained by comparing it with (7.155).

To find the zeros, let $m = N - 1$ and consider the Tschebyscheff polynomial of the mth degree. From (7.158) and (7.159),

$$T_m(x) = \cos(m \cos^{-1} x) = \cos(mw)$$

The nulls are given by

$$\cos(mw) = 0$$

or

$$w = w_i = \frac{(2i - 1)}{2m} \qquad i = 1, 2, \ldots, m \tag{7.173}$$

Since $x = \cos w$, the nulls of the pattern occur at values of x given by

$$x_i = \cos w_i \tag{7.174}$$

The corresponding position for the nulls on the unit circle will then be

$$x_i = x_0 \cos(\tfrac{1}{2}u_i)$$

or

$$u_i = 2\cos^{-1}(x_i/x_0) = 2\cos^{-1}[(\cos w_i)/x_0] \tag{7.175}$$

The correspondences between χ, u, and x are given by

$$u = \frac{2\pi d}{\lambda} \cos \chi + \alpha$$

$$x = x_0 \cos(\tfrac{1}{2}u)$$

As an example, consider a broadside array for which $\alpha = 0$. As χ varies from 0 to $\frac{1}{2}\pi$ through π, u varies from $2\pi d/\lambda$ through 0 to $-2\pi d/\lambda$, and x will vary from $x_0 \cos(\pi d/\lambda)$ to x_0 back to $x_0 \cos(-\pi d/\lambda) = x_0 \cos(\pi d/\lambda)$. If $d = \frac{1}{2}\lambda$, u will range from π through zero to $-\pi$, and x will range from 0 to x_0 to 0. If $d = \lambda$, u will range twice around the circle from 2π through 0 to -2π, (two main lobes) and x will range from $-x_0$ to x_0 and back to $-x_0$.

7.10.5 Example

Design a Tschebyscheff four-element broadside array having a spacing $d = \frac{1}{2}\lambda$ between elements. The side lobe to main beam level is to be 0.1.

Solution

$$\rho = \cosh^{-1} b = \cosh^{-1}(10.0) = 3.0$$
$$x_0 = \cosh(3.0/3) = 1.53$$

Nulls:

$$w_i = \frac{(2i-1)\pi}{2m} = \frac{(2i-1)}{6} \qquad i = 1, 2, 3$$

The following table is obtained:

i	w_i	$x_i = \cos w_i$	x_i/x_0	$u_i = 2\cos^{-1}(x_i/x_0)$	u_i(radians)
1	$\pi/6$	0.866	0.562	111.6°	1.95
2	$3\pi/6$	0	0	180°	π
3	$5\pi/6$	−0.866	−0.562	248.4°	4.34

The polynomial representing the array is

$$\begin{aligned}|f| &= |(z - \exp(j1.95))(z - \exp(j\pi))(z - \exp(j4.34))| \\ &= |z^3 + 1.736z^2 + 1.736z + 1|\end{aligned}$$

The required relative currents in the elements are therefore

$$1:1.736:1.736:1$$

7.11 PATTERN SYNTHESIS

7.11.1 Theory

In this section, we present a method of designing a linear array to produce a prescribed radiation pattern. The method is based on the idea that the array factor of a symmetrical array of an odd number of elements can be expressed in the form of a finite trigonemetric series. The coefficients of the series are related to the amplitudes and phases of the excitation currents. By choosing the coefficients to be the Fourier coefficients, the array factor will approximate the prescribed pattern in the least mean-squared sense.

Let the number of elements be $N = 2m + 1$, where m is a positive integer. Then, from (7.140),

$$|f| = |A_0 + A_1 z + \cdots + A_m z^m + A_{m+1} z^{m+1} + \cdots + A_{2m} z^{2m}| \tag{7.176}$$

where $z = \exp(ju)$.

After dividing (7.176) by z^m, which does not alter $|f|$, we have

$$|f| = |a_0 z^{-m} + A_1 z^{-m+1} + \cdots + A_{m-1} z^{m-1} + A_m + A_{m+1} z + \cdots + A_{2m} z^m| \tag{7.177}$$

We now require the array to possess the following symmetric properties:

(a) The currents on the corresponding elements on either side of the centre element be equal in magnitude.
(b) The phase of the left-side element shall lag that of the centre element by the same amount that the corresponding right-side element leads the centre element (or vice versa). Note, however, that deviations from progressive phaseshift are allowed.

Then we can write

$$\begin{aligned} A_m &= c_0 \\ A_{m-n} &= c_n - \mathrm{j}d_n \\ A_{m+n} &= c_n + \mathrm{j}d_n \end{aligned} \tag{7.178}$$

Using (7.178), the pair of terms

$$\begin{aligned} A_{m-n}z^{-n} + A_{m+n}z^{n} &= c_n(z^n + z^{-n}) + \mathrm{j}d_n(z^n - z^{-n}) \\ &= 2c_n \cos(nu) - 2b_n \sin(nu) \end{aligned}$$

since $z^n - \exp(\mathrm{j}nu)$. Thus

$$\begin{aligned} |f| &= 2\{\tfrac{1}{2}c_0 + c_1 \cos u + \cdots + c_m \cos(mu) - [d_1 \sin u + \cdots + d_m \sin(mu)]\} \\ &= 2\left(\tfrac{1}{2}c_0 + \sum_{n=1}^{m} [c_n \cos(nu) + (-d_n) \sin(nu)]\right) \end{aligned} \tag{7.179}$$

Equation (7.179) shows that a finite trigonometric series can be identified as the array factor of a symmetric array of odd elements. If a pattern specified as a function $f(u)$ is given, the series (7.179) can be used to approximate $f(u)$. If the coefficients c_0, c_n and $(-d_n)$ are chosen to be the Fourier coefficients, the approximation with the least mean-squared error is obtained, i.e. the mean-square difference between the desired and the approximate pattern for u from 0 to 2π is minimized. Obviously, if we have an infinite number of elements in the array, (7.179) will contain an infinite number of terms and the series will converge to the desired function.

7.11.2 Example

Design a linear array of five elements with half-wavelength spacing that will produce, in the least mean-square sense, the array factor in the range $0 < \chi < \pi$ described by

$$f(\chi) = \begin{cases} 1 & \pi/4 < \chi < 3\pi/4 \\ 0 & 0 \leqslant \chi < \dfrac{\pi}{4},\quad \dfrac{3\pi}{4} < \chi \leqslant \pi \end{cases} \tag{7.180}$$

Solution

The pattern is symmetrical about the line of the array $\chi = 0$. It is illustrated in Figure 7.32(a).

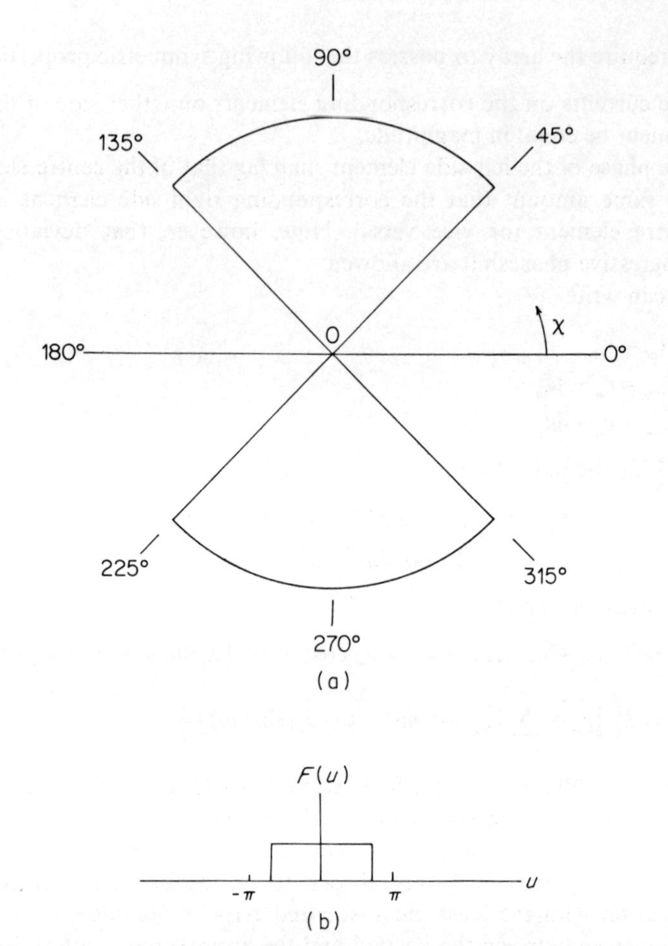

Figure 7.32 (a) The prescribed array factor in the example in section 7.11.2; (b) the corresponding function $F(u)$

Let us choose $\alpha = 0$. This would result in a broadside array, which is consistent with the given pattern. For $d = \frac{1}{2}\lambda$, we have

$$u = kd \cos \chi + \alpha = \pi \cos \chi$$

The corresponding function in the u variable is

$$F(u) = \begin{cases} 1 & -0.707\pi < u < 0.707\pi \\ 0 & -\pi < u < -0.707\pi,\ 0.707\pi < u < \pi \end{cases} \tag{7.181}$$

$F(u)$ is shown in Figure 7.32(b). The Fourier series expansion for $F(u)$ is

$$F(u) = 0.707 + \frac{2}{\pi} \sum_{n=1}^{\infty} \frac{1}{n} \sin(0.707n\pi) \cos(nu) \tag{7.182}$$

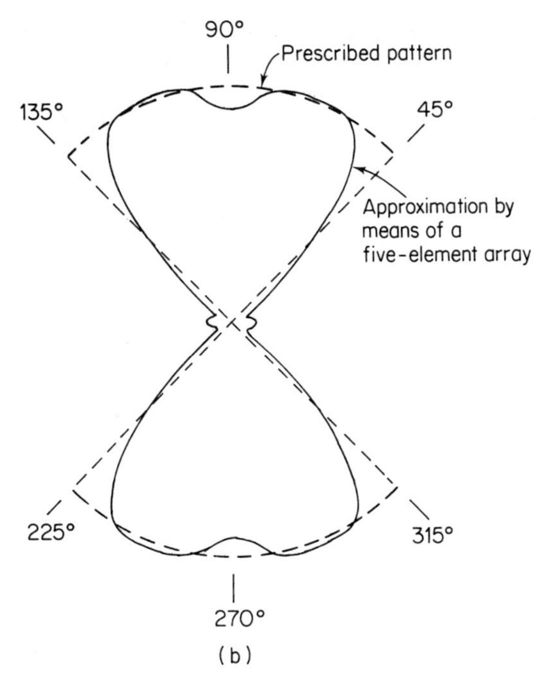

Figure 7.33 (a) A five-element array designed to approximate the prescribed pattern

$$f(\chi) = \begin{cases} 1 & \pi/4 < \chi < 3\pi/4 \\ 0 & \text{elsewhere} \end{cases}$$

in the least mean-square sense. (b) Comparison of the actual and the prescribed patterns

Thus

$$\begin{aligned} c_0 &= 0.707 \\ c_n &= (1/n\pi)\sin(0.707n\pi) \\ d_n &= 0 \end{aligned} \tag{7.183}$$

The pattern obtained using the value $n = 2$ is given by

$$|f| = |-0.154z^{-2} + 0.254z^{-1} + 0.707 + 0.254z - 0.154z^2| \tag{7.184}$$

This is a five-element array having the current ratios indicated in Figure 7.33(a) and an overall length of 2λ. The pattern produced is shown in Figure 7.33(b). The approximation can be improved by taking more terms in (7.182), i.e. using a greater number of elements.

7.12 WORKED EXAMPLES

Example 1

An array composed of two vertical half-wave dipoles with currents of equal magnitudes is placed with its axis (the array axis) along the east–west direction. Let the spacing be d and the phase difference of the currents be α.

(a) Determine α and d (in units of λ) such that the horizontal pattern has a maximum to the east and a null at an azimuth angle of 120° measured from the east.

(b) What is the condition on d such that the horizontal pattern has no null in any direction, while maintaining the maximum in the east direction.

Solution

(a) In the horizontal plane, let ϕ be the angle measured from the east. The array factor is

$$\begin{aligned} f &= 1 + \exp[\mathrm{j}(\alpha + kd\cos\phi)] \\ &= \exp[\mathrm{j}(\tfrac{1}{2}\alpha + \tfrac{1}{2}kd\cos\phi)]\{\exp[-\mathrm{j}(\tfrac{1}{2}\alpha + \tfrac{1}{2}kd\cos\phi)] \\ &\quad + \exp[\mathrm{j}(\tfrac{1}{2}\alpha + \tfrac{1}{2}kd\cos\phi)]\} \end{aligned}$$

$$|f| = 2\left|\cos\left(\frac{\pi d}{\lambda}\cos\phi + \frac{\alpha}{2}\right)\right|$$

The condition for the maximum to occur in the east direction ($\phi = 0$) is

$$\frac{\pi d}{\lambda} + \frac{\alpha}{2} = 0 \qquad \text{or} \qquad \frac{d}{\lambda} = -\frac{\alpha}{2\pi}$$

The condition for a null to occur at $\phi = 120°$ is

$$\frac{\pi d}{\lambda}\cos 120° + \frac{\alpha}{2} = \pm\frac{\pi}{2}$$

Solving the above equations for α and d, we obtain the result

$$\alpha = -120°$$

$$d = \lambda/3$$

(b) To maintain the maximum in the east, $d/\lambda = -\alpha/2\pi$. If a null is to occur at an angle ϕ, then

$$\frac{\pi d}{\lambda}\cos\phi + \frac{1}{2}(-2\pi)\frac{d}{\lambda} = -\frac{\pi}{2}$$

or

$$\cos\phi = 1 - \frac{\lambda}{2d}$$

If $d/\lambda < 1/4$, $|\cos\phi| > 1$ and this is not possible for real angles. Hence there will be no null in the pattern in the horizontal plane if the spacing is less than a quarter of a wavelength.

Example 2

It is desired to construct an array composed of half-wave dipoles oriented horizontally, all parallel to one another. The array is to be designed to produce a single main beam of width 20 ± 1 degrees in the broadside direction. Using uniform excitation of the elements, what should be the spacing of the antennas which would lead to the fewest number of dipoles consistent with the specifications?

Solution

For a uniform array of N elements with inter-element spacing d, the first zero of the array factor occurs at

$$\cos\chi = \lambda/Nd$$

where χ is measured from the array axis. If Θ is the angle measured from the broad side direction for which the first zero occurs, then the main-beam width is 2Θ. The relation between Θ, N, and d is

$$\cos(\tfrac{1}{2}\pi - \Theta) = \lambda/Nd$$

or

$$N = \frac{1}{\sin\Theta(d/\lambda)}$$

If the pattern is to have only one main beam, the spacing d must satisfy the condition $d/\lambda < 1$. Hence the minimum value of N is

$$N = \frac{1}{\sin\Theta}$$

Using the upper limit allowed for the main-beam width, $\Theta = 10.5^\circ$ and the above equation gives $N = 5.5$. Since N must be an integer, let us try $N = 6$ and $d/\lambda = 0.99$. Then

$$\sin\Theta = \frac{1}{0.99 \times 6} = 0.168$$

$$\Theta = 9.7^\circ$$

$$\text{main-beam width} = 19.4^\circ$$

Since this is within the specification, the choice $N = 6$, $d/\lambda = 0.99$ is an acceptable design.

Example 3

Consider a 15-element array of centre-fed z-directed half-wave dipoles, with centres lying symmetrically with respect to the origin along the positive and negative x-axis. The elements are equispaced with $d = 0.5\lambda$. The feed-point currents of the nth element are given by

$$I_n = I_0 \exp(jn\alpha)$$

where I_0 is the same for all elements.

(a) If α is set to the value corresponding to an endfire array, determine the width of the main beam.

(b) If a receiving antenna in the form of a half-wave dipole is placed with its centre at P($r = 100\lambda$, $\theta = 90°$, $\phi = 30°$), determine the value of α and the orientation of the receiving dipole such that the received signal is maximum.

(c) If five elements at the centre of the array are removed and the value of α is set to zero but I_0 remains unchanged, compare the magnitude of the far-zone electric field in the broadside direction ($\theta = 90°$, $\phi = 90°$) with the corresponding value for the full 15-element array excited by in-phase currents of amplitude I_0.

Solution

(a) From equation (7.78), the main-beam width of an endfire array is

$$4 \sin^{-1}\left(\sqrt{\frac{\lambda}{2Nd}}\right) = 4 \sin^{-1}\left(\sqrt{\frac{1}{15}}\right) = 60°$$

(b) The value of α should be such that the main beam of the array occurs at $\phi = 30°$. Hence

$$\alpha = -kd \cos 30° = -\frac{\sqrt{3}}{2}\pi \text{ radians}$$

Since the electric field from the array is parallel to the z-axis, so should be the axis of the receiving dipole in order for the received signal to be maximum.

(c) If five elements at the centre of the array are removed, the array factor is, from the principle of superposition,

$$f' = \frac{1 - \exp(j15u)}{1 - \exp(ju)} - \frac{1 - \exp(j5u)}{1 - \exp(ju)} = \frac{\exp(j5u) - \exp(j15u)}{1 - \exp(ju)}$$

where $u = kd \cos \chi = \pi \sin \theta \cos \phi$.

In the broadside direction, $u = 0$ and

$$f'(u \to 0) = \lim_{u \to 0} \frac{\cos 5u + j \sin 5u - \cos 15u - j \sin 15u}{1 - \cos u - j \sin u}$$

$$= \frac{1 + j5u - 1 - j15u}{-ju} = 10$$

For the full 15-element array, the array factor in the broadside direction is

$$f(u \to 0) = \lim_{u \to 0} \frac{1 - \exp(j15u)}{1 - \exp(ju)} = 15$$

Since the far-zone electric field is equal to the electric field due to a single element multiplied by the array factor, it follows that the ratio of the magnitude of the electric fields is the same as the ratio of the array factors, namely,

$$\frac{|E'|}{|E|} = \frac{f'(u \to 0)}{f(u \to 0)} = \frac{2}{3}$$

7.13 PROBLEMS

1. An antenna system consists of two quarter-wave monopoles above a perfect ground, with a spacing of a quarter-wavelength. The antennas are to be fed with currents that are equal in amplitude but 90° out of phase. If they are to radiate 10 kW power, what is the magnitude and phase of the required driving-point voltages for the two antennas. Assume that $Z_{11} = Z_{22} = 36 + j30$ and $Z_{12} = 20.5 - j14.5$ ohm.

2. (a) Using the identity (Whittaker and Watson, 1962)

$$\left(\frac{\sin(Nu/2)}{N \sin(u/2)}\right)^2 = \frac{1}{N} + \frac{2}{N^2} \sum_{m=1}^{N-1} (N-m)\cos(mu)$$

show that the directivity of a uniform array of N isotropic sources spaced a distance d apart with inter-element phaseshift α is given by

$$D = \frac{1}{\dfrac{1}{N} + \dfrac{2}{N^2} \displaystyle\sum_{m=1}^{N-1} \frac{(N-m)}{mkd} \sin(mkd)\cos(m\alpha)} \tag{7.185}$$

(b) For $N = 10$, plot a curve of directivity as a function of d/λ.

(c) What is the directivity when the spacing is equal to an odd multiple of $\lambda/2$.

3. Verify (7.83).

4. A uniform array has 100 elements and an inter-element spacing of a half-wavelength. Determine the width of the main beam if this array is designed to operate as (a) a broadside array; (b) an endfire array; (c) an array with the main-beam maximum occurring at an angle 45° with respect to the array axis.

5. The zeros of the array polynomials representing three four-element arrays are as follows: (a) $z_1 = \exp(j\pi/2)$, $z_2 = \exp(j\pi)$, and $z_3 = \exp(-j\pi/2)$ for array 1; (b) $z_1 = \exp(j\pi/3)$, $z_2 = \exp(j\pi)$, $z_3 = \exp(-j\pi/3)$ for array 2; (c) $z_1 = \exp(j2\pi/3)$, $z_2 = \exp(j\pi)$, $z_3 = \exp(-j2/\pi/3)$ for array 3. The spacing between the elements is a half-wavelength and the elements are excited in phase. Without carrying out the detailed calculations, determine which

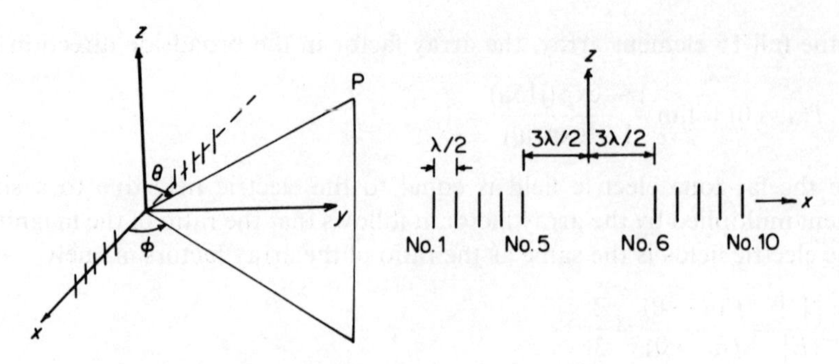

Figure 7.34 Geometry for problem 7

array has the narrowest beamwidth and which array has the smallest ratio of side lobe to main beam.

6. The zeros of the array polynomial representing a four-element linear array are: $z_1 = \exp(j2\pi/3)$, $z_2 = \exp(-j2\pi/3)$, $z_3 = \exp(j\pi)$. If the spacing between the elements is a half-wavelength and the elements are excited in phase, find (a) the beamwidth of the array, (b) the relative current amplitudes of the array.

7. A 10-element array of vertical centre-fed half-wave dipoles, with centres lying along the x-axis, is arranged in the manner shown in Figure 7.34. The spacing between elements no. 5 and no. 6 is 3λ, while the rest of the elements are spaced $\lambda/2$ apart. The feed-point currents of all the elements are equal in both amplitude and phase. If a receiving antenna in the form of a vertical half-wave dipole is placed at the point $P(r, \theta, \phi)$, obtain an expression for the open-circuit voltage induced at the terminals of the receiving dipole.

8. Find the array polynomial of a four-element array that is to radiate an optimum broadside pattern with a beamwidth between nulls of 120°. Use a spacing $d = \lambda/2$ between elements. Determine the ratio of main beam to side lobe and compare this with a four-element uniform array of the same spacing. What is the beamwidth of the latter?

9. (a) Design a linear array of five elements with half-wavelength spacings that will produce, in the least mean-square sense, the array factor in the range $0 < \chi < \pi$ described by

$$f(\chi) = \begin{cases} 1 & 0 < \chi \leqslant \pi/3 \\ 0 & \pi/3 < \chi \leqslant 2\pi/3 \\ 1 & 2\pi/3 < \chi \leqslant \pi \end{cases}$$

where χ is the angle subtended by the array axis and the line joining the array and the field point.

(b) Discuss whether a unique solution can be found if a spacing of less than a half-wavelength is used.

10. For a broadside array of N isotropic radiators spaced d apart such that $kd \ll 1$ but $Nkd \gg 1$, show that the directivity of the array is approximately equal to $2N(d/\lambda)$. [Hint: $\int_{-\infty}^{\infty} (\sin x/x)^2 \, dx = \pi$.]

11. For an ordinary endfire array of N isotropic radiators with spacing d such that $kd \ll 1$ but $Nkd \gg 1$, show that the directivity of the array is approximately equal to $4N(d/\lambda)$.

BIBLIOGRAPHY

Elliott, R. S. (1966). 'The theory of antenna arrays', in *Microwave Scanning Antennas*, Vol. II, *Array Theory and Practice* (ed. R. C. Hansen), Academic Press, New York, Chapter 1.

CHAPTER 8
The Uda–Yagi Antenna and the Corner Reflector

8.1 INTRODUCTION

In this chapter, we discuss two directive endfire antennas that are widely used in the VHF–UHF range, particularly in FM radio and television reception. These are the Uda–Yagi antenna and the corner reflector, both of which employ appropriately placed passive conductors to enhance the directivity of an active element.

The Uda–Yagi antenna is commonly known as the Yagi for short. In its simplest form, a Yagi consists of a resonant half-wave dipole which is directly fed and two passive or parasitic elements, as shown in Figure 8.1. Although the

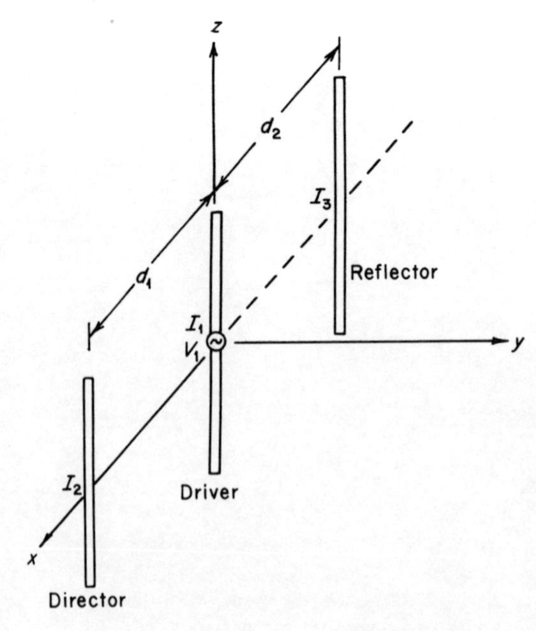

Figure 8.1 A three-element Yagi antenna

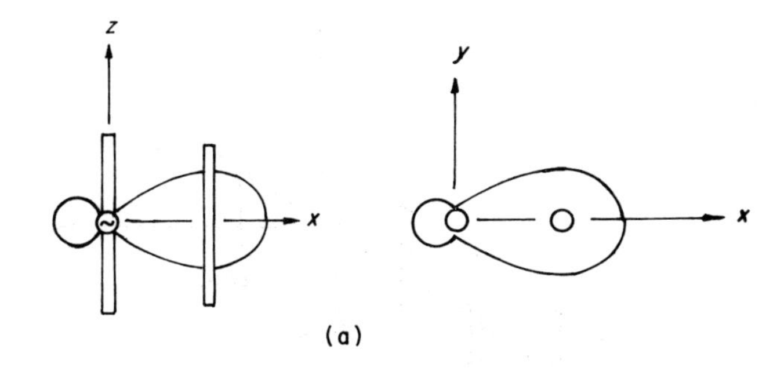

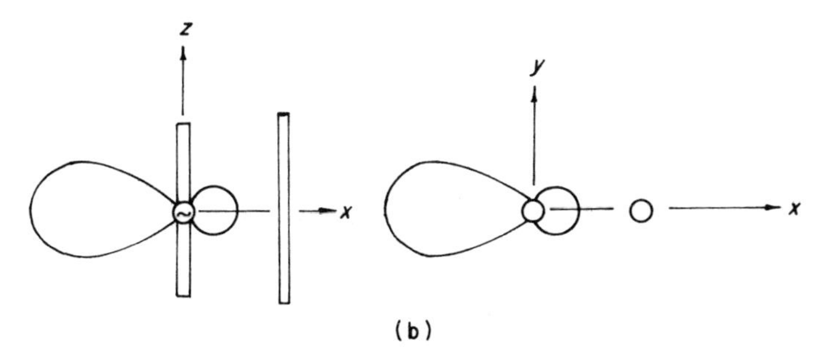

Figure 8.2 (a) A driven element and a director; (b) a driven element and a reflector

parasitic elements are not connected directly to a source, the radiation field of the active element induces currents on them, thereby causing them to radiate. By suitably choosing the length and position of a parasitic element, the phase of the induced current can be such that it has the effect of either enhancing the radiation in its own direction or in the direction of the driver. In the former case it is called a director (Figure 8.2a), and in the latter case it is called a reflector (Figure 8.2b). The three-element Yagi of Figure 8.1 consists of a driver, a director, and a reflector. The reflector is usually about 5% longer and the director 5% shorter than the driver. The spacings are of the order of 0.15λ to 0.25λ between the reflector and driver and also between the driver and director. The maximum directivity obtainable from such a Yagi is about 9 dB. Higher directivity can be obtained from a multi-element Yagi consisting of one reflector and several directors (Figure 8.3).

While a passive conductor in the form of a thin element can function as a reflector, it has the disadvantage of being highly sensitive to frequency. Moreover, radiation in the backlobe can be significant. On the other hand, a reflector in the form of a large metal sheet, such as that shown in Figure 8.4, is relatively insensitive to frequency and the backlobe radiation can be reduced to a negligible level. The directivity can be enhanced further by bending the large flat sheet

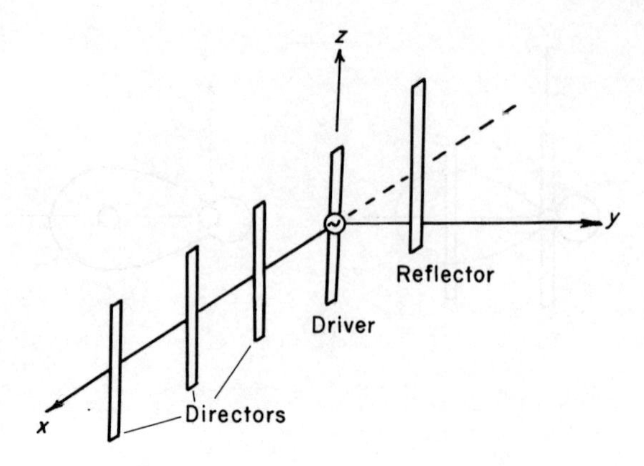

Figure 8.3 A multi-element Yagi antenna

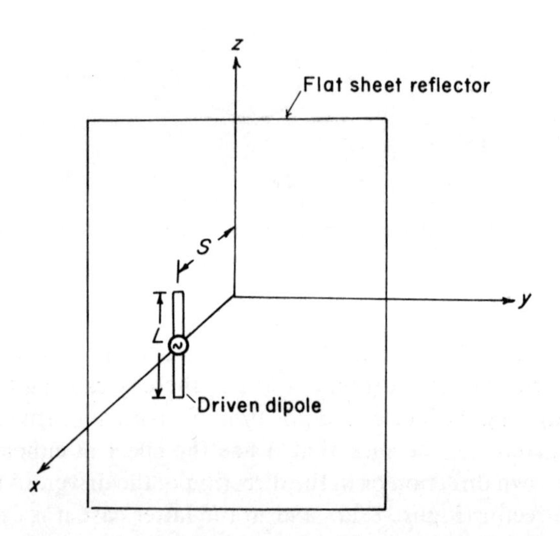

Figure 8.4 A driven element with a large flat sheet reflector

into two sheets intersecting at an angle, forming what is known as a corner reflector (Figure 8.5). Most practical corner reflectors have corner angles equal to $180°/N$, where N is a positive integer. For these configurations, the analyses of the corner reflector and the Yagi antenna are similar. They are both based on the concepts of mutual coupling and array theory developed in chapters 5 and 7.

In order to emphasize the physical principles involved, we shall assume the elements to be thin and neglect the effect of neighbouring elements or the conducting sheet on the current distribution, which is taken to be sinusoidal. Closed-form solutions are then obtained. Readers who have digested the introductory materials in this chapter and are interested in modern computing

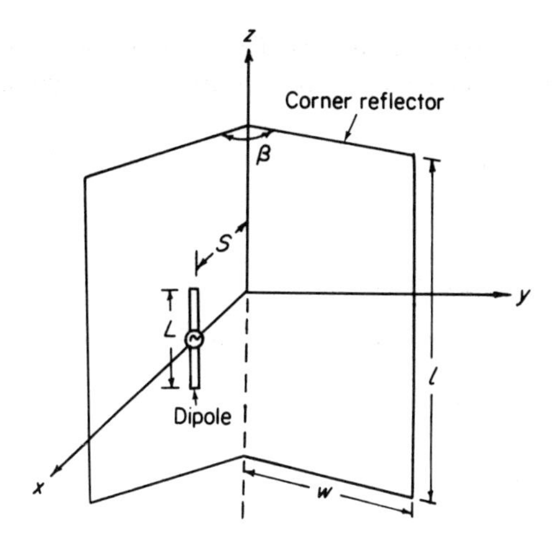

Figure 8.5 A corner reflector antenna

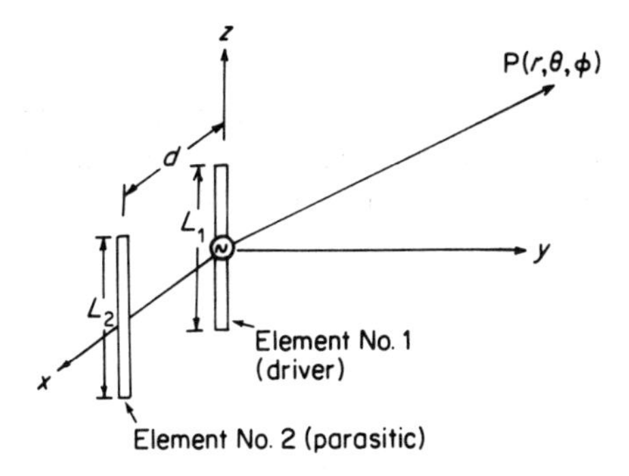

Figure 8.6 A driven element and a parasitic element

methods should consult the references by Thiele (1969, 1973), Cheng and Chen (1973), and Chen and Cheng (1975).

8.2 PARASITIC ELEMENT AS REFLECTOR AND DIRECTOR

Consider a two-element antenna consisting of a driven dipole and a parasitic element a distance d from it, as shown in Figure 8.6. The dipole is fed from a source via a transmission line. Let V_1 and I_1 be the voltage and current at the terminals of the driver. Since the parasitic element is not connected to a source, it is convenient to think of it as an element having a pair of terminals at the centre which is short-circuited. Let the current at these short-circuited terminals be I_2, which arises as a result of the near-field coupling with the driver. Writing the circuit relations for the voltages and currents at the terminals of elements 1 and 2, we have

$$V_1 = I_1 Z_{11} + I_2 Z_{12} \tag{8.1}$$

$$0 = I_2 Z_{22} + I_1 Z_{12} \tag{8.2}$$

Solving (8.2) for I_2, we obtain

$$\begin{aligned} I_2 &= -I_1 Z_{12}/Z_{22} = -I_1 |Z_{12}/Z_{22}| \exp[\mathrm{j}(\tau_{12} - \tau_{22})] \\ &= I_1 |Z_{12}/Z_{22}| \exp[\mathrm{j}(\pi + \tau_{12} - \tau_{22})] \end{aligned} \tag{8.3}$$

where

$$Z_{12} = R_{12} + \mathrm{j}X_{12}$$

is the mutual impedance between elements 1 and 2,

$$Z_{22} = R_{22} + \mathrm{j}X_{22}$$

is the self-impedance of the parasitic element,

$$\tau_{12} = \tan^{-1}(X_{12}/R_{12}) \tag{8.4}$$

$$\tau_{22} = \tan^{-1}(X_{22}/R_{22}) \tag{8.5}$$

Equation (8.3) shows that the induced current I_2 is dependent on I_1, the mutual impedance of the two elements, and the self-impedance of the parasitic element. By adjusting the position and length of the parasitic element, the phase of I_2 can either lag or lead that of I_1.

If we let

$$v = \pi + \tau_{12} - \tau_{22} \tag{8.6}$$

(8.3) becomes

$$I_2 = I_1 |Z_{12}/Z_{22}| \exp^{\mathrm{j}v} \tag{8.7}$$

For the purposes of calculating the far field, we shall assume that the principle of pattern multiplication developed in Chapter 7 can be applied to the present problem. Strictly speaking, it is valid only for arrays of identical elements, which is clearly not the case here since, in general, the two elements are not of equal length. However, in practice, the difference in length is small, so that use of the pattern multiplication principle will not result in any appreciable error and at the same time will greatly simplify the calculation.

The array factor of a two-element array is, from (7.41),

$$f(\psi_1) = I_1 + I_2 \exp(\mathrm{j}kd\psi_1) \tag{8.8}$$

where

$$\psi_1 = \cos\chi = \sin\theta\cos\phi$$

The far field of a dipole of length L is given by (3.87) and (3.88). By multiplying the individual pattern with the array factor, we obtain the following expressions for the electric and magnetic fields:

$$E_\theta(\chi) = \frac{\mathrm{j}\eta \exp(-\mathrm{j}kr)P(\theta)}{2\pi r \sin(\frac{1}{2}kL)} f(\cos\chi)$$

$$H_\phi(\chi) = E_\theta/\eta \tag{8.9}$$

where $f(\cos\chi)$ is given by (8.8) and

$$P(\theta) = \frac{\cos(\frac{1}{2}kL\cos\theta) - \cos(\frac{1}{2}kL)}{\sin\theta} \tag{8.10}$$

Equation (8.9) applies to any point in space. Let us simplify the discussion somewhat by considering the pattern in the H-plane where $\theta = 90^\circ$ and $\cos\chi = \cos\phi$. Then

$$E_\theta(\phi) = C[I_1 + I_2 \exp(\mathrm{j}kd\cos\phi)] \tag{8.11}$$

where

$$C = \frac{\mathrm{j}\eta I_1 \exp(-\mathrm{j}kr)[1 - \cos(\frac{1}{2}kL)]}{2\pi r \sin(\frac{1}{2}kL)} \tag{8.12}$$

Substituting (8.7) into (8.11), we obtain

$$E_\theta(\phi) = CI_1\{1 + |Z_{12}/Z_{22}| \exp[\mathrm{j}(v + kd\cos\phi)]\} \tag{8.13}$$

It is seen from (8.13) that maximum radiation occurs in the direction corresponding to

$$\cos\phi = -v/kd \tag{8.14}$$

For a given driver, v is dependent on the length and spacing of the parasitic element. If v is equal to kd, maximum radiation occurs in the $\phi = 180^\circ$ direction and the parasitic element is called a reflector since it appears to reflect radiation from the driver. On the other hand, if $v = -kd$, maximum radiation occurs in the $\phi = 0^\circ$ direction, and the parasitic element is called a director since it appears to direct radiation from the driver towards the director. The front-to-back (F/B) ratio is defined as

$$\mathrm{F/B} = |E_\theta(\phi = 180^\circ)|/|E_\theta(\phi = 0^\circ)| \tag{8.15}$$

If the lengths of the driven and parasitic elements are close to a half-wavelength, it can be verified from the equations developed in Chapter 5 that the phase of Z_{12} as a function of d/λ is quite insensitive to the lengths of the elements. Because of this insensitivity, it follows from (8.7) that the parameter v, at a given spacing, is governed primarily by the phase of Z_{22}. The phase of Z_{22} can be altered appreciably by making the length of the parasitic element a few per cent shorter or longer than that of the driver. The former usually yields a

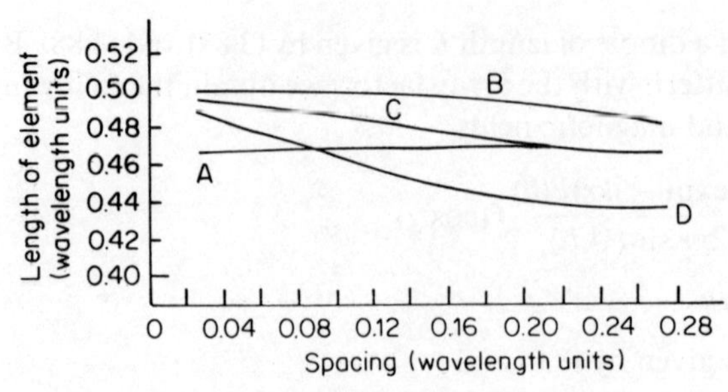

Figure 8.7 Required lengths of driven and parasitic elements. Thickness of conductors is assumed to be infinitesimal. Curve A is the length of the driven element when it is associated with a reflector. Curve B is the length of a reflector-type parasitic element for maximum gain. Curve C is the length of the driven element when it is associated with a director. Curve D is the length of a director-type parasitic element for maximum gain

director and the latter a reflector. Some quantitative results are shown in Figure 8.7. The H-plane power patterns for one particular set of parameters corresponding to a director and another set corresponding to a reflector are shown in Figure 8.8.

The driving-point or input impedance Z_1 of the driven element is

$$Z_1 = V_1/I_1 \tag{8.16}$$

On solving (8.1) and (8.2), we obtain

$$Z_1 = Z_{11} - \frac{Z_{12}^2}{Z_{22}} = Z_{11} - \frac{|Z_{12}|^2}{|Z_{22}|} \exp[\mathrm{j}(2\tau_{12} - \tau_{22})] \tag{8.17}$$

The real part of Z_1 is

$$R_1 = R_{11} - \frac{|Z_{12}|^2}{|Z_{22}|} \cos(2\tau_{12} - \tau_{22}) \tag{8.18}$$

Equation (8.18) shows that the input resistance of the driven element is reduced by the presence of the parasitic element.

Since the elements are closely spaced ($d < \frac{1}{4}\lambda$), we have seen in section 7.7.5 that the loss resistance is important and should be included in the analysis. Adding a term R_{1L} to account for the loss, (8.18) becomes

$$R_1 = R_{11} + R_{1L} - \frac{|Z_{12}|^2}{|Z_{22}|} \cos(2\tau_{12} - \tau_{22}) \tag{8.19}$$

For a power input W to the driven element,

$$I_1 = \sqrt{\frac{2W}{R_1}} = \sqrt{\frac{2W}{R_{11} + R_{1L} - (|Z_{12}|^2/|Z_{22}|)\cos(2\tau_{12} - \tau_{22})}} \tag{8.20}$$

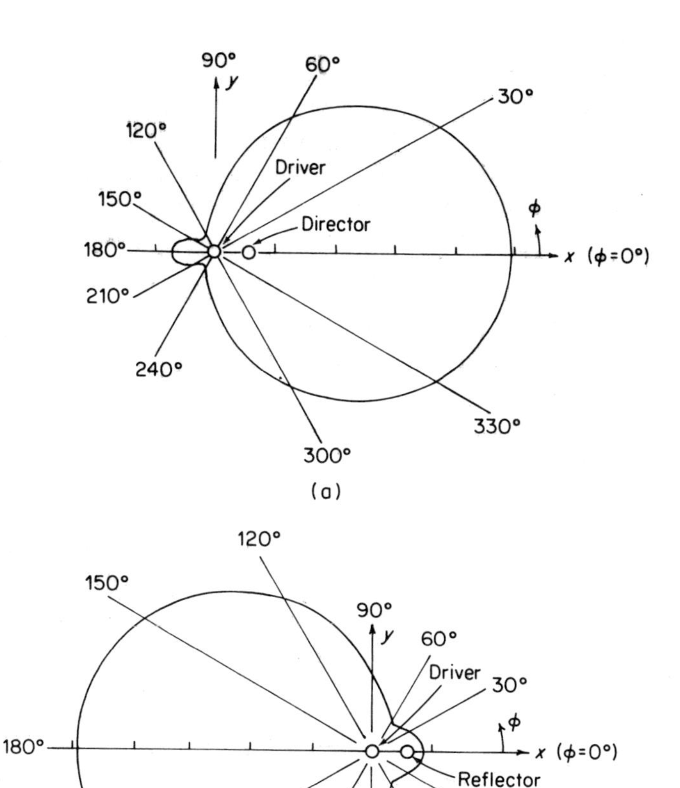

Figure 8.8 (a) H-plane power pattern of a driver and a director ($d/\lambda = 0.12$, $L_1/\lambda = 0.482$, $L_2/\lambda = 0.450$). (b) H-plane power pattern of a driver and a reflector ($d/\lambda = 0.16$, $L_1/\lambda = 0.454$, $L_2/\lambda = 0.500$) (Source: Robert S. Elliott, *Antenna Theory and Design*, © 1981, p. 372. Reprinted by permission of Prentice-Hall, Inc., Englewood Cliffs, N. J.)

Substitution of (8.20) for I_1 into (8.13) yields

$$E_\theta(\phi) = C\sqrt{\left(\frac{2W}{R_{11} + R_{1L} - (|Z_{12}|^2/|Z_{22}|)\cos(2\tau_{12} - \tau_{22})}\right)} \times \left(1 + \frac{|Z_{12}|}{|Z_{22}|}\exp[j(v + kd\cos\phi)]\right) \tag{8.21}$$

For a power input W to a single $\frac{1}{2}\lambda$ element, the electric field at the same distance is

$$E_{\lambda/2}(\phi) = C(\lambda/2)\sqrt{[2W/(73 + R_{0L})]} \tag{8.22}$$

where R_{0L} is the loss resistance of a single $\frac{1}{2}\lambda$ element.

If the lengths of the driven and the parasitic elements are not much different from $\lambda/2$, the constant C in (8.21) is approximately equal to $C(\lambda/2)$. Moreover, R_{1L} can be set equal to R_{0L} without introducing any significant error. The gain in field intensity as a function of ϕ of the array with respect to a single $\frac{1}{2}\lambda$ dipole with the same power input is therefore

$$G(\phi) = \sqrt{\left(\frac{73 + R_{1L}}{R_{11} + R_{1L} - (|Z_{12}|^2/|Z_{22}|)\cos(2\tau_{12} - \tau_{22})}\right)} \times \left(1 + \frac{|Z_{12}|}{|Z_{22}|}\exp[\mathrm{j}(v + kd\cos\phi)]\right) \quad (8.23)$$

The maximum power gain (square of the gain in field intensity) is typically about 5 dB for the two-element arrangement.

The analysis following equation (8.10) is for the H-plane. It is clear that, by putting $\phi = 0°$ in (8.9), a similar analysis can be carried out for the E-plane.

8.3 THE THREE-ELEMENT UDA–YAGI ANTENNA

As mentioned in section 8.2, a single driven element in the form of a half-wave dipole, together with a parasitic element acting as a reflector or director, will produce an endfire beam with a gain of about 5 dB. By combining a reflector and a director to form a three-element array, additional enhancement of the gain in the endfire direction can be obtained. This idea can be developed further by adding more directors and reflectors. However, by the very nature of a reflector, if a reflector element is placed adjacent to a driven element, another element placed beyond it will not be appreciably excited. Hence little improvement in gain is obtained by using more than one reflector. An arrangement consisting of a driven element, a reflector, and one or more directors is called a Uda–Yagi antenna, or simply a Yagi. In this section, the three-element Yagi will be analysed in some detail.

Consider the three-element Yagi shown in Figure 8.1. Let the voltage and current at the terminals of the driven element be V_1 and I_1 respectively. The parasitic elements can be regarded as having a pair of terminals at the centre which are short-circuited. The voltages at these terminals are therefore zero while the currents, denoted by I_2 and I_3, arise from near-field coupling with the driver and with each other. The circuit relations are therefore

$$V_1 = I_1 Z_{11} + I_2 Z_{12} + I_3 Z_{13} \quad (8.24)$$

$$0 = I_1 Z_{12} + I_2 Z_{22} + I_2 Z_{23} \quad (8.25)$$

$$0 = I_1 Z_{13} + I_2 Z_{23} + I_3 Z_{33} \quad (8.26)$$

Solving (8.25) and (8.26) for I_2 and I_3, we have

$$I_2 = \frac{Z_{13}Z_{23} - Z_{12}Z_{33}}{Z_{22}Z_{33} - Z_{23}^2} I_1 \quad (8.27)$$

$$I_3 = \frac{Z_{12}Z_{23} - Z_{13}Z_{22}}{Z_{22}Z_{33} - Z_{23}^2} I_1 \tag{8.28}$$

As in section 8.2, we assume that the lengths of the three elements differ only slightly so that the principle of pattern multiplication can be used to calculate the far field. The electric field is then given by equation (8.9), where the array factor $f(\cos\chi)$ in the present case is

$$f(\cos\chi) = I_1 + I_2 \exp(jkd_1 \cos\chi) + I_3 \exp(-jkd_2 \cos\chi) \tag{8.29}$$

For the H-plane where $\theta = 90°$, we have

$$E_\theta(\phi) = C[I_1 + I_2 \exp(jkd_1 \cos\phi) + I_3 \exp(-jkd_2 \cos\phi)] \tag{8.30}$$

where C is given by (8.12).

On letting

$$Z_{13}Z_{23} - Z_{12}Z_{33} = |Z_{n2}| \exp(j\tau_{n2}) \tag{8.31}$$

$$Z_{12}Z_{23} - Z_{13}Z_{22} = |Z_{n3}| \exp(j\tau_{n3}) \tag{8.32}$$

$$Z_{22}Z_{33} - Z_{23}^2 = |Z_d| \exp(j\tau_d) \tag{8.33}$$

the currents induced on the parasitic elements can be written as

$$I_2 = I_1 |Z_{n2}/Z_d| \exp[j(\tau_{n2} - \tau_d)] \tag{8.34}$$

$$I_3 = I_1 |Z_{n3}/Z_d| \exp[j(\tau_{n3} - \tau_d)] \tag{8.35}$$

Substitution of (8.34) and (8.35) into (8.30) yields

$$E_\theta(\phi) = CI_1 \{1 + |Z_{n2}/Z_d| \exp[j(\tau_{n2} - \tau_d + kd_1 \cos\phi)] + |Z_{n3}/Z_d| \exp[j(\tau_{n3} - \tau_d - kd_2 \cos\phi)]\} \tag{8.36}$$

The driving-point or input impedance is given by $Z_1 = V_1/I_1$. From (8.24)–(8.26), we obtain

$$Z_1 = Z_{11} + Z_{12}|Z_{n2}/Z_d| \exp[j(\tau_{n2} - \tau_d)] + Z_{13}|Z_{n3}/Z_d| \exp[j(\tau_{n3} - \tau_d)] \tag{8.37}$$

The real part of Z_1 is

$$R_1 = R_{11} + \left|\frac{Z_{12}Z_{n2}}{Z_d}\right| \cos(\tau_{12} + \tau_{n2} - \tau_d) + \left|\frac{Z_{13}Z_{n3}}{Z_d}\right| \cos(\tau_{13} + \tau_{n3} - \tau_d) \tag{8.38}$$

If an effective loss resistance R_{1L} is added to (8.38), we have

$$R_1 = R_{11} + R_{1L} + \left|\frac{Z_{12}Z_{n2}}{Z_d}\right| \cos(\tau_{12} + \tau_{n2} - \tau_d) + \left|\frac{Z_{13}Z_{n3}}{Z_d}\right| \cos(\tau_{13} + \tau_{n3} - \tau_d) \tag{8.39}$$

For a power input W to the driven element,

$$I_1 = \sqrt{(2W/R_1)} \tag{8.40}$$

Using (8.40), (8.37), and (8.22), we obtain the gain in field intensity as a function of ϕ of the array with respect to a single $\frac{1}{2}\lambda$ dipole:

$$G(\phi) = \sqrt{\left(\frac{73 + R_{0L}}{R_1}\right)}\left(1 + \left|\frac{Z_{n2}}{Z_d}\right| \exp[j(\tau_{n2} - \tau_d + kd_1 \cos\phi)] + \left|\frac{Z_{n3}}{Z_d}\right| \exp[j(\tau_{n3} - \tau_d - kd_2 \cos\phi)]\right) \tag{8.41}$$

A three-element Yagi can be designed by assuming that the presence of the director does not significantly affect the proper length of the reflector and vice versa. One can then use the data of section 8.2 to determine the spacings and lengths. Once these are determined, the mutual impedance can be calculated and the formulae developed in this section to find the input impedance, the pattern, and the gain in field intensity in the H-plane. Similar formulae, of course, can be readily derived for the E-plane or any other plane without difficulty. The H-plane power pattern for a particular set of parameters is illustrated in Figure 8.9. The directivity of this three-element Yagi is 7.5 dB above that of a single dipole.

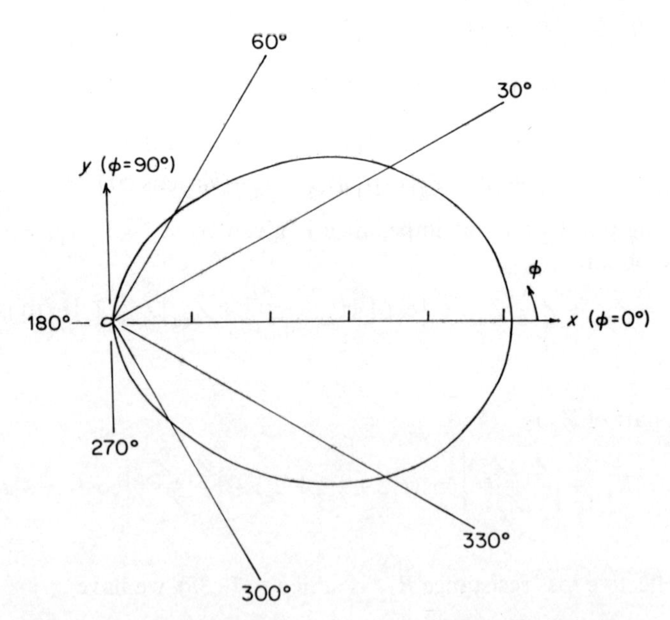

Figure 8.9 H-plane pattern of a three-element Yagi antenna (driver length = 0.475λ, director length = 0.450λ, reflector length = 0.500λ, inter-element spacing = 0.20λ) (Source: Elliott, *Antenna Theory and Design*, © 1981, p. 375. Reprinted by permission of Prentice-Hall, Inc., Englewood Cliffs, N. J.)

8.4 SOME DESIGN DATA FOR THE YAGI ANTENNA

The maximum gain obtainable from a three-element Yagi is about 9.8 dB. While adding more reflectors provides little improvement since little radiation can reach beyond the first reflector to excite them, increasing the number of directors does lead to further enhancement in gain. The method of analysis of a multi-element Yagi is similar to that used in sections 8.2 and 8.3. If the lengths and spacings of the elements are known, then, by solving the circuit equations, the currents induced on the parasitic elements can be obtained in terms of the mutual impedances and the current on the driver. Once the currents are known, the pattern, gain, and input impedance of the array are determined in the usual manner.

Using this method, Green (1966) has performed extensive numerical computations on Yagis with three to ten elements and his results provide useful data for the design of such Yagis. More recent data have been obtained by Stutzman and Thiele (1981), who used the numerical method known as the method of moments, which can take into accout the mutual effect on the current distributions. Their results for conductor diameters equal to 0.005λ, corresponding to $L/2a \simeq 100$, are shown in Table 8.1. Since the characterstics obtained for the diameters equal to 0.0025λ, 0.01λ, and 0.02λ are similar, the data can be used for the design of Yagis using commonly available materials. The bandwidth of a Yagi so designed is typically 2%. The narrow bandwidth is the principal disadvantage of the Yagi antenna.

Two features about the data shown in Table 8.1 are worthy of note. First, the half-power beamwidth (HP) in the E-plane is smaller than in the H-plane. The reason for this is as follows. The patterns in both planes are approximately given by the product of the array factor and the element pattern. The former is the same for both planes while the latter is approximately that of a half-wave dipole since all elements are about a half-wave long. In the E-plane, the element pattern is a figure of eight with a half-power beamwidth of 78°, which is obviously narrower than the isotropic element pattern in the H-plane. The second feature to note is that the input resistance of the Yagi antenna is typically much smaller than the 73 ohm of a half-wave dipole. This is the result of the loading effect of the parasitic elements, which we already alluded to in section 8.2. The input resistance can be stepped up by using a folded dipole instead of a simple dipole as the driven element. As discussed in section 3.13, a half-wavelength folded dipole has a radiation resistance of 292 ohm, which is four times that of a simple dipole.

8.5 APPLICATION OF THE YAGI ANTENNA

The Yagi antenna, because of the simplicity of its feed-system design, low cost, light weight, and relatively high gain, is one of the most popular receiving antennas in use in the VHF–UHF frequency range. This covers the FM frequencies (88–108 MHz), the low (54–88 MHz) and high (174–216 MHz) VHF TV bands and the UHF TV band (470–890 MHz). It is usual to use a three-

Table 8.1 Characteristics of equally spaced Uda–Yagi antennas (Source: W. L. Stutzman and G. A. Thiele, (1981), *Antenna Theory and Design*, Wiley, New York. Reproduced by permission of John Wiley & Sons Inc.)

Number of elements,	Spacing (wave lengths)	Element lengths (wavelengths)			Gain (dB)	Front-to-back ratio (dB)	Input impedance (ohm)	*H*-plane		*E*-plane	
		Reflector	Driver	Director				HP_H (deg)	SLL_H (dB)	HP_E (deg)	SLL_E (dB)
3	0.25	0.479	0.453	0.451	9.4	5.6	22.3 + j15.0	84	−11.0	66	−34.5
4	0.15	0.486	0.459	0.453	9.7	8.2	36.7 + j9.6	84	−11.6	66	−22.8
4	0.20	0.503	0.474	0.463	9.3	7.5	5.6 + j20.7	64	−5.2	54	−25.4
4	0.25	0.486	0.463	0.456	10.4	6.0	10.3 + j23.5	60	−5.8	52	−15.8
4	0.30	0.475	0.453	0.446	10.7	5.2	25.8 + j23.2	64	−7.3	56	−18.5
5	0.15	0.505	0.476	0.456	10.0	13.1	9.6 + j13.0	76	−8.9	62	−23.2
5	0.20	0.486	0.462	0.449	11.0	9.4	18.4 + j17.6	68	−8.4	58	−18.7
5	0.25	0.477	0.451	0.442	11.0	7.4	53.3 + j6.2	66	−8.1	58	−19.1
5	0.30	0.482	0.459	0.451	9.3	2.9	19.3 + j39.4	42	−3.3	40	−9.5
6	0.20	0.482	0.456	0.437	11.2	9.2	51.3 − j1.9	68	−9.0	58	−20.0
6	0.25	0.484	0.459	0.446	11.9	9.4	23.2 + j21.0	56	−7.1	50	−13.8
6	0.30	0.472	0.449	0.437	11.6	6.7	61.2 + j7.7	56	−7.4	52	−14.8
7	0.20	0.489	0.463	0.444	11.8	12.6	20.6 + j16.8	58	−7.4	52	−14.1
7	0.25	0.477	0.454	0.434	12.0	8.7	57.2 + j1.9	58	−8.1	52	−15.4
7	0.30	0.475	0.455	0.439	12.7	8.7	35.9 + j21.7	50	−7.3	46	−12.6

Conductor diameter = 0.005λ
HP = half-power beamwidth
SLL = Side-lobe level

element Yagi for the lower VHF TV and FM bands and a five- or six-element Yagi for the higher VHF TV band. In the UHF TV band, because of the greater propagation attenuation, higher values of gain are required and a 10- to 12-element Yagi is usually employed. At these frequencies, the length of the elements are relatively short and more than 10 elements do not present a practical problem.

Several general properties, derived partly from practical experience and partly from theoretical computations, are useful guides for the design of Yagis. These are briefly stated below. For more detailed information, the reader is referred to Jasik (1961), the *ARRL Antenna Book*, and Viezbicke (1968).

(a) Close spacings between the elements will result in a higher front-to-back ratio with a broader main beam while wider spacings give a sharper main beam but more and larger minor lobes. Wider spacing also yields greater bandwidth than closer spacing.

(b) In general, the gain of the antenna drops off less rapidly when the reflector length is increased beyond the optimum value than it does for a corresponding decrease below the optimum value. The opposite is true of a director. Consequently, if the reflector is made longer and the directors shorter than their optimum values, the antenna would be less dependent on the exact frequency at which it is operated. This is because in increase above the design frequency has the same effect as increasing the length of both types of parasitic elements, while a decrease in frequency has the same effect as shortening them. By making the directors slightly shorter and the reflectors slightly longer, there will be a greater spread between the upper and lower frequencies at which the gain starts to show a rapid decrease. This offers a simple method of achieving greater bandwidth at the expense of lower peak again.

(c) The radiation pattern of a Yagi antenna is almost independent of the length of its driven element. The length and the construction of the driven element are determined mainly by the impedance characteristics. As a general rule, greater bandwidth is obtained by using an element with a fairly large ratio of diameter to length. An example is the folded dipole, which, in addition to stepping up the input impedance, also leads to a broader bandwidth. This is based on the well known fact that a strictly parallel combination of two closely spaced dipoles is effectively equivalent to a fat dipole whose radius is the geometrical mean value of the radius of the individual dipole and the spacing between them. Another useful element based on the same principle is the fan dipole, which is essentially a parallel combination of two or more ordinary dipoles in which at one end they are connected together and at the other end spread out like a fan. In VHF TV Yagis covering both the high and low band, the fan dipole is usually tilted forward by approximately 33° as illustrated in Figure 8.10. The effect of the tilt is to restore the forward radiation in the high band, when the electrical length of the element increases to more

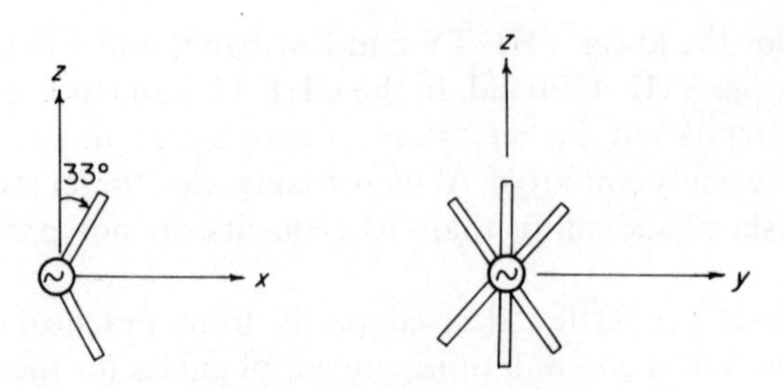

Figure 8.10 A tilted fan dipole

than 1.2 times the wave length. At such electrical lengths, the pattern of an untilted dipole would have a small forward radiation and large side lobes.

(d) It is possible to replace the thin-element reflector by a screen-type corner reflector made of grid wires. While this makes the antenna more bulky, it has the advantage of improving both the gain and the bandwidth. The theory of the corner reflector will be discussed in the next several sections.

8.6 FLAT SHEET REFLECTOR

As shown in section 8.2, a thin parasitic element can be used as a reflector. The condition on the element length and its distance from the driver is given by the relation $v = kd$, namely,

$$\pi + \tan^{-1}(X_{12}/R_{12}) - \tan^{-1}(X_{22}/R_{22}) = kd \tag{8.42}$$

Since the mutual impedance, the self-impedance, and the term kd are all frequency-dependent, the thin parasitic element is highly sensitive to frequency. Indeed, if sufficiently out of tune, it may cease to function as a reflector. If the element is replaced by a large flat conducting sheet, it is clear that it would always function as a reflector. Detailed analysis also shows that the electrical properties of a flat sheet reflector are relatively insensitive to small frequency changes compared to the case of a thin-element reflector.

Figure 8.4 shows a dipole antenna of length L placed at a distance S from a perfectly conducting plane sheet of metal, with the axis of the dipole parallel to the sheet. Assuming the sheet to be infinite in extent, the problem is easily handled by the methods of images. As discussed in section 3.14, the boundary condition on the conducting sheet can be satisfied by an image dipole placed at a distance S on the opposite side of the sheet carrying a current of equal magnitude but opposite phase. As for as calculating the fields in the space to the right of the sheet is concerned, the original problem is equivalent to that of two out-of-phase doublets, as shown in Figure 8.11. Let the current in the dipole be I_1. The current in the image will then be $-I_1$. The array factor is

$$f(\cos\chi) = I_1\,[\exp(-jkS\cos\chi) - \exp(jkS\cos\chi)] = -2jI_1\sin(kS\cos\chi) \tag{8.43}$$

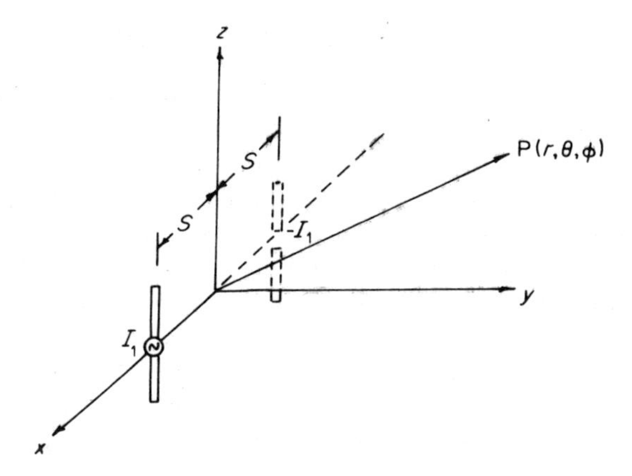

Figure 8.11 Replacing the flat sheet reflector by an image current

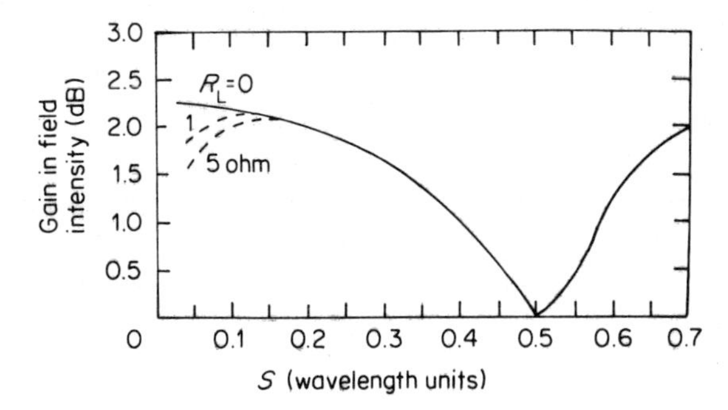

Figure 8.12 The gain in field intensity in the direction $\theta = 90°$, $\phi = 0°$ as a function of spacing S for a half-wave dipole in front of a flat sheet reflector (Source: Kraus (1950), *Antennas*. Reproduced by permission of McGraw-Hill)

The far field of the antenna is equal to the array factor multiplied by the individual pattern. It is again of the form given by (8.9), except that in the present case $f(\cos\chi)$ is given by (8.43).

The gain in field intensity over a half-wave dipole can be calculated in the manner of sections 8.2 and 8.3. In the plane perpendicular to the dipole axis, where $\chi = \phi$, the result is

$$|G(\phi)| = 2\sqrt{\left(\frac{R_{11} + R_L}{R_{11} + R_L - R_{12}}\right)}|\sin(kS\cos\phi)| \tag{8.44}$$

For the case of a half-wavelength driven element, the gain at $\phi = 0°$ as a function of spacing S is shown in Figure 8.12 for assumed loss resistances $R_L = 0$, 1, and 5 ohm. It is seen that, if $R_L = 1$ ohm, a spacing of 0.125λ yields the maximum gain.

8.7 THE CORNER REFLECTOR

If we bend the flat sheet reflector in the middle so that it becomes two flat sheets intersecting at an angle β, the resultant structure is called a corner reflector. The flat sheet reflector can be considered as a corner reflector with $\beta = 180°$.

The geometry of the corner reflector antenna is shown in Figure 8.5. For analysis purposes, the two conducting sheets are assumed to be perfectly conducting and infinite in extent. Even so, the problem is complicated for the general case of an arbitrary corner angle β. However, if β is restricted to $180°/N$ where N is any positive integer, which covers most of the corner reflectors used in practice, the problem can be conveniently handled by the method of images. In this approach, the effects of the conducting planes are accounted for by an appropriate number of suitably placed images of the proper signs. For the case in which the dipole is placed on the bisector of the corner angle and oriented parallel to the apex of the reflector, there are $(2N - 1)$ images. The actual current element and its images are spaced uniformly on a circle of radius S, with the adjacent elements alternating in polarity. This is shown in Figure 8.13.

To calculate the far field of the driven dipole and its $(2N - 1)$ images, one can follow either of two approaches. The first is to divide the N-element circular array into $\frac{1}{2}N$ two-element arrays. Each of the two-element arrays is formed by the two elements on the opposite side of the circle. The total far field is obtained by applying the principle of pattern multiplication and superposition. For small corner angles (large N), this approach leads to very cumbersome algebra, especially if one is interested in the radiation field at an arbitrary point in space, not just in the two principal planes.

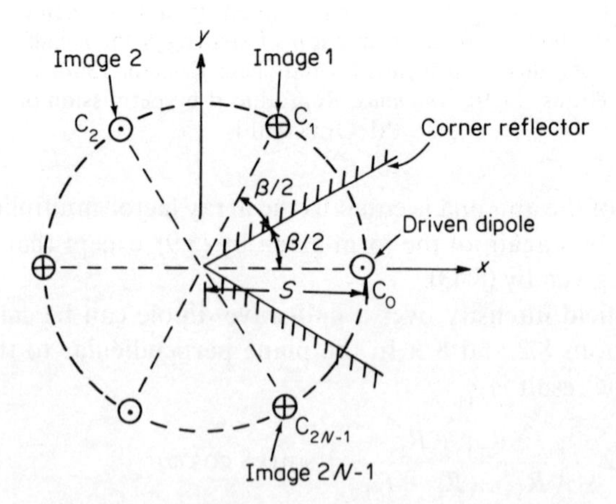

Figure 8.13 The $(2N - 1)$ images of the driven dipole in a corner reflector of corner angle $\beta = 180°/N$ are uniformly spaced in a circle of radius S, with the adjacent elements alternating in polarity

In the second approach, one simply adds up the radiation from the N vertical dipoles directly. It is somewhat surprising that this direct method enables one to express the far field in a compact form, regardless of the value of N. For this reason, we shall use the second approach instead of the first.

To be specific, let us take the driven element to be a half-wave dipole; it will be evident that the analysis can easily be extended to dipoles of any length. For a z-directed half-wave dipole situated at the origin, the electric field at any point $P(r, \theta, \phi)$ in the far zone is given by

$$\mathbf{E}(r) = \mathrm{j}CI\frac{\cos(\frac{1}{2}\pi\cos\theta)}{\sin\theta}\hat{\boldsymbol{\theta}} \tag{8.45}$$

where

$$C = \frac{\eta\exp(-\mathrm{j}kr)}{2\pi r} \tag{8.46}$$

and I is the current at the centre. Suppose now that the dipole is situated at a distance S from the origin. The line joining the origin and the centre of the dipole makes an angle ψ with the x-axis, as shown in Figure 8.14. As far as the far-zone radiation is concerned, the only difference is that the phase is now advanced by an amount kD, where D is the path difference. Hence

$$\mathbf{E}(r) = \mathrm{j}CI\frac{\cos(\frac{1}{2}\pi\cos\theta)}{\sin\theta}\exp(\mathrm{j}kD)\hat{\boldsymbol{\theta}} \tag{8.47}$$

The path difference

$$D = S\sin\theta\cos(\psi - \phi) \tag{8.48}$$

To prove (8.48), let $\hat{\mathbf{p}}$ and $\hat{\mathbf{s}}$ be two unit vectors in the directions OP and OC respectively and let γ be the angle between them. Then

$$D = S\cos\gamma \tag{8.49}$$

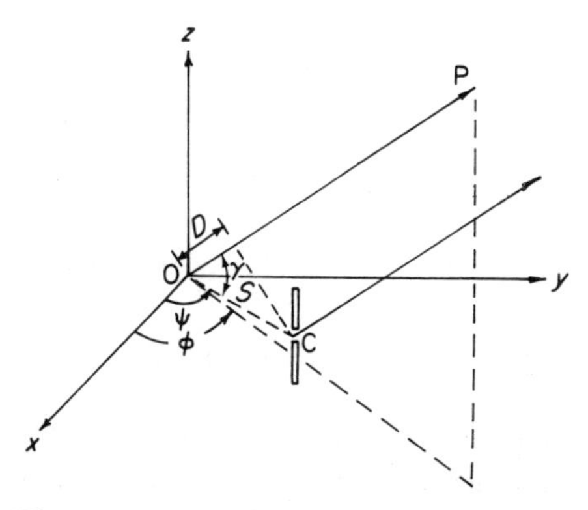

Figure 8.14 A z-directed dipole at a distance S from the origin

where

$$\cos \gamma = \hat{\mathbf{p}} \cdot \hat{\mathbf{s}} \tag{8.50}$$

But

$$\hat{\mathbf{p}} = \sin\theta \cos\phi \, \hat{\mathbf{x}} + \sin\theta \sin\phi \, \hat{\mathbf{y}} + \cos\theta \, \hat{\mathbf{z}} \tag{8.51}$$

$$\hat{\mathbf{s}} = \cos\psi \, \hat{\mathbf{x}} + \sin\psi \, \hat{\mathbf{y}} \tag{8.52}$$

Hence

$$\cos\gamma = \sin\theta\cos\phi\cos\psi + \sin\theta\sin\phi\sin\psi = \sin\theta\cos(\psi - \phi)$$

and (8.48) is proved.

For the corner reflector antenna of corner angle $\beta = 180°/N$, we have $2N$ elements equispaced on a circle of radius S, with the currents in adjacent elements alternating in polarity. As shown in Figure 8.13, we denote each element's centre by a point C_i, with C_0 being the driven dipole's centre. Applying (8.47) to each of the elements, we obtain the following compact formula for the electric field at any point in the far zone:

$$\mathbf{E}(\theta, \phi) = \mathrm{j}CI \frac{\cos(\frac{1}{2}\pi\cos\theta)}{\sin\theta} \sum_{i=0}^{2N-1} (-1)^i \exp(\mathrm{j}kD_i)\hat{\theta} \tag{8.53}$$

where

$$D_i = S\sin\theta\cos(\psi_i - \phi) \tag{8.54}$$

and

$$\psi_i = i\beta \tag{8.55}$$

The far-zone magnetic field is related to $\mathbf{E}$ by the usual formula, namely, $\mathbf{H} = H_\phi \hat{\boldsymbol{\phi}}$ where $H_\phi = E_\theta/\eta$.

The far field in the two principal planes, namely, azimuth and elevation, can be obtained by putting $\theta = 90°$ and $\phi = 0°$ in (8.53) respectively. Thus we have

$$\mathbf{E}(90°, \phi) = \mathrm{j}CI \sum_{i=0}^{2N-1} (-1)^i \exp[\mathrm{j}kS\cos(i\beta - \phi)]\hat{\boldsymbol{\theta}} \tag{8.56}$$

$$\mathbf{E}(\theta, 0°) = \mathrm{j}CI \frac{\cos(\frac{1}{2}\pi\cos\theta)}{\sin\theta} \sum_{i=0}^{2N-1} (-1)^i \exp[\mathrm{j}kS\sin\theta\cos(i\beta)]\hat{\boldsymbol{\theta}} \tag{8.57}$$

In the broadside direction ($\theta = 90°$, $\phi = 0°$), (8.56) and (8.57) reduce to

$$\mathbf{E}(90°, 0°) = \mathrm{j}CI \sum_{i=0}^{2N-1} (-1)^i \exp[\mathrm{j}kS\cos(i\beta)]\hat{\boldsymbol{\theta}} \tag{8.58}$$

For a given corner angle β, the series in (8.56) and (8.57) can be expanded. If desired, the exponential functions can be converted to trigonometric functions. However, the trigonometric forms quickly become long and unwieldy for corner angles less than 60° ($N > 3$).

The gain in field intensity over a half-wave dipole in free space with the same power input can be computed in the usual manner. In the horizontal plane, the expressions for corner angles of 90° and 60° are given below:

(a) $\beta = 90°$ (square corner reflector, $N = 2$)

$$|G(\phi)| = 2\sqrt{\left(\frac{73 + R_L}{73 + R_L + R_{02} - 2R_{01}}\right)}|\cos(kS\cos\phi) - \cos(kS\sin\phi)| \tag{8.59}$$

(b) $\beta = 60°$ ($N = 3$)

$$|G(\phi)| = 2\sqrt{\left(\frac{73 + R_L}{73 + R_L + 2R_{02} - 2R_{01} - 2R_{03}}\right)}|\sin(kS\cos\phi) - \sin[kS\cos(60° - \phi)] - \sin[kS\cos(60° + \phi)]| \tag{8.60}$$

In deriving (8.59) and (8.60), the driven element is designated with subscript 0 and the images are numbered in an anticlockwise order, as shown in Figure 8.13. R_L is the loss resistance of the half-wave dipole and $R_{0i}(i = 1, \ldots, 2N - 1)$ is the real part of the mutual impedance between the driven element and image i.

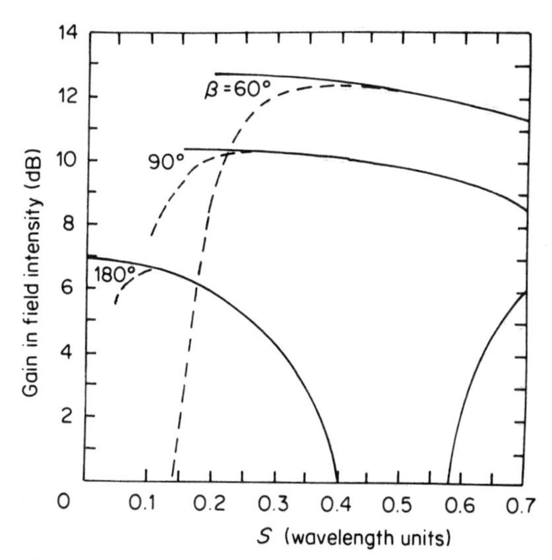

Figure 8.15 Gain in field intensity of a corner reflector antenna over a half-wave dipole in the direction $\theta = 90°$, $\phi = 0°$ for corner angles 180°, 90°, and 60°. Gain is in decibels and is shown for zero loss resistance (full curves) and for an assumed loss resistance of 1 ohm (broken curves) (Source: Kraus (1950), *Antennas*. Reproduced by permission of McGraw-Hill)

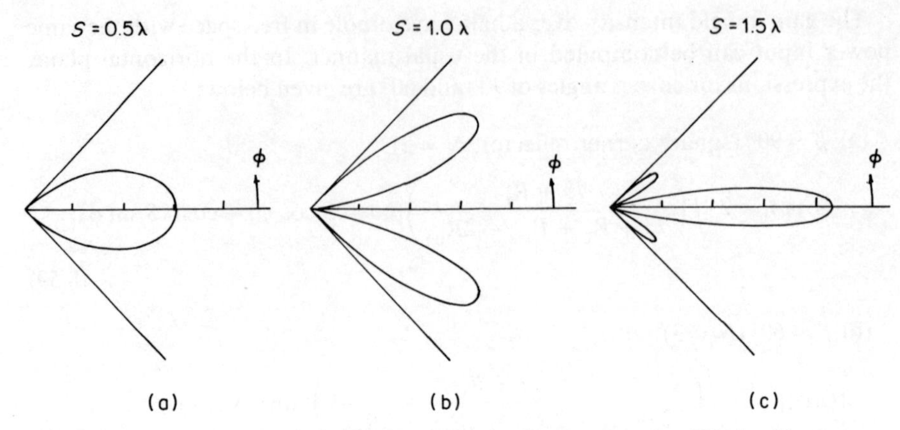

Figure 8.16 Patterns of gain in field intensity in the direction $\theta = 90°$, $\phi = 0°$ of square corner reflector antennas with dipole-to-apex spacing of (a) 0.5λ, (b) 1.0λ, and (c) 1.5λ (Source: Kraus (1950), *Antennas*. Reproduced by permission of McGraw-Hill)

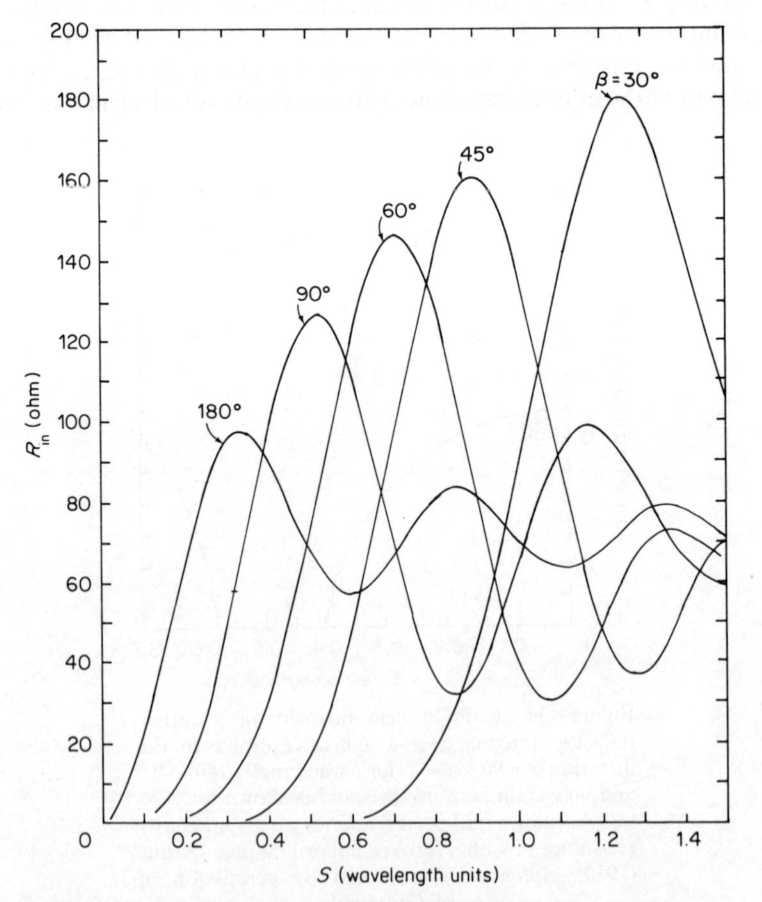

Figure 8.17 Driving-point resistance of corner reflector antenna (Source: Ng and Lee (1982), *IEE Proc.*, 129, Pt H, 11–17. Reproduced by permission of the Institute of Electrical Engineers, England)

Figure 8.15 shows the gain in field intensity versus spacing in the direction $\phi = 0$ for $\beta = 180°$, $90°$, and $60°$. Two curves are shown for each corner angle. The full curve in each case is computed for $R_L = 0$ while the broken curve is for an assumed loss resistance $\mathbf{R}_L = 1$ ohm.

The pattern in the horizontal plane for a square ($\beta = 90°$) corner reflector is shown in Figure 8.16 for three values of the dipole-to-corner spacing S. For $S = 0.5\lambda$, the pattern is single-lobed. A two-lobed pattern results if S is increased to 1.0λ. If S is further increased to 1.5λ, the pattern has a major lobe in the $\phi = 0$ direction but with minor lobes present.

The input or driving-point impedance of the antenna is given by

$$Z_{in} = R_{in} + jX_{in} \equiv V(0)/I(0)$$

where $V(0)$ and $I(0)$ are the voltage and current at the terminals of the driven dipole. Since the currents in the elements are equal in magnitude, we have

$$Z_{in} = Z_{00} + \sum_{i=1}^{2N-1} Z_{0i}(-1)^i \tag{8.61}$$

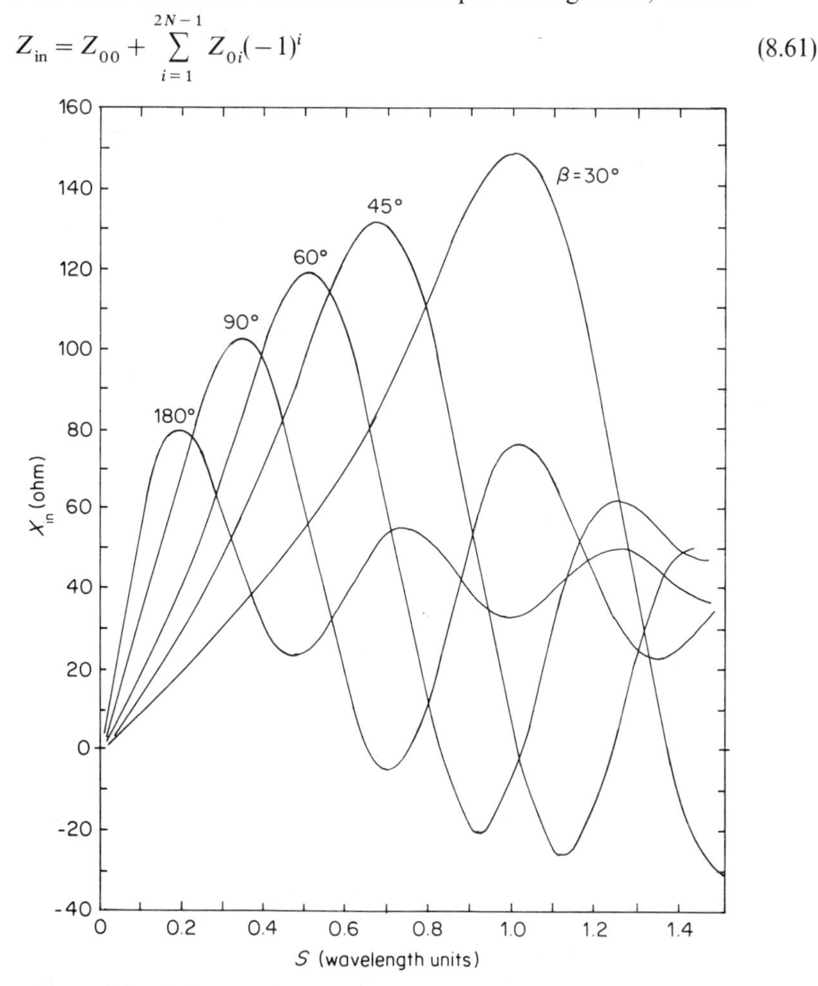

Figure 8.18 Driving-point reactance of corner reflector antenna (Source: Ng and Lee (1982), *IEE Proc.*, 129, Pt H, 11–17. Reproduced by permission of the Institute of Electrical Engineers, England)

where Z_{00} is the self-impedance of the driven dipole and Z_{0i} is the mutual impedance between the driven dipole and image i. In writing (8.61) we have neglected the copper loss of the dipole.

For a given corner angle $\beta = 180°/N$, the mutual impedances can be computed in the manner discussed in Chapter 4. The variation of the driving-point resistance, R_{in}, as a function of dipole-to-corner spacing (S) is shown in Figure 8.17. The corresponding variation of the driving-point reactance, X_{in}, is shown in Figure 8.18.

In the above analysis using the method of images, it has been assumed that each of the reflecting sheets making up the corner reflector is semi-infinite in extent. In practice, it has been found that, referring to Figure 8.5, if the dimension l is about 20% longer than the driven element and the dimension w is at least equal to twice the dipole-to-corner spacing S, the effect of finite sides is negligible (Kraus, 1950).

To reduce wind resistance, the solid reflector is often replaced by a grid of parallel conductors, as shown in Figure 8.19. The supporting member may be either a conductor or an insulator. Experimentally, it has been found that, for a 90° corner angle, the spacing d between the reflector conductors should be equal to or less than 0.1λ in order to simulate the performance of a solid reflector (Kraus, 1950). Quantitative theoretical justification of this assertion and that of the preceding paragraph concerning the dimensions l and w do not appear to be available in the literature.

8.8 CORNER REFLECTOR WITH TILTED DIPOLE

8.8.1 General Comments

In our discussion of the corner reflector antenna in section 8.7, the driven dipole is placed on the bisector of the corner angle and oriented parallel to the apex of the reflector. The radiation produced is linearly polarized, i.e. the locus of the

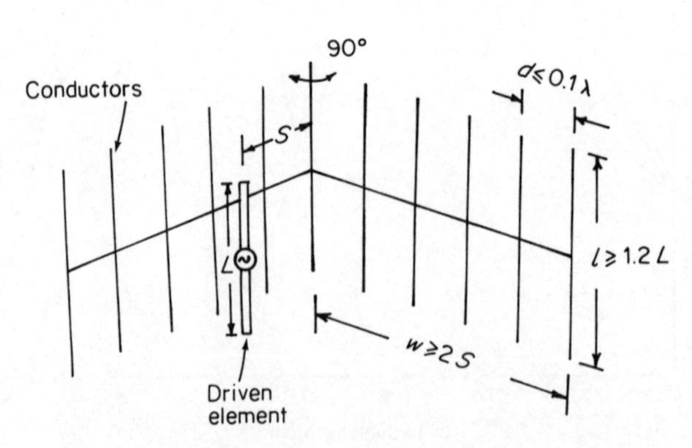

Figure 8.19 Grid-type reflector

electric field vector always lies along a straight line as time varies. This is because the driven dipole and its images are all pointed in the same direction. Indeed, all the antennas we have discussed so far produced linearly polarized radiation, since they have been limited to single dipoles or arrays of similarly oriented dipoles.

There are applications in which elliptically or circularly polarized radiation is preferred over linearly polarized radiation. For example, in communications with satellites and space vehicles, an electromagnetic wave must traverse the Earth's ionosphere, which is that part of the Earth's upper atmosphere where an appreciable number of the molecules are ionized. It extends from about 60 km upwards. Since it is immersed in the Earth's magnetic field, the propagation characteristics of an electromagnetic wave are dependent on the direction which the wave vector makes with the magnetic field. The ionosphere is therefore an anisotropic medium. As a linearly polarized electromagnetic wave propagates in such an anisotropic medium, the direction of its electric field vector rotates, the amount of rotation being dependent on the electron density, the magnetic field strength, and the distance travelled. This phenomenon is known as Faraday rotation. Since the electron density is a variable quantity, depending on conditions prevailing in the ionosphere at the time, it is difficult to predict the total amount of rotation and hence the direction of the electric field at the end of the journey. If, on arriving at the satellite or spacecraft, the electric field vector is prependicular to the receiving antenna, there will be no signal received. This problem can be avoided if the electromagnetic wave transmitted is circularly polarized. For such a wave, its electric field vector rotates in a circle and there is always a time-averaged signal at the receiving antenna.

As another example, in radar applications, circular polarization is found to minimize the 'clutter' echoes received from raindrops, in relation to larger targets such as an aircraft. For more detailed discussion of this phenomenon, the reader is referred to Kerr (1965).

There are many ways of making a circularly polarized antenna. Some of these will be discussed in Chapter 9. The objective of this section is to show that a corner reflector antenna excited by a tilted dipole can serve as a circularly polarized antenna. Before analysing it in detail, let us briefly review the mathematical description of elliptically polarized waves.

8.8.2 Elliptical Polarization

Let u_1, u_2, u_3 be the mutually perpendicular axes of a right-handed coordinate system. Consider a plane electromagnetic wave propagating along the $\hat{\mathbf{u}}_3$ direction. The electric and magnetic fields will then lie in the u_1-u_2 plane. The polarization of the wave refers to the shape of the locus traced out by the electric field vector. The electric field vector can be resolved into two perpendicular components along $\hat{\mathbf{u}}_1$ and $\hat{\mathbf{u}}_2$. If these components are in phase, the resultant vector always lie along a straight line and the wave is said to be linearly polarized. If

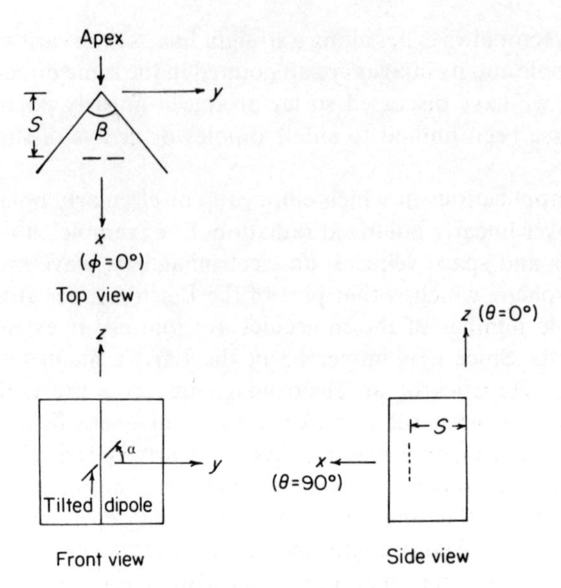

Figure 8.20 Geometry of the corner reflector antenna with tilted dipole

they are in phase quadrature, the tip of the resultant vector describes an ellipse and the wave is said to be elliptically polarized. An anticlockwise rotation of the vector corresponds to left elliptical polarization while a clockwise rotation corresponds to right elliptical polarization. When the major and minor axes of the ellipse are equal, we have circular polarization. Mathematically, let A and B be positive real numbers. If

$$\mathbf{E} = (A\hat{\mathbf{u}}_1 + B\hat{\mathbf{u}}_2)E_0 \tag{8.62}$$

the polarization is linear. If

$$\mathbf{E} = (A\hat{\mathbf{u}}_1 \pm \mathrm{j}B\hat{\mathbf{u}}_2)E_0 \tag{8.63}$$

the positive (negative) sign corresponds to left (right) elliptical polarization, respectively. If $A = B$ in (8.63), the polarization is circular.

It is easy to show qualitatively that elliptically polarized radiation can be generated by tilting the axis of the driven dipole in a corner reflector so that it is no longer parallel to the apex, as shown in Figure 8.20. For this purpose, let the corner angle β be 90° and consider the radiation in the broadside direction ($\theta = 90°$, $\phi = 0°$). To use the method of images, we resolve the current flowing in the tilted dipole into horizontal and vertical components:

$$\mathbf{I} = I_{\mathrm{H}}\hat{\mathbf{y}} + I_{\mathrm{V}}\hat{\mathbf{z}} = I(\cos\alpha)\hat{\mathbf{y}} + I(\sin\alpha)\hat{\mathbf{z}} \tag{8.64}$$

The images of the vertical component I_{V} and the horizontal component I_{H} are shown in Figure 8.21(a) and (b) respectively. At any point in the far zone in the broadside direction, I_{V} and its images will contribute a vertical component

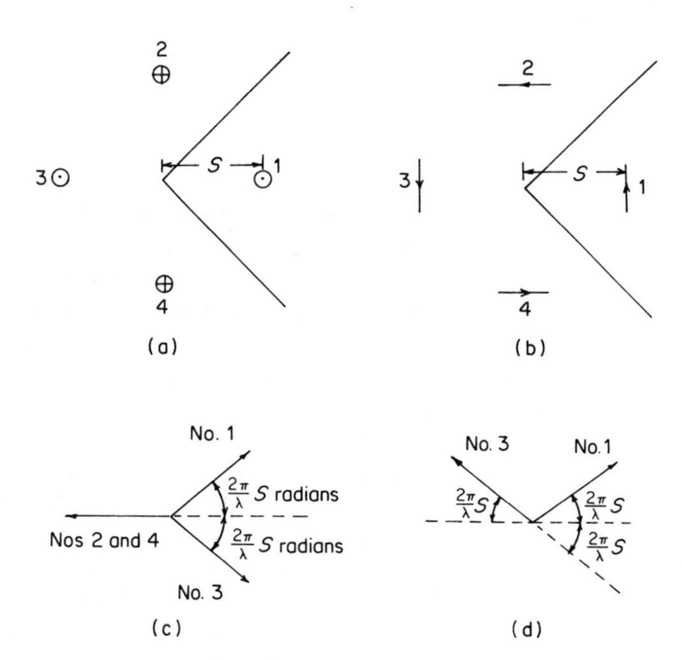

Figure 8.21 (a) The vertical current component and its images of a square corner reflector excited by a tilted dipole; (b) the horizontal current component and its images; (c) phase relations of the vertical electric field components; (d) phase relations of the horizontal electric field components

to the electric field while I_H and its images will contribute a horizontal component. Let us examine the phase relationships of the various contributions. Taking the apex to be the point of phase reference, the phasor diagram for the vertical electric field components are shown in Figure 8.21(c). It is seen that the resultant vector is 180° out of phase with respect to the reference for all values of the dipole-to-corner spacing S.

The phasor diagram for the horizontal electric field components is shown in Figure 8.21(d). Note that there is no contribution from images 2 and 4 because their radiation patterns have a null in the broadside direction. The resultant vector is in phase quadrature with respect to the reference. It follows that, in the broadside direction, the far-zone electric field has a horizontal component and a vertical component which are 90° out of phase. Hence the polarization is elliptical. The ellipticity depends on the tilt angle α and the dipole-to-apex spacing S. It is reasonable to expect that, by suitable arrangement of these two parameters, circular polarization can be obtained.

Phasor diagrams similar to Figure 8.21 can be constructed for other corner angles. If $\beta = 60°$, it is easy to show that the vertical and horizontal components of the electric field in the broadside direction is in phase and the polarization is linear. In general, the two components are in phase quadrature only for β equal to submultiples of 90°.

We now turn to a quantitative analysis of the corner reflector with tilted dipole.

8.8.3 Far-Field Formulae

Consider again the corner reflector antenna with tilted dipole ($\alpha \neq 90°$) shown in Figure 8.20. As in the untilted case ($\alpha = 90°$) treated in section 8.7, we restrict β to $180°/N$ so that the method of images can be used to analyse the operation of the antenna. For both of the horizontal and vertical current components I_H and I_V respectively, the effect of the conducting planes can be taken care of by $(2N-1)$ images. The actual current element and its images are spaced uniformly on a circle of radius S. Their orientations are illustrated in Figure 8.22. For the vertical currents, the adjacent elements alternate in polarity (Figure 8.22a). For the horizontal currents, they flow as in a continuous loop (Figure 8.22b).

We shall assume the driven dipole to be of infinitesimal length dl (Hertzian dipole). We do this partly for the sake of obtaining relatively simple formulae and partly because the results for the far field are similar to those of dipole lengths up to a half-wavelength. However, it will be aparent that the approach to be used can be extended to any dipole length in a straightforward manner.

In order to find the radiation field at any point in the far zone, it is necessary to add the contributions from the vertical and horizontal current components and their images. The former has already been obtained in section 8.7 for half-wave dipoles. Modifying (8.53) to Hertzian dipoles, we have

$$\mathbf{E}_V(\theta, \phi) = 0.5jCI_V \sin\theta \sum_{i=0}^{2N-1} (-1)^i \exp(jkD_i)\hat{\boldsymbol{\theta}} \tag{8.65}$$

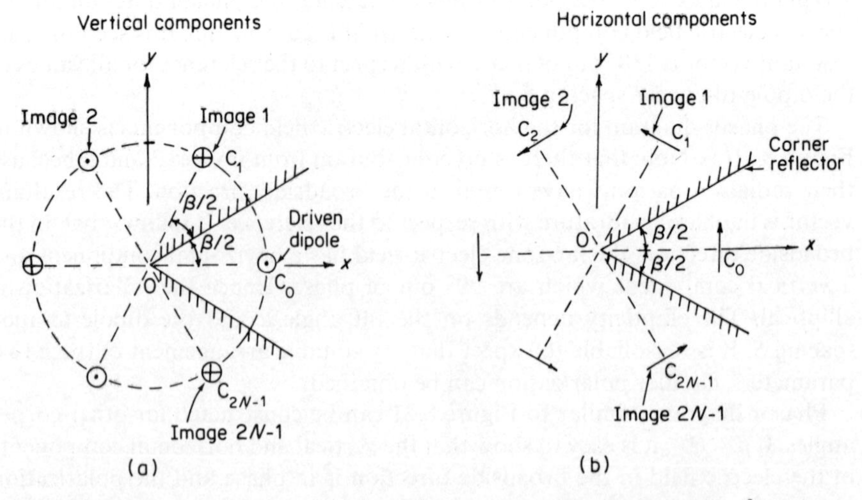

Figure 8.22 Images of the vertical and horizontal current components of a corner reflector excited by a tilted dipole

where

$$C = \frac{\eta \mathrm{d}l}{\lambda r} \exp(-\mathrm{j}kr) \tag{8.66}$$

and D_i is given by (8.54).

The radiation from the horizontal current component is more complicated, since each of its images is pointed in a different direction, as is evident from Figure 8.22(b). We first consider a Hertzian dipole $I_{\mathrm{H}}\mathrm{d}l$ which is situated at the origin and oriented parallel to the x-axis, as shown in Figure 8.23(a). Referring to example 2 in section 3.17, the field at any point $P(r, \theta, \phi)$ in the far zone is given by

$$\mathbf{E}(r) = 0.5\mathrm{j}CI_{\mathrm{H}}(-\hat{\boldsymbol{\theta}} \cos\theta \cos\phi + \hat{\boldsymbol{\phi}} \sin\phi) \tag{8.67}$$

Suppose the dipole is oriented such that it makes an angle ζ with respect to the x-axis while still situated at the origin, as shown in Figure 8.23(b). Introducing

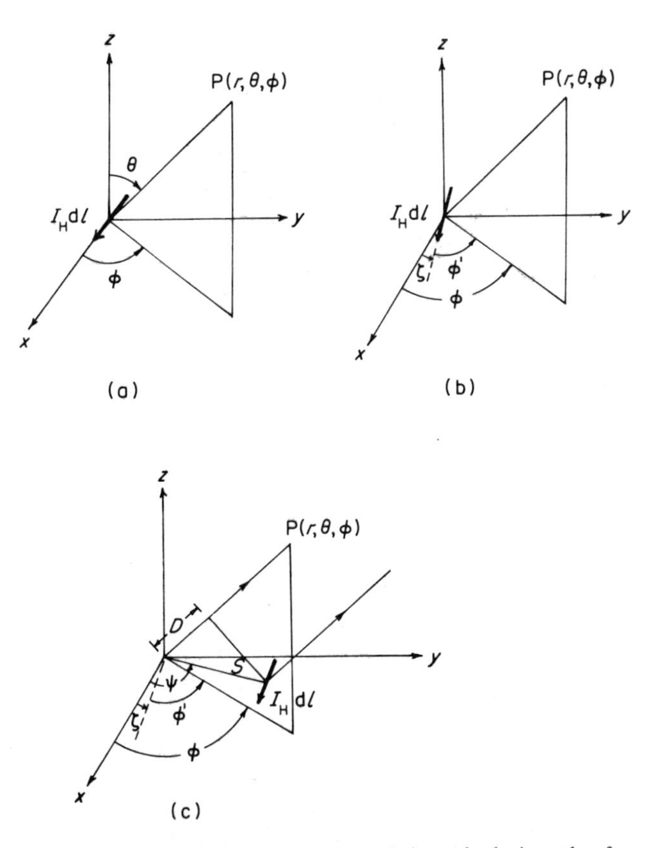

Figure 8.23 Coordinate system used in calculating the far field due to the horizontal component

the variable

$$\phi' = \phi - \zeta \tag{8.68}$$

(8.67) becomes

$$\begin{aligned}\mathbf{E}(r) &= 0.5\mathrm{j}CI_{\mathrm{H}}(-\hat{\boldsymbol{\theta}} \cos\theta \cos\phi' + \hat{\boldsymbol{\phi}} \sin\phi') \\ &= 0.5\mathrm{j}CI_{\mathrm{H}}[-\hat{\boldsymbol{\theta}} \cos\theta \cos(\phi - \zeta) + \hat{\boldsymbol{\phi}} \sin(\phi - \zeta)]\end{aligned} \tag{8.69}$$

Let us now consider the situation shown in Figure 8.23(c). The dipole, in addition to making an angle ζ with the x-axis, is at a distance S from the origin. The line joining the origin and the centre of the dipoles makes an angle ψ with the x-axis. From Figure 8.23(c), it is clear that for this case,

$$\mathbf{E}(r) = 0.5\mathrm{j}CI_{\mathrm{H}} \exp(\mathrm{j}kD)[-\hat{\boldsymbol{\theta}} \cos\theta \cos(\phi - \zeta) + \hat{\boldsymbol{\phi}} \sin(\phi - \zeta)] \tag{8.70}$$

where D is given by (8.48).

Suppose now that we have in the x–y plane an array of $2N$ Hertizan dipoles, all of which are of strength $I_{\mathrm{H}}\mathrm{d}l$ and at a distance S from the origin. Let the orientation of each be specified by the angles ζ_i and ψ_i. Then the far field due to the array is

$$\mathbf{E}(r) = 0.5\mathrm{j}CI_{\mathrm{H}} \sum_{i=0}^{2N-1} \exp(\mathrm{j}kD_i)[-\hat{\boldsymbol{\theta}} \cos\theta \cos(\phi - \zeta_i) + \hat{\boldsymbol{\phi}} \sin(\phi - \zeta_i)] \tag{8.71}$$

where D_i is given by (8.54).

To use (8.71) in our problem, let us define the variables in the equation by the geometry of the images and the driven element of the corner reflector antenna. As shown in Figure 8.22, we denote each current element's centre by a point C_i, with C_0 being the driven dipole's centre. We have

$$\beta = 180°/N$$

$$\angle \mathrm{C}_0\mathrm{O}\mathrm{C}_i = i\beta \qquad i = 0, 1, 2, \ldots, (2N - 1)$$

For the horizontal current elements, it is clear that

$$\zeta_i = 90° + i\beta \tag{8.72}$$

$$\psi_i = i\beta \tag{8.73}$$

If we denote the far-zone electric field due to the horizontal current component by $\mathbf{E}_{\mathrm{H}}(\theta, \phi)$, then from (8.71),

$$\mathbf{E}_{\mathrm{H}}(\theta, \phi) = 0.5\mathrm{j}CI_{\mathrm{H}} \sum_{i=0}^{2N-1} \exp(\mathrm{j}kD_i)[-\hat{\boldsymbol{\theta}} \cos\theta \cos(\phi - \zeta_i) + \hat{\boldsymbol{\phi}} \sin(\phi - \zeta_i)] \tag{8.74}$$

The total electric field is given by the sum of $\mathbf{E}_{\mathrm{V}}$ and $\mathbf{E}_{\mathrm{H}}$. It contains two perpendicular components, E_θ and E_ϕ, which in general are not in phase. The far-zone magnetic field is related to $\mathbf{E}$ by the usual formulae, namely, $H_\phi = E_\theta/\eta$, $H_\theta = -E_\phi/\eta$.

The electric field in the two principal planes, namely, azimuth and elevation, can be obtained by putting $\theta = 90^\circ$ and $\phi = 0^\circ$ in (8.65) and (8.74) respectively. Thus we have (problem 12 in section 8.10)

$$\mathbf{E}(90^\circ, \phi) = \mathbf{E}_{\mathrm{H}}(90^\circ, \phi) + \mathbf{E}_{\mathrm{V}}(90^\circ, \phi) = E_{\mathrm{H}}(90^\circ, \phi)\hat{\boldsymbol{\phi}} + E_{\mathrm{V}}(90^\circ, \phi)\hat{\boldsymbol{\theta}} \tag{8.75}$$

$$\mathbf{E}(\theta, 0) = \mathbf{E}_{\mathrm{H}}(\theta, 0) + \mathbf{E}_{\mathrm{V}}(\theta, 0) = E_{\mathrm{H}}(\theta, 0)\hat{\boldsymbol{\phi}} + E_{\mathrm{V}}(\theta, 0)\hat{\boldsymbol{\theta}} \tag{8.76}$$

where

$$E_{\mathrm{H}}(90^\circ, \phi) = 0.5\mathrm{j}CI_{\mathrm{H}} \sum_{i=0}^{2N-1} \sin(\phi - \zeta_i)\exp[\mathrm{j}kS\cos(i\beta - \phi)] \tag{8.77}$$

$$E_{\mathrm{V}}(90^\circ, \phi) = 0.5\mathrm{j}CI_{\mathrm{V}} \sum_{i=0}^{2N-1} (-1)^i \exp[\mathrm{j}kS\cos(i\beta - \phi)] \tag{8.78}$$

$$E_{\mathrm{H}}(\theta, 0) = -0.5\mathrm{j}CI_{\mathrm{H}} \sum_{i=0}^{2N-1} \sin\zeta_i \exp[\mathrm{j}kS\sin\theta\cos(i\beta)] \tag{8.79}$$

$$E_{\mathrm{V}}(\theta, 0) = 0.5\mathrm{j}CI_{\mathrm{V}} \sum_{i=0}^{2N-1} (-1)^i \sin\theta \exp[\mathrm{j}kS\sin\theta\cos(i\beta)] \tag{8.80}$$

8.8.4 Condition for Circular Polarization

Equations (8.75) and (8.76) show that, in the azimuth and elevation planes, the far field due to the horizontal current component is along the $\hat{\boldsymbol{\phi}}$-direction while that due to the vertical current component is along the $\hat{\boldsymbol{\theta}}$-direction. We will now show that these two perpendicular field components are in phase quadrature and can be adjusted to have equal magnitudes for corner angles equal to sub-multiples of 90°, resulting in circularly polarized radiation.

Azimuth $(x\text{–}y)$ *plane*

From (8.77),

$$\begin{aligned}
E_{\mathrm{H}}(90^\circ, \phi) &= 0.5\mathrm{j}CI_{\mathrm{H}} \sum_{i=0}^{2N-1} \sin(\phi - 90^\circ - i\beta)\exp[\mathrm{j}kS\cos(i\beta - \phi)] \\
&= -0.5\mathrm{j}CI_{\mathrm{H}}\Bigg(\sum_{i=0}^{N-1} \cos(i\beta - \phi)\exp[\mathrm{j}kS\cos(i\beta - \phi)] \\
&\qquad + \sum_{i=N}^{2N-1} \cos(i\beta - \phi)\exp[\mathrm{j}kS\cos(i\beta - \phi)]\Bigg) \\
&= -0.5\mathrm{j}CI_{\mathrm{H}}\Bigg(\sum_{i=0}^{N-1} \cos(i\beta - \phi)\exp[\mathrm{j}kS\cos(i\beta - \phi)] \\
&\qquad - \sum_{k=0}^{N-1} \cos(k\beta - \phi)\exp[-\mathrm{j}kS\cos(k\beta - \phi)]\Bigg) \\
&= CI\cos\alpha\Bigg(\sum_{i=0}^{N-1} \cos(i\beta - \phi)\sin[kS\cos(i\beta - \phi)]\Bigg)
\end{aligned} \tag{8.81}$$

From (8.78),

$$E_V(90°, \phi) = 0.5jCI \sin\alpha\left(\sum_{i=0}^{N-1} (-1)^i \exp[jkS\cos(i\beta - \phi)] + \sum_{i=N}^{2N-1} (-1)^i \exp[jkS\cos(i\beta - \phi)]\right)$$

$$= \begin{cases} jCI \sin\alpha \sum_{i=0}^{N-1} (-1)^i \cos[kS\cos(i\beta - \phi)] & \text{if } N = \text{even} \quad (8.82a) \\ -CI \sin\alpha \sum_{i=0}^{N-1} (-1)^i \sin[kS\cos(i\beta - \phi)] & \text{if } N = \text{odd} \quad (8.82b) \end{cases}$$

It follows from (8.81) and (8.82) that the two perpendicular field components are in phase for odd N and in phase quadrature for even N. Circularly polarized radiation is therefore not possible for odd N but it can be obtained for even N if the condition

$$\cot\alpha \sum_{i=0}^{N-1} \frac{\sin[kS\cos(i\beta - \phi)]\cos(i\beta - \phi)}{(-1)^i \cos[kS\cos(i\beta - \phi)]} = \pm 1 \tag{8.83}$$

is satisfied, where the positive and negative signs refer to right- and left-hand rotation, respectively. Given a corner angle that is a submultiple of 90°, we can adjust the dipole-to-apex spacing S and the tilt angle α according to (8.83) to yield circular polarization in a given direction ϕ.

Elevation (x–z) plane

Starting with (8.79) and (8.80) and manipulating them in a similar manner, we find that circular polarization in the elevation plane can be obtained if N is even and the parameters satisfy the equation

$$\frac{\cot\alpha}{\sin\theta} \sum_{i=0}^{N-1} \frac{\sin[kS\sin\theta\cos(i\beta)]\cos(i\beta)}{(-1)^i \cos[kS\sin\theta\cos(i\beta)]} = \pm 1 \tag{8.84}$$

Broadside direction

The condition for circular polarization in the broadside direction ($\theta = 90°$, $\phi = 0°$) is of considerable interest. This can be obtained by either putting $\phi = 0°$ in (8.83) or $\theta = 90°$ in (8.84). For $\beta = 90°$, 45°, and 30°, we have the following results:

$$\beta = 90°: \quad \frac{\cot\alpha \sin(kS)}{\cos(kS) - 1} = \pm 1 \tag{8.85}$$

$$\beta = 45°: \quad \frac{\cot\alpha[\sin(kS) + \sqrt{2}\sin(kS/\sqrt{2})]}{1 + \cos(kS) - 2\cos(kS/\sqrt{2})} = \pm 1 \tag{8.86}$$

$$\beta = 30^\circ: \qquad \frac{\cot\alpha[\sin(kS) + \sqrt{3}\sin(kS\sqrt{3}/2) + \sin(0.5\,kS)]}{\cos(kS) - 2\cos(kS\sqrt{3}/2) + 2\cos(0.5\,kS) - 1} = \pm 1$$

(8.87)

It is interesting to note that, for $\beta = 90^\circ$, the relation between α and S for circular polarization in the broadside direction is a linear one. This can be seen from solving (8.85) for α. Taking the positive sign, we have

$$\begin{aligned}\alpha &= \tan^{-1}\left(\frac{\sin(kS)}{\cos(kS) - 1}\right) = \tan^{-1}\left(\frac{2\sin(0.5kS)\cos(0.5kS)}{-2\sin^2(0.5kS)}\right) \\ &= -\tan^{-1}[\cot(0.5\,kS)] = \tan^{-1}[\tan(\tfrac{1}{2}\pi + 0.5kS)] \\ &= \tfrac{1}{2}\pi(1 + 2S/\lambda) \text{ radians} \qquad (8.88)\end{aligned}$$

Equation (8.88) shows that α varies linearly with S, increasing at a rate of π radians per wavelength.

For information on the driving-point impedance of the corner reflector with tilted dipole, the reader is referred to Ng and Lee (1982). The materials of sections 8.8.3 and 8.8.4 also follow the treatment of this paper, and are reprinted with the permission of IEE, London.

8.9 WORKED EXAMPLES

Example 1

Figure 8.24 shows a system of transmitting and receiving antennas, each of which is in the far zone of the other. The transmitting antenna consists of a vertical 1.5λ driven dipole and a vertical 1.5λ parasitic element placed a fraction of a wavelength from it. The centre of the former is at the origin and the latter is at $(d, 0, 0)$. The receiving antenna is a vertical half-wave dipole, the centre of which is at the point $P(r, \theta, \phi)$ with respect to the coordinate system of the transmitter. Let the current at the centre of the driven dipole be I_1.

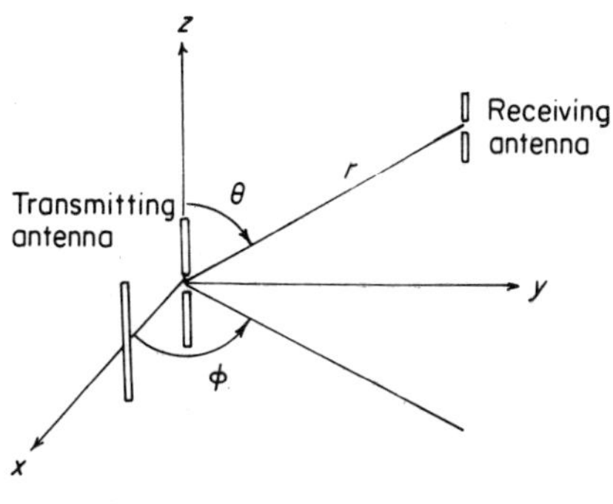

Figure 8.24 Geometry for worked example 1

(a) Obtain an expression for the electric field at the point P in terms of I_1, d, and the self- and mutual impedances of the elements.
(b) Obtain an expression for the open-circuit voltage induced at the terminals of the receiving dipole.

Solution

(a) Let I_2 be the current induced at the centre of the parasitic element. Since $0 = I_1 Z_{12} + I_2 Z_{22}$, we have

$$I_2 = -I_1 Z_{12}/Z_{22}$$

The array factor is

$$f = I_1 + I_2 \exp(jkd \sin\theta \cos\phi)$$
$$= I_1\left(1 - \frac{Z_{12}}{Z_{22}} \exp(jkd \sin\theta \cos\phi)\right)$$

The electric field at point P is $\mathbf{E} = E_\theta \hat{\boldsymbol{\theta}}$ where

$$E_\theta = (\text{field of } 1.5\lambda \text{ dipole}) \times (\text{array factor})$$
$$= \frac{jI_1\eta \exp(-jkr)\cos(\frac{3}{2}\pi \cos\theta)}{2\pi r \sin\theta}\left(1 - \frac{Z_{12}}{Z_{22}} \exp(jkd \sin\theta \cos\phi)\right)$$

(b) $|V_{oc}| = |h_e(\theta')||E_\theta|$ where $\theta' = 180° - \theta$ and $h_e(\theta')$ is the effective length of the receiving antenna evaluated at θ'. For a half-wave dipole.

$$h_e(\theta') = \frac{2\cos(\frac{1}{2}\pi \cos\theta')}{k \sin\theta'} = -\frac{2\cos(\frac{1}{2}\pi \cos\theta)}{k \sin\theta}$$

Let

$$Z_{12}/Z_{22} = |Z_{12}/Z_{22}| \exp[j(\tau_{12} - \tau_{22})]$$
$$u = \tau_{12} - \tau_{22} + kd \sin\theta \cos\phi$$

Then

$$|E_\theta| = \frac{I_1\eta \cos(\frac{3}{2}\pi \cos\theta)}{2\pi r \sin\theta}\sqrt{\left[\left(1 - \left|\frac{Z_{12}}{Z_{22}}\right| \cos u\right)^2 + \left(\left|\frac{Z_{12}}{Z_{22}}\right| \sin u\right)^2\right]}$$

$$V_{oc} = \frac{I_1\eta \cos(\frac{3}{2}\pi \cos\theta)\cos(\frac{1}{2}\pi \cos\theta)}{r\pi \sin^2\theta}\sqrt{\left[\left(1 - \left|\frac{Z_{12}}{Z_{22}}\right| \cos u\right)^2 + \left(\left|\frac{Z_{12}}{Z_{22}}\right| \sin u\right)^2\right]}$$

Example 2

An infinitesimally thin centre-fed half-wave dipole with feed-point current $I(0)$ is placed a distance S from a flat sheet reflector. Assume the reflector to be a perfect conductor occupying the whole x–z plane and the axis of the dipole is along the z-direction.

(a) For a point $P(r, \theta, \phi)$ in the far zone, find the electric field and compare its magnitude and phase with the case when the reflector is removed but the position and the current of the dipole remain the same.

(b) If $S = 0.4\lambda$ and $I(0) = 2$ A r.m.s., find the power delivered to the half-wave dipole and compare it with the case when the reflector is removed. Assume the dipole to have negligible copper loss.

(c) The system (dipole + reflector) is now used as a receiving antenna. For the case $S = 0.4\lambda$, determine the value of the load impedance which, when connected to the dipole terminals, will yield the maximum power transfer.

Solution

(a) The effect of the flat sheet reflector can be simulated by an image $\frac{1}{2}\lambda$ dipole with a feed-point current $-I(0)$. The equivalent problem is shown in Figure 8.25. For a point $P(r, \theta, \phi)$ in the far zone, let α be the angle between OP and the y-axis.

$$r_1 \simeq r + S\cos\alpha = r + S\sin\theta\sin\phi$$
$$r_2 \simeq r - S\cos\alpha = r - S\sin\theta\sin\phi$$

The electric field $\mathbf{E} = E_\theta\hat{\boldsymbol{\theta}}$ where

$$E_\theta = \frac{\mathrm{j}I(0)\exp(-\mathrm{j}kr_2)}{2\pi r_2}\frac{\cos(\frac{1}{2}\pi\cos\theta)}{\sin\theta} - \frac{\mathrm{j}I(0)\exp(-\mathrm{j}kr_1)}{2\pi r_1}\frac{\cos(\frac{1}{2}\pi\cos\theta)}{\sin\theta}$$
$$= \frac{I(0)\cos(\frac{1}{2}\pi\cos\theta)}{\pi r\sin\theta}\sin(kS\sin\theta\cos\phi)\exp[\mathrm{j}(\pi - kr)]$$

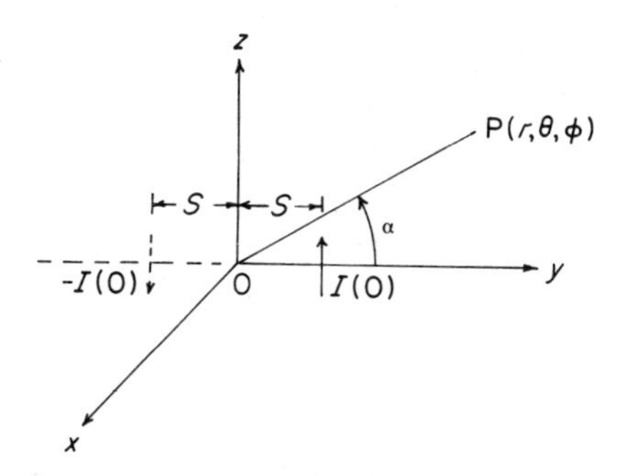

Figure 8.25 Geometry for worked example 2

For the case when the reflector is removed,

$$E'_\theta = \frac{jI(0)\exp(-jkr)}{2\pi r}\frac{\cos(\frac{1}{2}\pi\cos\theta)}{\sin\theta}\exp(jkS\sin\theta\sin\phi)$$

$$= \frac{I(0)\cos(\frac{1}{2}\pi\cos\theta)}{2\pi r\sin\theta}\exp[j(\tfrac{1}{2}\pi + kS\sin\theta\sin\phi - kr)]$$

Thus

$$|E_\theta|/|E'_\theta| = 2\sin(kS\sin\theta\sin\phi)$$

and

$$\frac{\text{phase of } E_\theta}{\text{phase of } E'_\theta} = \frac{\pi - kr}{\frac{1}{2}\pi + kS\sin\theta\sin\phi - kr}$$

Note that, if $kS \ll 1$, $|E_\theta|/|E'_\theta| \simeq 2kS\sin\theta\sin\phi \ll 1$ and the phase of E_θ leads that of E'_θ by approximately $\frac{1}{2}\pi$.

(b) $Z_{in} = Z_{11} - Z_{12}$

$$Z_{11} = 73 + j43.5$$

For two side-by-side $\frac{1}{2}\lambda$ dipoles with spacing $2S = 0.8\lambda$, $Z_{12} = -17.5 + j11$ ohm from Figure 5.5.

Hence $Z_{in} = 90.5 + j32.5$ ohm and $R_{in} = 90.5$ ohm, so $W = I^2(0)R_{in} = 3.62$ W If the reflector is removed,

$$W' = I^2(0)R_{in} = 4 \times 10^{-2}(73) = 2.92 \text{ W}$$

(c) The load impedance must be the complex conjugate of the input impedance of the antenna for maximum power transfer. Hence $Z_L = 90.5 - j32.5$ ohm.

Example 3

(a) A square corner reflector is fed by a half-wave dipole spaced 0.3λ from the corner and lying on the bisector of the corner angle, as shown in Figure 8.26(a). If the current at the feed point is 2 A r.m.s., find the power delivered to the half-wave dipole if it has negligible power loss.
(b) Repeat (a) if the half-wave dipole is spaced 0.3λ from the corner but displaced from the bisector of the corner angle by 15°, as shown in Figure 8.26(b).
(c) For the corner reflector described in (b), obtain an expression for the gain in field intensity in the x–y plane over a $\frac{1}{2}\lambda$ dipole.

Solution

(a) The images of a square corner reflector are shown in Figure 8.27(a). The circuit equation for the driven element (no. 1) is

$$V_1 = I_1 Z_{11} - I_1 Z_{12} + I_1 Z_{13} - I_1 Z_{14}$$

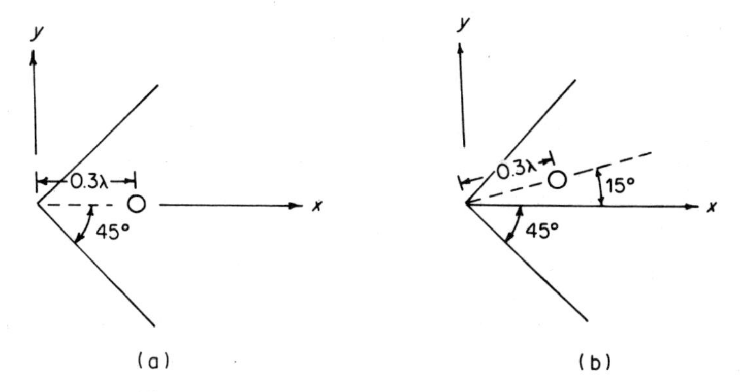

Figure 8.26 Geometry for worked example 3

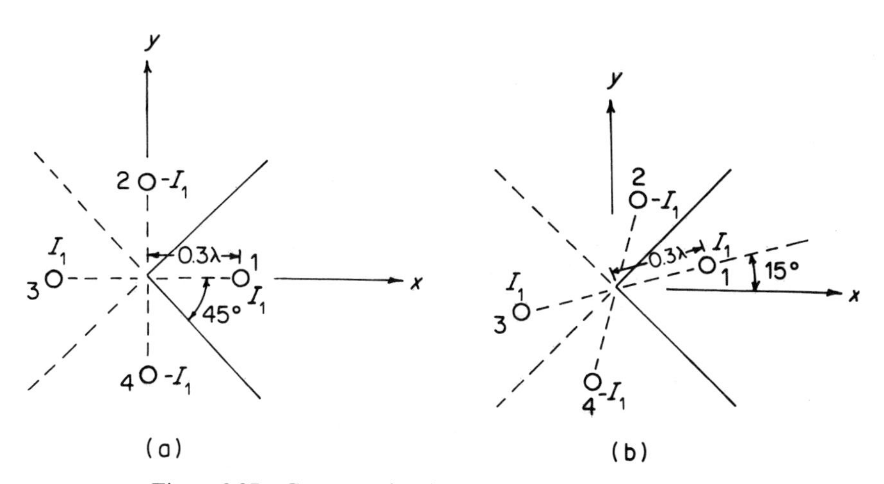

Figure 8.27 Geometry for the images in worked example 3

The input impedance

$$Z_1 = V_1/I_1 = Z_{11} - 2Z_{12} + Z_{13}$$

since $Z_{12} = Z_{14}$.

$$R_1 = \text{Re}\, Z_1 = R_{11} - 2R_{12} + R_{13}$$

$$R_{11} = 73$$

Spacing between elements 1 and 3 $= 0.6\lambda$
Spacing between elements 1 and 2 $= 0.3 \times \sqrt{2} = 0.42\lambda$
From Figure 5.5, $R_{13} = -23$ ohm, $R_{12} = 6$ ohm
Hence $R_1 = 38$ ohm

$$W = I_1^2 R_1 = 4 \times 38 = 152 \text{ W}$$

(b) The images are now shown in Figure 8.27(b).

$$R_1 = R_{11} + R_{13} - R_{12} - R_{14}$$

Spacing between elements 1 and 2 = 0.3λ, $R_{12} = 29$ ohm
Spacing between elements 1 and 3 = 0.6λ, $R_{13} = -23$ ohm
Spacing between elements 1 and 4 = $0.3 \times \sqrt{3} = 0.52\lambda$, $R_{14} = -13$ ohm
$R_1 = 34$ ohm

$$W = I_1^2 R_1 = 136 \text{ W}$$

(c) The far field is due to the contributions from the driven dipole and its images. Adding the pairs 1 and 3 and 2 and 4, we have, in the x–y plane,

$$|E(\phi)| = (\eta/\pi r)I_1 |\cos[kS\cos(\phi - 15^\circ)] - \cos[kS\cos(75^\circ - \phi)]|$$

For a power input W,

$$I_1 = \sqrt{[W/(R_{11} + R_{13} - R_{12} - R_{14})]}$$

For a half-wave dipole,

$$E_{\lambda/2}(\phi) = (\tfrac{1}{2}\eta/\pi r)\sqrt{(W/R_{11})}$$

Gain in field intensity

$$|G(\phi)| = |E(\phi)|/|E_{\lambda/2}(\phi)|$$

$$= 2\sqrt{\left(\frac{R_{11}}{R_{11} + R_{13} - R_{12} - R_{14}}\right)}|\cos[kS\cos(\phi - 15^\circ)] - \cos[kS\cos(75^\circ - \phi)]|$$

8.10 PROBLEMS

1. The statement after equation (8.15) reads: 'If the lengths of the driven and parasitic elements are close to a half-wavelength, it can be verified from the equations developed in Chapter 5 that the phase of Z_{12} as a function of d/λ is quite insensitive to the lengths of the elements.' To verify this statement for a specific case, let the length of the driven dipole (L_1) be 0.475λ. For a spacing of 0.20λ, obtain Z_{12} for $L_2 = 0.450\lambda$, 0.475λ, and 0.5λ. Repeat for another spacing of 0.30λ and comment on your results.

2. In the arrangement of Figure 8.6, $L_1 = 0.453\lambda$, $L_2 = 0.479\lambda$, $d = 0.25\lambda$.
   (a) Obtain an approximate expression for the far-zone electric field at P(r, 90°, 0°) using the principle of pattern multiplication.
   (b) Obtain an exact expression (without using the pattern multiplication principle) for the far-zone electric field at P(r, 90°, 0°) and compare its magnitude with that of (a) and comment on your result.

3. In the arrangement of Figure 8.6, $L_1 = 0.453\lambda$, $L_2 = 0.479\lambda$, $d = 0.25\lambda$. The diameters of the conductors are equal to 0.005λ. Find the input resistance

at the terminals of the driven dipole and compare it with the value when the parasitic element is removed.

4. In the driver and reflector arrangement of problem 3, a director of length 0.451λ is added at a distance of 0.25λ from the driver to form a three-element Yagi. Find the input resistance of the antenna and compare it with the value obtained in problem 3.

5. In the arrangement of problem 3, a second reflector of length 0.479λ is placed at a distance of 0.25λ from the first reflector (0.50λ from the driver). Find the current induced in the second reflector and compare it with the current induced in the first reflector. Comment on the result you obtain.

6. (a) The gain in field intensity in the H-plane as given by equation (8.23) is a complex number. What is the meaning of the gain being complex?
   (b) Derive an expression for the gain in field intensity in the E-plane of the driver and parasitic element combination of Figure 8.6 over a half-wave dipole.

7. Derive an expression for the gain in field intensity in the E-plane of a three-element Yagi over a half-wave dipole.

8. Verify equations (8.59) and (8.60).

9. Consider the 60° corner reflector with a tilted dipole. In the broadside direction, construct the phasor diagram showing the phase relationships of the contributions to the electric field due to the horizontal and vertical current components and their images. Show that the polarization is linear.

10. A half-wave dipole is placed parallel to an infinite flat ground plane and at a distance $d = 0.4\lambda$ above it, as shown in Figure 8.28. Let E_0 be the magnitude of the electric field measured at a point A in the far zone in the zenith direction. The ground plane is now removed but the same input power W is supplied to the isolated half-wave dipole. The electric field strength at point A is found to change to a value E'_0. Find the ratio E_0/E'_0. Assume the conductivities of the dipole and the ground plane to be infinite.

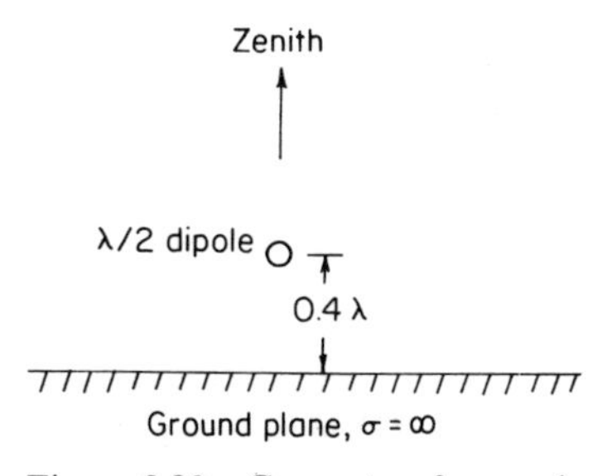

Figure 8.28 Geometry for problem 10

11. In problem 10, the ground plane, instead of being removed, is bent upwards into two semi-infinite flat sheets intersecting at an angle 90°, with the apex directly under the dipole. Let E'_0 be the electric field strength measured at point A for the same input power W to the dipole. Find the ratio E_0/E'_0.
12. Verify equations (8.76) and (8.79).

BIBLIOGRAPHY

ARRL Antenna Book (1974). The American Radio Relay League, Newington, CT, Chapters 4 and 11.

Jasik, H. (ed.) (1961). *Antenna Engineering Handbook*, McGraw-Hill, New York, Chapters 5 and 24.

Kraus, J. (1950). *Antennas*, McGraw-Hill, New York, Chapter 12.

Walkinshaw, W. (1946). 'Theoretical treatment of short Yagi aerials', *J. IEE (London), Pt III*, **93**, 598.

Woodward, Jr, O. M. (1957). 'A circularly-polarized corner reflector antenna', *IRE Trans. Antennas and Propagation*, **AP-5**, 290–296.

CHAPTER 9
The Turnstile Antenna, the Loop Antenna, the Helical Antenna, and Frequency-independent Antennas

9.1 INTRODUCTION

In the previous chapters, we developed the basic concepts of antenna theory by concentrating on the linear dipole. We first considered the dipole as a transmitting antenna and as a receiving antenna. We then considered an array of similarly oriented dipoles as well as the effect of passive conductors on the radiation from dipoles. By using the linear dipole antenna as the basic element, the fundamental concepts of reciprocity, pattern multiplication, mutual coupling, and image theory were introduced and described in a simple yet quantitative manner, using only junior-level mathematics and electromagnetic theory. It turns out that these concepts are also applicable to other forms of radiating elements, both of the wire type and the aperture type.

In this chapter, we study several other basic radiators of the wire type. The first is the cross-dipole, which consists of two dipoles arranged in the form of a cross. The second is the loop antenna and the third is the helical antenna. The chapter ends with an introduction to frequency-independent antennas.

9.2 THE CROSS-DIPOLE (TURNSTILE ANTENNA)

A cross-dipole, also known as a turnstile antenna, consists of two perpendicularly oriented dipoles fed with currents that are equal in magnitude but 90° out of phase. It is not a linear array in the sense discussed in Chapter 7 because the two elements are not similarly oriented, with the result that the principle of pattern multiplication cannot be applied. Moreover, although each of the elements is a linear dipole, the composite antenna is not strictly linear since it consists of conductors not lying along a straight line.

The cross-dipole has two principal characteristics. The first is that the radiation is in general elliptically polarized. The advantage of using non-linear polarization has been discussed in section 8.8.1 in connection with the corner reflector with tilted dipole. The second characteristic of the cross-dipole is that its three-dimensional radiation pattern is nearly omnidirectional. We shall show these features by first considering Hertzian dipoles ($dl \ll \lambda$) for simplicity. The calculation will then be extended to the case of half-wave elements.

9.2.1 Hertzian Dipoles

Polarization Characteristics

Consider a cross-dipole consisting of a z-directed Hertizan dipole $I_z dl$ (A) and an x-directed Hertzian dipole $I_x dl$ (B), as shown in Figure 9.1. In the far zone, the electric field of dipole A is given by equation (3.16):

$$\mathbf{E}_A(r) = \frac{jk\eta I_z dl}{4\pi r} \exp(-jkr) \sin\theta \,\hat{\boldsymbol{\theta}} \tag{9.1}$$

The electric field of dipole B is given by equation (3.129):

$$\mathbf{E}_B(r) = \frac{jk\eta I_x dl}{4\pi r} \exp(-jkr)(-\hat{\boldsymbol{\theta}} \cos\theta \cos\phi + \hat{\boldsymbol{\phi}} \sin\phi) \tag{9.2}$$

In the usual operation of the antenna, the dipoles are fed with currents that are equal in magnitude but 90° out of phase. Let I_x lead I_z by 90°. Then

$$I_z = I_0 \tag{9.3}$$

$$I_x = jI_0 \tag{9.4}$$

where I_0 is a real number.

The electric field of the cross-dipole is given by the vectorial sum of $\mathbf{E}_A$ and $\mathbf{E}_B$. Using (9.3) and (9.4), we have

$$\mathbf{E} = \mathbf{E}_A + \mathbf{E}_B = \frac{jk\eta \exp(-jkr)}{4\pi r} I_0 dl[\hat{\boldsymbol{\theta}}(\sin\theta - j\cos\theta\cos\phi) + \hat{\boldsymbol{\phi}}\, j \sin\phi] \tag{9.5}$$

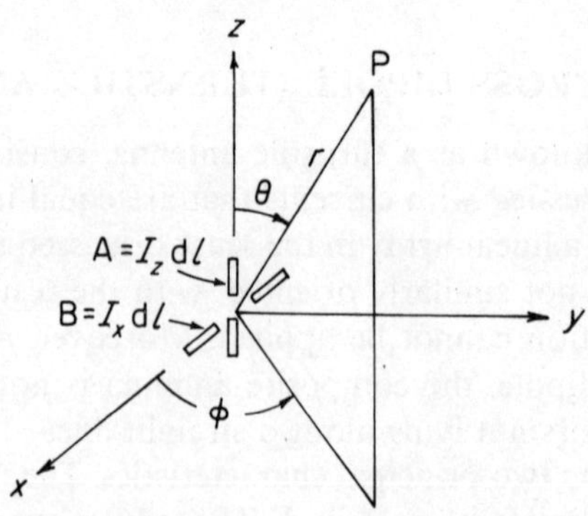

Figure 9.1 A cross-dipole (turnstile antenna) of two perpendicular infinitesimal current elements

Equation (9.5) shows that the far field consists of two perpendicular components. The polarization depends on the direction specified by θ and ϕ. Let us consider the three coordinate planes.

(1) *x–z plane.* In the plane containing the dipoles, i.e. the x–z plane, $\phi = 0$ and (9.5) becomes

$$\mathbf{E}(r) = \frac{\mathrm{j}k\eta \exp(-\mathrm{j}kr)}{4\pi r} I_0 \mathrm{d}l(\sin\theta - \mathrm{j}\cos\theta)\hat{\boldsymbol{\theta}} \tag{9.6}$$

Since the electric field has only the $\hat{\boldsymbol{\theta}}$-component, the radiation is linearly polarized.

(2) *x–y plane.* In the x–y plane, which is perpendicular to the plane of the dipole, $\theta = 90°$ and (9.5) becomes

$$\mathbf{E}(r) = \frac{\mathrm{j}k\eta \exp(-\mathrm{j}kr)}{4\pi r} I_0 \mathrm{d}l(\hat{\boldsymbol{\theta}} + \hat{\boldsymbol{\phi}}\,\mathrm{j}\sin\phi) \tag{9.7}$$

Referring to the discussion in section 8.8.2, we see from (9.7) that the polarization is left-elliptical with respect to the radial ($\hat{\mathbf{r}}$) direction for $0 < \phi < \pi$. If the thumb of the left hand points in the radial direction, the fingers curl in the direction of rotation of the instantaneous electric vector. For $\phi = 0$ and $\phi = \pi$, $\mathbf{E}(r)$ is linearly polarized. For $\phi = \pi/2$, the θ- and ϕ-components are equal and in phase quadrature and the polarization is left-circular. In the range $\pi < \phi < 2\pi$, the sign of the ϕ-component is reversed and so is the sense of polarization (direction of electric vector rotation). Similarly, if the current in dipole B lags instead of leads that of dipole A by 90°, we replace j by $-$j in (9.7) and the sense of polarization is reversed.

Thus, as the direction is varied from $\phi = 0$ to $\phi = \pi/2$, the polarization starts out to be linear, then acquires ellipticity which decreases as ϕ increases. At $\phi = \pi/2$, the polarization becomes circular. This phenomenon repeats itself in the other three quadrants.

(3) *y–z plane.* In the y–z plane, $\phi = 90°$ and (9.5) becomes

$$\mathbf{E}(r) = \frac{\mathrm{j}k\eta \exp(-\mathrm{j}kr)}{4\pi r} I_0 \mathrm{d}l(\hat{\boldsymbol{\theta}}\sin\theta + \mathrm{j}\,\hat{\boldsymbol{\phi}}) \tag{9.8}$$

The situation in the y–z plane is similar to that in the x–z plane. As θ varies from 0 to $\pi/2$, the polarization starts out to be linear, then acquires ellipticity which decreases as θ increases. The polarization becomes circular at $\theta = \pi/2$. This phenomenon repeats itself in the other three quadrants.

To summarize, in the directions lying in the plane of the dipoles, the electric field is linearly polarized. In the broadside directions ($\theta = 90°$, $\phi = \pm 90°$), the electric field is circularly polarized. In all other directions, it is elliptically polarized.

Pattern

The magnitude of the electric field of the cross-dipole is given by

$$|\mathbf{E}| = \sqrt{(|E_\theta|^2 + |E_\phi|^2)} \tag{9.9}$$

where

$$E_\theta = \frac{\eta k I_0 \mathrm{d}l}{4\pi r}(\sin^2\theta + \cos^2\theta\cos^2\phi)^{1/2} \tag{9.10}$$

$$E_\phi = \frac{\eta k I_0 \mathrm{d}l}{4\pi r}\sin\phi \tag{9.11}$$

In the three coordinate planes, we have the following results:

x–z plane:

$$|\mathbf{E}| = \frac{\eta k I_0 \mathrm{d}l}{4\pi r} \tag{9.12}$$

x–y plane:

$$|\mathbf{E}| = \frac{\eta k I_0 \mathrm{d}l}{4\pi r}(1 + \sin^2\phi)^{1/2} \tag{9.13}$$

y–z plane:

$$|\mathbf{E}| = \frac{\eta k I_0 \mathrm{d}l}{4\pi r}(1 + \sin^2\theta)^{1/2} \tag{9.14}$$

Equation (9.12) shows that the pattern in the x–z plane is an isotropic one (independent of θ). This is to be contrasted with the $\sin\theta$ variation of a single z-directed dipole, as illustrated in Figure 9.2(a). Equations (9.13) and (9.14) show that the patterns in the x–y and y–z planes are the same. (Figure 9.2(b)) For a single z-directed dipole, the pattern in the x–y plane is a circle while that is the y–z plane varies as $\sin\theta$.

A sketch of the three-dimensional pattern of the cross-dipole is shown in Figure 9.2(c). Since there is radiation in all directions, such a pattern can be said to be nearly omnidirectional in three dimensions.

Radiation Intensity and Directive Gain

The far-zone magnetic field of the cross-dipole is given by

$$\mathbf{H}(r) = H_\theta\hat{\boldsymbol{\theta}} + H_\phi\hat{\boldsymbol{\phi}} \tag{9.15}$$

where H_θ and H_ϕ are related to E_θ band E_ϕ by

$$H_\phi = E_\theta/\eta \tag{9.16}$$

$$H_\theta = -E_\phi/\eta \tag{9.17}$$

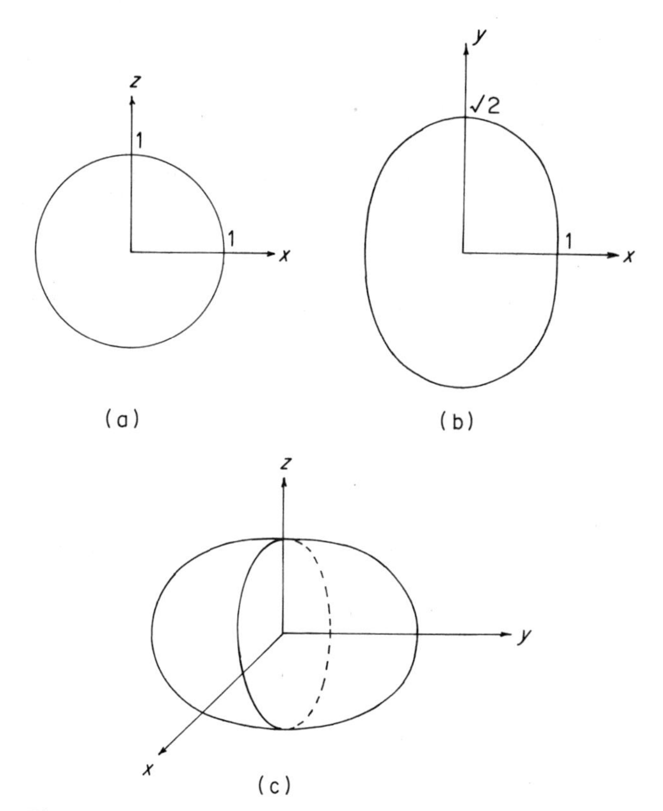

Figure 9.2 The patterns of a turnstile antenna of infinitesimal current elements

From (9.15) we obtain

$$H_\theta = \frac{kI_0\mathrm{d}l}{4\pi r}\exp(-\mathrm{j}kr)\sin\phi \tag{9.18}$$

$$H_\phi = -\frac{\mathrm{j}kI_0\mathrm{d}l}{4\pi r}\exp(-\mathrm{j}kr)(\sin\theta - \mathrm{j}\cos\theta\cos\phi) \tag{9.19}$$

The radiation intensity

$$\begin{aligned} U(\theta,\phi) &= r^2\hat{\mathbf{r}}\cdot\tfrac{1}{2}\,\mathrm{Re}[\mathbf{E}(r)\times\mathbf{H}^*(r)] = r^2\frac{|\mathbf{E}|^2}{2\eta} \\ &= \frac{(I_0\mathrm{d}l)^2\eta k^2}{32\pi^2}(1+\sin^2\phi\sin^2\theta) \end{aligned} \tag{9.20}$$

where use has been made of (9.9)–(9.11).

The time-averaged power radiated is given by

$$W = \int_0^{2\pi}\int_0^{\pi}\frac{(I_0\mathrm{d}l)^2\eta k^2}{32\pi^2}(1+\sin^2\phi\sin^2\theta)\sin\theta\,\mathrm{d}\theta\,\mathrm{d}\phi = \frac{(I_0\mathrm{d}l)^2\eta k^2}{6\pi} \tag{9.21}$$

The directive gain

$$g(\theta, \phi) = \frac{4\pi U(\theta, \phi)}{W} = \tfrac{3}{4}(1 + \sin^2\phi \sin^2\theta) \tag{9.22}$$

The directivity is $D = 1.5$ occurring in the broadside direction ($\theta = 90°$, $\phi = 90°$).

Effect of Phase Change

It is of interest to consider the case when the phase between the two dipoles is not 90°. Let

$$I_z = I_0 \tag{9.23}$$

$$I_x = I_0\, e^{j\delta} \tag{9.24}$$

where I_0 and δ are real numbers. Then it is readily shown that

$$\mathbf{E}(r) = \frac{jk\eta I_0 dl}{4\pi r}\exp(-jkr)[\hat{\boldsymbol{\theta}}(\sin\theta - e^{i\delta}\cos\theta\cos\phi) + \hat{\boldsymbol{\phi}}(e^{i\delta}\sin\phi)] \tag{9.25}$$

In the three coordinate planes, we have:

x–z plane:

$$\mathbf{E}(r) = \frac{jk\eta I_0 dl}{4\pi r}\exp(-jkr)[\hat{\boldsymbol{\theta}}(\sin\theta - \cos\theta\cos\delta - j\sin\delta\cos\theta)] \tag{9.26}$$

x–y plane:

$$\mathbf{E}(r) = \frac{jk\eta I_0 dl}{4\pi r}\exp(-jkr)(\hat{\boldsymbol{\theta}} + \hat{\boldsymbol{\phi}}\sin\phi\exp(j\delta)) \tag{9.27}$$

y–z plane:

$$\mathbf{E}(r) = \frac{jk\eta I_0 dl}{4\pi r}\exp(-jkr)(\hat{\boldsymbol{\theta}}\sin\theta + \hat{\boldsymbol{\phi}}\exp(j\delta)) \tag{9.28}$$

On examining (9.26)–(9.28), we arrive at the following conclusions:

(a) For $\delta \neq 90°$, the patterns in the x–y and y–z planes are the same as those for $\delta = 90°$. However, the pattern in the x–z plane depends on the value of δ. Figure 9.3 shows the patterns for $\delta = 0$, $\pi/6$, $\pi/3$, and $\pi/2$. It is seen that in these planes the pattern is no longer isotropic for values of δ different from $\pi/2$.

(b) For $\delta \neq 90°$, circular polarization in the broadside direction can no longer be obtained.

From the above considerations, it is clear why the cross-dipole is usually operated with a 90° phase difference in the currents.

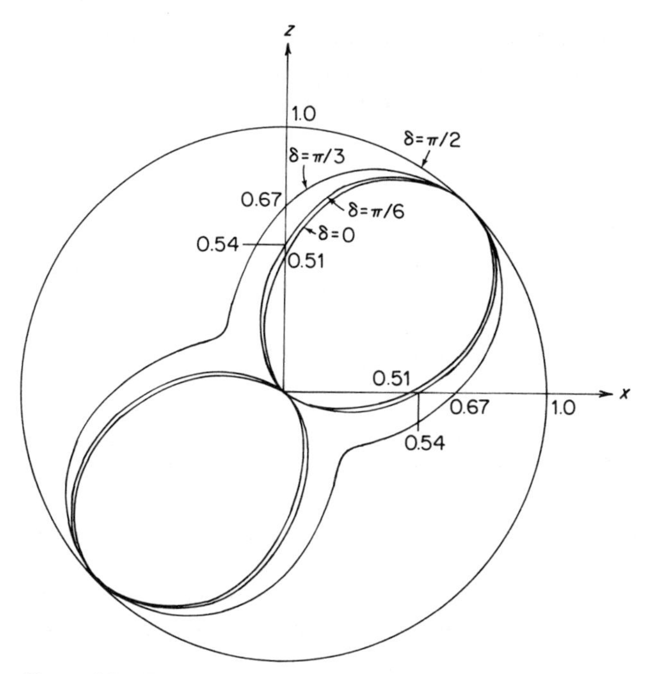

Figure 9.3 Effect of phase change on the pattern in the x–z plane

9.2.2 Half-wave Elements

In section 9.2.1, we have established the main features of the cross-dipole by considering the lengths of each dipole to be infinitesimal. In practice, the turnstile antenna is implemented by using half-wave dipoles operating in time and space quadrature. One way of feeding the cross-dipole is shown in Figure 9.4.

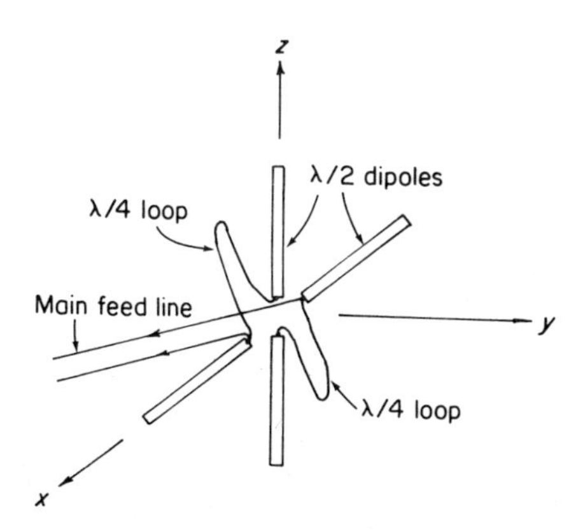

Figure 9.4 Turnstile antenna of half-wave dipoles

By using two quarter-wave loops of transmission line between the dipole terminals, a phase quadrature between the currents is obtained.

For the case $I_z = I_0$, $I_x = I_0 \exp(j\pi/2)$, the far-zone electric field is readily found to be

$$\mathbf{E}(r) = \frac{j\eta I_0}{2\pi r}\exp(-jkr)$$

$$\times\left[\hat{\theta}\left(\frac{\cos(\frac{1}{2}\pi\cos\theta)}{\sin\theta} - j\cos\theta\cos\phi\frac{\cos(\frac{1}{2}\pi\sin\theta\cos\phi)}{(1-\sin^2\theta\cos^2\phi)}\right)\right.$$

$$\left. + \hat{\phi}\, j\sin\phi\frac{\cos(\frac{1}{2}\pi\sin\theta\cos\phi)}{(1-\sin^2\theta\cos^2\phi)}\right] \tag{9.29}$$

Equation (9.29) can be studied with regard to polarization characteristics and relative field patterns in the manner of section 9.2.1. It is found that the polarization characteristics are similar to those of a cross-dipole of infinitesimal current elements. In particular, the polarization is left-circular at points along the positive y-axis and is right-circular along the negative y-axis.

The relative field pattern in the x–z plane is shown in Figure 9.5(a). Note that while the case of infinitesimal elements is a circle, the case of half-wave dipoles is no longer a circle. However, the pattern is still nearly isotropic in this plane.

The relative field pattern in the x–y or y–z plane is shown in Figure 9.5(b). The shape here is not much different from that of infinitesimal elements, except that it is slightly narrower.

The radiation intensity is readily found to be

$$U(\theta,\phi) = \frac{I_0^2\eta}{8\pi^2}\left(\frac{\cos^2(\frac{1}{2}\pi\cos\theta)}{\sin^2\theta} + \frac{\cos^2(\frac{1}{2}\pi\sin\theta\cos\phi)}{1-\sin^2\theta\cos^2\phi}\right) \tag{9.30}$$

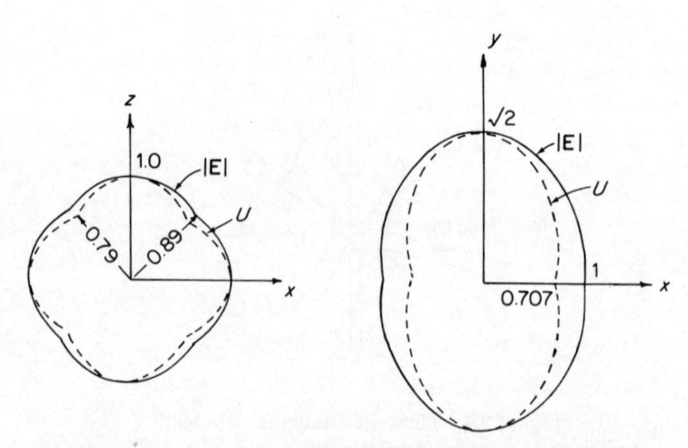

Figure 9.5 Relative patterns of electric field E and radiation intensity U of a turnstile antenna of half-wave dipoles

The time-averaged power is

$$W = \int_0^{2\pi} \int_0^{\pi} U \sin\theta \, d\theta \, d\phi$$

$$= \frac{I_0^2 \eta}{8\pi^2} \int_0^{2\pi} \int_0^{\pi} \left(\frac{\cos^2(\frac{1}{2}\pi\cos\theta)}{\sin^2\theta} + \frac{\cos^2(\frac{1}{2}\pi\sin\theta\cos\phi)}{1 - \sin^2\theta\cos^2\phi} \right) \sin\theta \, d\theta \, d\phi \quad (9.31)$$

The first term in the integral can be identified as $\pi \, \mathrm{Cin}(2\pi)$ and is equal to 7.658. The second integral does not appear to be expressible in terms of tabulated functions. However, its value can be obtained numerically and is found to be also 7.658. Thus

$$W = \frac{I_0^2 \eta}{8\pi^2}(15.316) = 0.194 I_0^2 \eta \text{ W} \quad (9.32)$$

The directive gain

$$g(\theta, \phi) = \frac{4\pi U(\theta, \phi)}{W} = 0.82 \left(\frac{\cos^2(\frac{1}{2}\pi\cos\theta)}{\sin^2\theta} + \frac{\cos^2(\frac{1}{2}\pi\sin\theta\cos\phi)}{1 - \sin^2\theta\cos^2\phi} \right) \quad (9.33)$$

The directivity

$$D = g(\theta, \phi)\big|_{\max} = 1.64$$

and its value is the same as that of a single half-wave dipole.

9.2.3 Applications of the Turnstile Antenna

The characteristics of the turnstile antenna or cross-dipole have been utilized in several areas of applications, as discussed below.

(a) The fact that the turnstile antenna radiates circularly polarized waves in the broadside direction makes it a useful antenna for use in situations where circular polarization is preferred over linear polarization. These include communication with space vehicles above the Earth's ionosphere and in radar applications. The former has already been discussed in section 8.2. A discussion of the latter will be given in section 9.3.

(b) In communication with an unstabilized spacecraft, three-dimensional omnidirectionality is desirable. While a turly isotropic antenna is not possible to obtain even in theory, reference to Figure 9.2(c) shows that the turnstile antenna does radiate in all directions. This property enables the antenna to be used successfully on artificial satellites that are not large in comparison with wavelength. The diameters of the satellites can be up to a few feet for frequencies up to about 150 MHz ($\lambda = 2$ m or about 6ft). Each dipole is mounted so that each half projects radially from the surface of the satellite, with the body of the satellite occupying the space between the two halves of the dipoles.

(c) It was shown in section 9.2.1 that, in the plane of the dipoles (x–z plane), the far-zone electric field is linearly polarized along the θ-direction and the pattern is a circle. If this plane is made to coincide with the horizontal plane, a horizontally polarized antenna with an omnidirectional pattern in the horizontal plane is obtained. This property is utilized in TV broadcasting. At the frequencies of TV broadcasting (see section 8.5), horizontal polarization has been adopted as standard. This choice was made to maximize signal-to-noise ratios, as it was found that the majority of man-made noises are predominantly vertically polarized. The second requirement in entertainment broadcasting is omnidirectionality in the horizontal plane. Both requirements can be met by using a turnstile antenna. Note that a vertical dipole alone will satisfy the second requirement but not the first while a horizontal dipole alone will satisfy the first requirement but not the second. If more gain is required, an array of turnstiles stacked vertically can be used.

9.3 COMPLEX EFFECTIVE LENGTH AND THE RECEPTION OF ELLIPTICALLY POLARIZED WAVES

In Chapter 3, it was shown that the far field of a finite-length dipole directed along the z-axis is linearly polarized along the $\hat{\boldsymbol{\theta}}$-direction. The effective length of the dipole, $h_e(\theta)$, was introduced and it was defined by equation (3.53). In terms of $h_e(\theta)$, the electric field is given by (3.54) and (3.55) as

$$\mathbf{E} = \frac{\mathrm{j}\eta k I(0)}{4\pi r} \exp(-\mathrm{j}kr) h_e(\theta)\, \hat{\boldsymbol{\theta}}$$

In dealing with antennas that radiate elliptically polarized waves, for example the turnstile antenna and the corner reflector with tilted dipole, it is convenient to think of the elliptically polarized radiation as being generated by two equivalent dipoles, one producing the θ-component and the other producing the ϕ-component. Each of the equivalent dipoles is assigned an appropriate effective length which takes care of not only the relative magnitudes but also the relative phases of the two perpendicular components. The vectorial sum of the effective lengths of the two equivalent dipoles is called the complex effective length of the transmitting antenna. Thus, for an antenna that in general radiates elliptically polarized waves, we write the far-zone electric field in the form

$$\mathbf{E} = \frac{\mathrm{j}k\eta I(0)}{4\pi r} \exp(-\mathrm{j}kr)\mathbf{h} \tag{9.34}$$

The quantity $\mathbf{h}$ may have both θ- and ϕ-components and both may be complex. However, one of these components (say the θ-component) can be made a positive real quantity by a suitable choice of the origin for time. Then

$$\mathbf{h} = h_\theta \hat{\boldsymbol{\theta}} + h_\phi \hat{\boldsymbol{\phi}} \tag{9.35}$$

where h_θ is real and

$$h_\phi = |h_\phi| \exp(\mathrm{j}\delta_1) \tag{9.36}$$

The quantity **h** has the dimension of length and is called the complex effective length of the antenna. For example, for the cross-dipole of infinitesimal elements, the radiation in the x–y plane is

$$\mathbf{E}(\theta = \pi/2) = \frac{jk\eta I}{4\pi r} \exp(-jkr)\, dl(\hat{\boldsymbol{\theta}} + j \sin\phi\, \hat{\boldsymbol{\phi}}) \tag{9.37}$$

Thus

$$\mathbf{h}(\theta = \pi/2) = dl(\hat{\boldsymbol{\theta}} + j \sin\phi\, \hat{\boldsymbol{\phi}}) \tag{9.38}$$

In section 4.2, it was shown that the open-circuit voltage developed across the terminals of a linearly polarized antenna in the field of an incident wave of linear polarization is given by

$$V_{oc} = -h_e(\theta)E$$

where E is the magnitude of the electric field in the incident plane. We now wish to consider an elliptically polarized antenna (one producing elliptically polarized radiation) in the field of an incident wave of elliptical polarization.

In general, an incident plane wave can be written as

$$\mathbf{E}_0 = (E_{0\theta}\hat{\boldsymbol{\theta}} + E_{0\phi}\hat{\boldsymbol{\phi}}) \exp(jkr) \tag{9.39}$$

In (9.39), the exponent is positive because the wave is an incoming wave. By a suitable choice of origin in time, it is possible to make $E_{0\theta}$ real and positive. Then

$$E_{0\phi} = |E_{0\phi}| \exp(-j\delta_2) \tag{9.40}$$

The signs are chosen so that, when δ_1 and δ_2 are in the same quadrant, the two equations (9.36) and (9.40) represent the same direction of rotation of the electric vector. The difference in sign required to do this arises from the difference in signs of the exponentials representing outgoing and incoming waves respectively.

The plane wave $E_{0\theta}$ will produce a terminal voltage

$$V_1 = -h_\theta E_{0\theta} \tag{9.41}$$

The plane wave $E_{0\phi}$ will produce a terminal voltage

$$V_2 = -h_\phi E_{0\phi} = -|h_\phi||E_{0\phi}| \exp[j(\delta_1 - \delta_2)] \tag{9.42}$$

The total open-circuit terminal voltage is the sum of V_1 and V_2:

$$V_{oc} = V_1 + V_2 = -(h_\theta E_{0\theta} + h_\phi E_{0\phi}) = -\mathbf{h}\cdot\mathbf{E}_0 \tag{9.43}$$

or

$$V_{oc} = -\{h_\theta E_{0\theta} + |h_\phi||E_{0\phi}| \exp[j(\delta_1 - \delta_2)]\} \tag{9.44}$$

As an example, consider an antenna which if transmitting would radiate circularly polarized waves in a given direction, and is used to receive circularly polarized waves from the same direction. Then

$$h_\theta = |h_\phi| = h \tag{9.45}$$

$$E_\theta = |E_{0\phi}| = E_0 \tag{9.46}$$

Hence

$$V_{oc} = -hE_0\{1 + \exp[j(\delta_1 - \delta_2)]\} \tag{9.47}$$

If the antenna and the incident wave are characterized by the same direction of rotation, then

$$\delta_1 = \delta_2 = \pm 90° \tag{9.48}$$

and

$$V_{oc} = -2hE \tag{9.49}$$

On the other hand, if the rotations are in opposite directions,

$$\delta_1 = -\delta_2 = \pm 90° \tag{9.50}$$

and

$$V_{oc} = 0 \tag{9.51}$$

Equations (9.50) and (9.51) show that a circularly polarized antenna is 'blind' to an incident circularly polarized radiation of the opposite or 'wrong' sense. This property has been utilized in radar applications to minimize the 'clutter' echoes received from raindrops, in relation to the echoes from larger targets such as aircraft. The quantitative explanation involves advanced scattering theory and is beyond the scope of this book. The qualitative explanation is as follows. Raindrops, being substantially spherical, qualify as symmetrical reflectors. A circularly polarized antenna is unable to see its own image in a symmetrical reflector. Therefore, if the radar antenna is circularly polarized, the echo from a symmetrical target such as a spherical raindrop will be circularly polarized with the wrong sense to be accepted by the antenna. The echo from a composite target such as an aircraft will have scrambled polarization and will usually contain a polarization component to which the circularly polarized radar antenna can respond.

9.4 MEASUREMENT OF POLARIZATION

We have introduced two antennas that radiate elliptically polarized waves, namely, the corner reflector with tilted dipole and the turnstile antenna. It is appropriate at this point to describe how the polarization of an incoming electromagnetic wave can be measured. We shall limit our discussion to the polarization pattern method, which is one of the simplest methods used for measuring the polarization characteristics of a wave.

Let $-\hat{\mathbf{r}}$ be the direction of the wave normal of an approaching electromagnetic wave. In the polarization pattern method, a linearly polarized antenna, usually a dipole, mounted so that its axis can be rotated in the θ–ϕ plane, is used as the receiving antenna (Figure 9.6a). For each orientation of the axis, the received signal voltage is proportional to the maximum of the projection of $\mathbf{E}$ along the antenna. The resultant pattern is known as the polarization

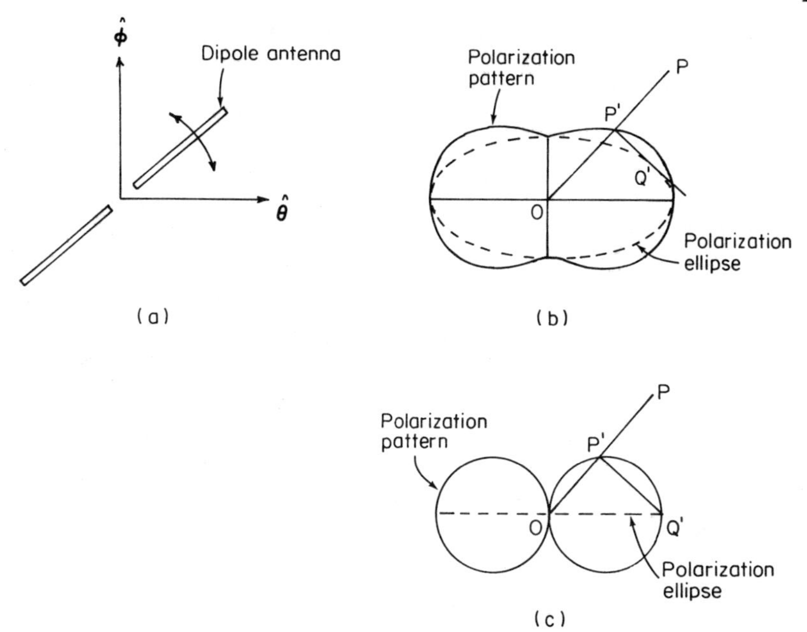

Figure 9.6 (a) A rotatable dipole antenna for measuring the polarization of an incoming wave; (b) relation of the pattern of the received signal voltage (polarization pattern) to the polarization ellipse of the incoming wave; (c) the polarization pattern becomes a figure of eight when the incoming wave is linearly polarized

pattern. Its relation with the polarization ellipse of the incident wave is illustrated in Figure 9.6(b). For a given orientation of the receiving antenna, such as OP, the received signal voltage is proportional to OP′; the line P′Q′ being both tangent to the ellipse and perpendicular to OP. For an elliptically polarized wave, the polarization pattern has two maxima and two minima. It becomes a figure of eight, with two maxima and two nulls if the incoming wave is linearly polarized (Figure 9.6c). The ratio of the maximum to the minimum of the polarization pattern yields the polarization ratio. This ratio is called the axial ratio if measurement is made in the main beam of the transmitting antenna.

The polarization pattern does not provide information about the direction of rotation of **E** of the incoming wave. This can be obtained by comparing the signal voltages received by two circularly polarized antennas, one left-handed and the other right-handed. As is evident from the discussion in section 9.3, the antenna with the larger response would be the one with the same sense of polarization as the incoming wave. In this connection, it is interesting to note that a single turnstile antenna (Figure 9.4) can serve as the two circularly polarized receiving antennas since the two senses of polarization can be obtained by connecting the main feed line first to the terminals of the z-directed dipole and then to the terminals of the x-directed dipole.

9.5 THE LOOP ANTENNA

Antennas having the configuration of a wire loop are much used in practice. The loop can take on many forms, with the most common ones being circular and rectangular. We consider below the analysis of the circular loop.

9.5.1 Current Distribution

The circular loop antenna, in its simplest form, consists of a single turn of wire. A current can be made to flow in the loop by breaking it at some point and connecting the terminals of a transmission line at the gap of the loop, as shown in Figure 9.7. The current distribution around the loop, when driven by a delta-function generator across the gap, can be determined by solving a boundary value problem similar to the case of the linear dipole antenna (section 3.8). The details of this problem are more complicated than those of the linear antenna and will not be taken up here. The interested reader is referred to the article by King in Collin and Zucker (1969).

As in the case of the linear antenna, the current distribution is governed by an integral equation. For a loop lying in the x–y plane, the solution is given in the form of a cosine Fourier series in the angular variable ϕ:

$$I(\phi) = -\frac{\mathrm{j}V}{\eta\pi}\left(B_0 + 2\sum_{n=1}^{\infty} B_n \cos(n\phi)\right) \tag{9.52}$$

In (9.52), V is the applied voltage across the gap. The coefficients B_0 and B_n are complex numbers that are functions of ka and the parameter

$$\Omega \equiv 2\ln(2\pi a/b) \tag{9.53}$$

where a is the radius of the loop and b is the radius of the wire. Curves of B_0, B_1, and B_2 are given in Figure 9.8 for $\Omega = 12$ and $\Omega = 8$.

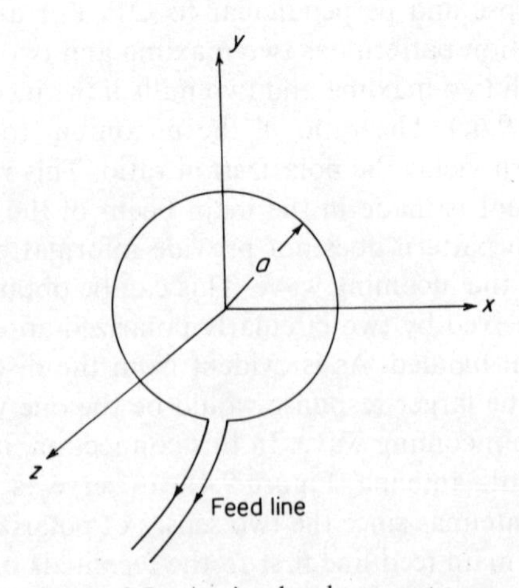

Figure 9.7 A circular loop antenna

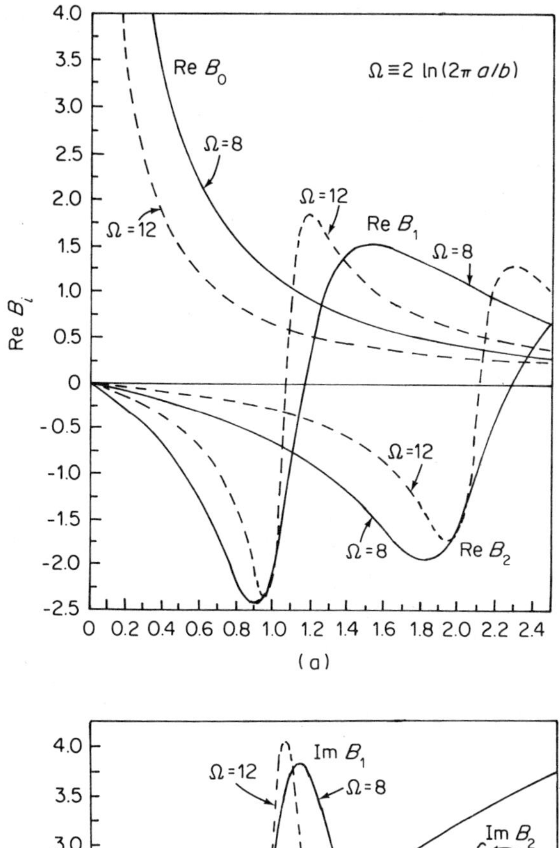

Figure 9.8 (a) Real parts of the functions B_0, B_1, and B_2; (b) imaginary parts of the functions B_0, B_1, and B_2 (Source: Collin and Zucker (1969), *Antenna theory*, Pt 1. Reproduced by permission of McGraw-Hill)

For $ka \leqslant 0.2$, we see from Figure 9.8 that the constant term in (9.52) dominates and the current can be taken to be independent of ϕ. This will be referred to as the case of the small circular loop. The expression for I is

$$I = I_0 = -\frac{jVB_0}{\eta\pi} \tag{9.54}$$

For $0.2 < ka \leqslant 1$, the $n = 1$ term in (9.54) needs to be considered while terms with $n > 1$ can be neglected. The current is approximately

$$I(\phi) = I_0 + I_1(\phi) \tag{9.55}$$

where I_0 is given by (9.54) and

$$I_1(\phi) = -\frac{2jVB_1}{\eta\pi}\cos\phi \tag{9.56}$$

9.5.2 Radiation from a Small Circular Loop

Let us consider a circular loop of radius a, with $ka \leqslant 0.2$, lying in the x–y plane, as shown in Figure 9.9. It carries a constant current I_0 of angular frequency ω around the circumference. Since there is cylindrical symmetry, the electromagnetic fields will be independent of ϕ. Consequently, there will be no loss of generality if we take the field point P to lie in the $\phi = 0$ plane so that its coordinate is $P(r, \theta, 0)$. Consider an element of current lying at the point $P'(a, \frac{1}{2}\pi, \phi')$ subtending an angle $d\phi'$. The length of the current element is $a\,d\phi'$. Let $\hat{\boldsymbol{\phi}}'$ be the unit vector tangent to the element. Then

$$\hat{\boldsymbol{\phi}}' = -\hat{\mathbf{x}}\sin\phi' + \hat{\mathbf{y}}\cos\phi' \tag{9.57}$$

For $\phi = 0$,

$$\hat{\mathbf{y}} = \hat{\boldsymbol{\phi}} \tag{9.58}$$

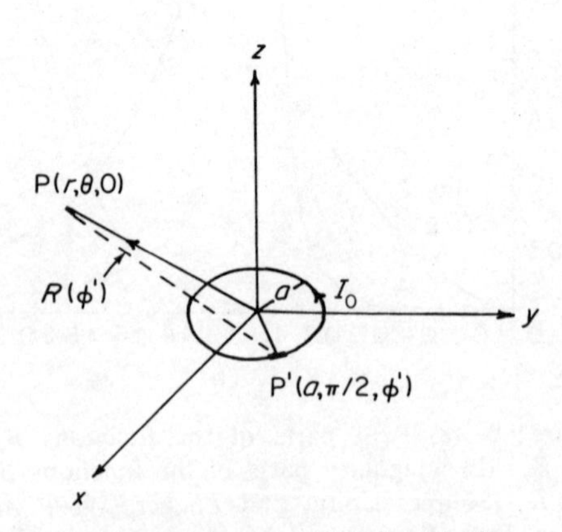

Figure 9.9 Geometry for the calculation of the radiation from a small circular loop

and

$$\hat{\boldsymbol{\phi}}' = -\hat{\mathbf{x}}\sin\phi' + \hat{\boldsymbol{\phi}}\cos\phi' \tag{9.59}$$

The contribution to the vector potential at P due to the current element at P′ is

$$\mathrm{d}\mathbf{A}(\mathbf{r}) = \frac{\mu I_0 a\,\mathrm{d}\phi'}{4\pi R(\phi')}\exp[-\mathrm{j}kR(\phi')]\hat{\boldsymbol{\phi}}' \tag{9.60}$$

where

$$\begin{aligned} R(\phi') &= [(r\sin\theta - a\cos\phi')^2 + a^2\sin^2\phi' + r^2\cos^2\theta]^{1/2} \\ &= [r^2 + a^2 - 2ar\sin\theta\cos\phi']^{1/2} \end{aligned} \tag{9.61}$$

The total vector potential at P is

$$\mathbf{A}(\mathbf{r}) = \frac{\mu I_0 a}{4\pi}\int_0^{2\pi}\frac{\exp[-\mathrm{j}kR(\phi')]}{R(\phi')}\hat{\boldsymbol{\phi}}'\,\mathrm{d}\phi' \tag{9.62}$$

Note that $\hat{\boldsymbol{\phi}}'$ is a function of ϕ' and hence cannot be taken outside the integral. Substituting (9.57) into (9.62), we have

$$\mathbf{A}(\mathbf{r}) = \hat{\mathbf{x}}\,A_x(r) + \hat{\phi}\,A_\phi(r) \tag{9.63}$$

where

$$A_x(r) = -\frac{\mu I_0 a}{4\pi}\int_0^{2\pi}\sin\phi'\frac{\exp[-\mathrm{j}kR(\phi')]}{R(\phi')}\mathrm{d}\phi' \tag{9.64}$$

and

$$A_\phi(r) = \frac{\mu I_0 a}{4\pi}\int_0^{2\pi}\cos\phi'\frac{\exp[-\mathrm{j}kR(\phi')]}{R(\phi')}\mathrm{d}\phi' \tag{9.65}$$

Equation (9.64) can be written as

$$A_x(r) = -\frac{\mu I_0 a}{4\pi}\left(\int_0^{\pi}\sin\phi'\frac{\exp[-\mathrm{j}kR(\phi')]}{R(\phi')}\mathrm{d}\phi' + \int_\pi^{2\pi}\sin\phi'\frac{\exp[-\mathrm{j}kR(\phi')]}{R(\phi')}\mathrm{d}\phi'\right)$$

In the second integral, let $u = 2\pi - \phi'$. Then $\mathrm{d}\phi'\sin\phi' = \mathrm{d}u\sin u$ and from (9.61), $R(\phi') = R(2\pi - u) = R(u)$. Hence

$$A_x(r) = -\frac{\mu I_0 a}{4\pi}\left(\int_0^{\pi}\sin\phi'\frac{\exp[-\mathrm{j}kR(\phi')]}{R(\phi')}\mathrm{d}\phi' + \int_\pi^{0}\sin u\frac{\exp[-\mathrm{j}kR(u)]}{R(u)}\mathrm{d}u\right) = 0 \tag{9.66}$$

and

$$\mathbf{A} = A_\phi \hat{\boldsymbol{\phi}} \tag{9.67}$$

where A_ϕ is given by (9.65). To simplify matters, let us evaluate A_ϕ at points such that the inequality

$$a/r \ll 1 \tag{9.68}$$

is satisfied. Then

$$\begin{aligned} R(\phi') &= r[1 - 2(a/r)\sin\theta\cos\phi' + (a/r)^2]^{1/2} \\ &\simeq r[1 - (a/r)\sin\theta\cos\phi'] \end{aligned} \tag{9.69}$$

where we retain only terms linear in a/r. Hence

$$\frac{1}{R(\phi')} \simeq \frac{1}{r}\left(1 + \frac{a}{r}\sin\theta\cos\phi'\right) \tag{9.70}$$

On using (9.70), (9.65) becomes

$$\begin{aligned} A_\phi(r) = \frac{\mu I_0 a}{4\pi r}\exp(-jkr)\int_0^{2\pi} &\cos\phi' \exp(jka\sin\theta\cos\phi') \\ &\times\left(1 + \frac{a}{r}\sin\theta\cos\phi'\right)\mathrm{d}\phi' \end{aligned} \tag{9.71}$$

Since $ka \ll 1$ for a small circular loop,

$$\exp(jka\sin\theta\cos\phi') \simeq 1 + jka\sin\theta\cos\phi' \tag{9.72}$$

and

$$\begin{aligned} A_\phi(r) &= \frac{\mu I_0 a\exp(-jkr)}{4\pi r}\int_0^{2\pi}\cos\phi' \\ &\quad\times\left(1 + jka\sin\theta\cos\phi' + \frac{a}{r}\sin\theta\cos\phi'\right)\mathrm{d}\phi' \\ &= \mu I_0 \pi a^2\frac{\exp(-jkr)}{4\pi r}jk\sin\theta\left(1 + \frac{1}{jkr}\right) \end{aligned} \tag{9.73}$$

Carrying out the operations indicated by equations (3.4) and (2.24), we obtain the following expressions for the electromagnetic fields:

$$E_r = E_\theta = H_\phi = 0$$

$$H_\theta(r) = -I_0\pi a^2\frac{\exp(-jkr)}{4\pi r}k^2\sin\theta\left(1 + \frac{1}{jkr} - \frac{1}{k^2r^2}\right) \tag{9.74}$$

$$H_r(r) = I_0\pi a^2\frac{\exp(-jkr)}{2\pi r^2}kj\sin\theta\left(1 + \frac{1}{jkr}\right) \tag{9.75}$$

$$E_\phi(r) = -j\omega\mu I_0\pi a^2\frac{\exp(-jkr)}{4\pi r}jk\sin\theta\left(1 + \frac{1}{jkr}\right) \tag{9.76}$$

Equations (9.74)–(9.76) can be simplified under the near-field and far-field approximations.

Near field: $kr \ll 1$ (but $r \gg a$)

$$H_r(r) = \frac{I_0 \pi a^2}{2\pi r^3} \cos\theta \tag{9.77}$$

$$H_\theta(r) = \frac{I_0 \pi a^2}{4\pi r^3} \sin\theta \tag{9.78}$$

$$E_\phi(r) = -\mathrm{j}\omega\mu \frac{I_0 \pi a^2}{4\pi r^2} \sin\theta \tag{9.79}$$

Far field: $kr \gg 1$

$$H_\theta(r) = -I_0 \pi a^2 \frac{\exp(-\mathrm{j}kr)}{4\pi r} k^2 \sin\theta = -\frac{I_0 \pi a^2}{\lambda^2} \pi \sin\theta \frac{\exp(-\mathrm{j}kr)}{r} \tag{9.80}$$

$$E_\phi(r) = I_0 \pi a^2 \frac{\exp(-\mathrm{j}kr)}{4\pi r} \eta k^2 \sin\theta = -\eta H_\theta \tag{9.81}$$

$$H_r \ll H_\theta \qquad \text{and can be neglected} \tag{9.82}$$

Comparing (9.80) and (9.81) with the far field of a Hertzian dipole given by (3.87) and (3.88), we see that the pattern of the loop has exactly the same shape as that of the Hertzian dipole oriented with its axis perpendicular to the plane of the loop. However, the directions of the electric and magnetic fields are interchanged relative to those of the Hertzian dipole. The polarization is linear but perpendicular to that of the corresponding electric dipole. A loop with its axis horizontal radiates maximum field intensity in the plane of the loop, but in the horizontal directions the polarization is vertical rather than horizontal. If the plane of the loop is horizontal, the radiation pattern in the horizontal plane is a circle, like that of a vertical Hertzian dipole, but the polarization is horizontal. Because the pattern has the same shape as the Hertzian dipole, the directivity is the same and so is the half-power beamwidth. They are equal to 1.5 and 90° respectively.

The radiation intensity U, the time-averaged radiated power W, the radiation resistance R_r, and the directive gain g are readily found to be:

$$U(\theta) = \frac{(I_0 \pi a^2)^2}{32\pi^2} \eta k^4 \sin^2\theta \tag{9.83}$$

$$W = \frac{(I_0 \pi a^2)^2}{12\pi} \eta k^4 \tag{9.84}$$

$$R_\mathrm{r} = 20(k^2 \pi a^2)^2 = 31\,171(\pi a^2/\lambda^2)^2 \text{ ohm} \tag{9.85}$$

$$g = 1.5 \sin^2\theta \tag{9.86}$$

On comparing (9.81) with (9.34), the effective length

$$\mathbf{h} = (\pi a^2) k \sin\theta\, \hat{\boldsymbol{\phi}} \tag{9.87}$$

The above formulae for the small circular loop turn out to apply also to a loop of any shape as long as its dimensions are sufficiently small compared to the wavelength so that the current can be assumed to be uniform. This is usually taken to mean that the loop perimeter is less than about one-fifth of a wavelength. Under such circumstances, one simply replaces the area of the circular loop, πa^2, by the area of the other loops, A_L, which may be rectangular, triangular, or even irregular in shape.

For the case of the small circular loop, the loss resistance is

$$R_L = aR_s/b \tag{9.88}$$

where b is the wire radius and R_s is the surface resistance ($R_s \simeq \sqrt{(\omega\mu/2\sigma)}$). The radiation efficiency is then given by

$$e = \frac{R_r}{R_r + R_L} = \frac{10(k^2\pi a^2)^2}{10(k^2\pi a^2)^2 + (a/b)R_s} \tag{9.89}$$

As an example, let $f = 1$ MHz (typical AM brodcast frequency) and consider a loop made of AWG 12 copper wire ($b = 1$ mm). If $a = 15$ cm, then $R_r = 2 \times 10^{-8}$ ohm, $R_L = 4.65 \times 10^{-2}$ ohm, and $e = 4.3 \times 10^{-7}$ or 4.3×10^{-5}%.

The radiation resistance can be increased by using more than one turn of wire. For a loop of N turns, each carrying a current I_0, the radiation resistance is

$$R_r = 20(Nk^2\pi a^2)^2 = 31\,171(NA_L/\lambda^2)^2 \text{ ohm} \tag{9.90}$$

However, the loss resistance also goes up by a factor of approximately N. Moreover, the total length of wire must be small compared to the wavelength if the current is to remain uniform so that the small-loop behaviour is to apply. This limits the value of N and, as a result, the efficiencies of small loops are quite low and they are seldom used as transmitting antennas. The main applications of small loops are as receiving antennas and directional finders, as discussed in the next section.

9.5.3 Applications of the Small Loop Antenna

Receiving Antenna

Although the small loop antenna has low radiation efficiency and therefore is not usually used as a transmitting antenna, its relative compactness lends itself to use as a receiving antenna in low-frequency devices such as AM receivers. In this application, the terminals of the loop are connected to a very high-impedance input circuit so that the quantity of primary interest is the voltage induced in the loop rather than the power radiated. This quantity is maximum when the wave normal and the polarization of the incoming wave are in the plane of the loop. Its value is

$$V_{oc} = -|h(\theta = 90°)||E| = -kNA_L E \text{ V} \tag{9.91}$$

where E is the magnitude of the incident electric field. The induced voltage can be enhanced by winding the loop turns around a core of ferromagnetic material such as ferrite. If μ_{eff} is the effective permeability of the core, the flux linkage of the loop will be increased by the factor μ_{eff}/μ_0 over an air core. As a result, the voltage induced at the terminals of the loop will also be increased by the same factor. The quantity μ_{eff} is less than the permeability of the core material. They are related by

$$\mu_{eff} = \frac{\mu}{1 + D(\mu - 1)} \tag{9.92}$$

where D is the demagnetization factor. This factor has been determined experimentally as a function of core length-to-diameter ratio and the results are shown in Figure 9.10.

Although a loop antenna with a ferromagnetic core is usually used for reception, it should be noted that the radiation resistance of such an antenna when used as a transmitter will be larger than the case with an air core by the factor $(\mu_{eff}/\mu_0)^2$. This follows since the radiation resistance is proportional to the square of the far-zone electric field, which in turn is proportional to the effective length. By the reciprocity theorem, the effective length of a transmitting antenna is the same as the effective length of the antenna when functioning as a receiver and therefore increase by the factor (μ_{eff}/μ_0) since it is linearly proportional to the induced voltage.

Directional Finder

Small loops are often used as directional finders when the incoming signals are vertically polarized. In this application, the plane of the loop is oriented

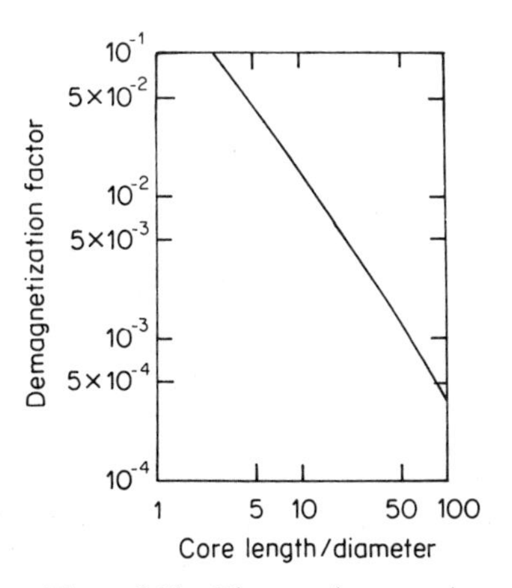

Figure 9.10 The demagnetization factor as a function of core length-to-diameter ratio (Source: Wolff (1966), *Antenna Analysis*. Reproduced by permission of E. A. Wolff)

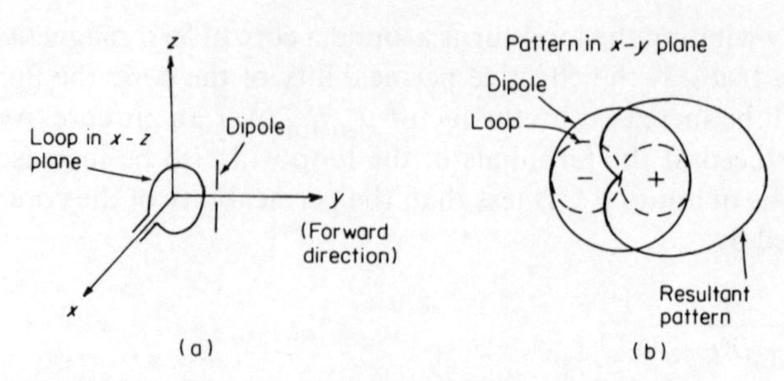

Figure 9.11 (a) A loop and a dipole used in direction finding; (b) patterns in the x–y plane

vertically. To find the direction of the incoming wave, the axis of the loop is rotated in the horizontal plane until a null is obtained. The null is very sharp so that the direction of the signal can be determined. However, there is a two-fold ambiguity in the direction since a null exists in the pattern on both sides of the loop. To resolve this ambiguity, the loop is usually used in conjunction with a vertical dipole, as shown in Figure 9.11(a). The pattern of the latter is a circle while that of the former is a figure of eight, with the forward radiation 180° out of phase with the backward radiation. If the loop and dipole signal amplitudes are equal and the two are added, the backward radiation is largely cancelled out and the forward radiation increased, producing a cardioid pattern. This is illustrated in Figure 9.11(b). Thus, if the loop is rotated from one null position, say 0° towards 90°, the signal will decrease. On the other hand, if rotation in the same direction starts from 180°, the signal will increase. This is illustrated in Figure 9.12. In this way, one can determine both the direct and reciprocal bearings of the incoming signal.

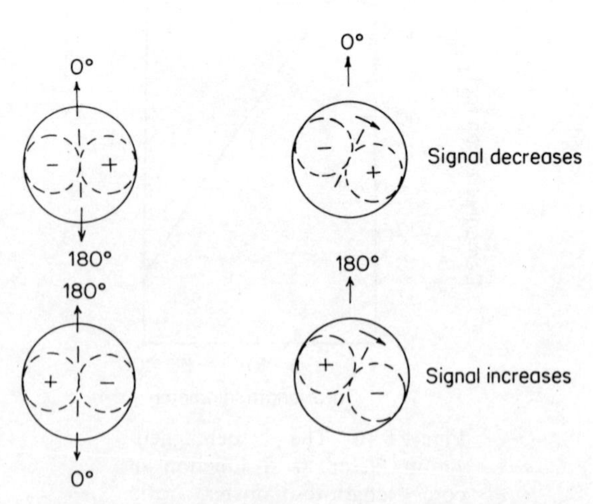

Figure 9.12 Determination of the direct and reciprocal bearings of an incoming wave by rotating the loop

9.5.4 Radiation from a Circular Loop with $0.2 < ka \leqslant 1$

For $0.2 < ka \leqslant 1$, the current in the loop is given by (9.55), consisting of a constant term I_0 and a term I_1 which varies as $\cos\phi$. The former is referred to as the circulating 'zero' mode and the latter the 'first' dipole mode. The radiation from the zero mode has been obtained in section 9.5.2. The far-zone electric field has only the ϕ-component and is given by

$$E_{\phi 0} = I_0 \pi a^2 \frac{\exp(-\mathrm{j}kr)}{4\pi r} \eta k^2 \sin\theta \tag{9.93}$$

The radiation from the first dipole mode can be calculated as follows.

Since I_1 is a function of ϕ, the problem no longer has cylindrical symmetry and the field point P cannot be chosen to lie on the x–z plane. Let its spherical coordinates be r, θ, ϕ. Restricting ourselves to field points in the far zone, the distance between P and the source point $\mathrm{P}'(a, \frac{1}{2}\pi, \phi')$ is

$$R(\phi') \simeq r - a\sin\theta\cos(\phi - \phi') \tag{9.94}$$

The x- and y-components of the vector potential $\mathbf{A}_1$ due to current I_1 are obtained by replacing I_0 by I_1 in (9.64) and (9.65):

$$A_{x1} = -\frac{\mu a}{4\pi}\int_0^{2\pi} I_1(\phi')\sin\phi' \frac{\exp[-\mathrm{j}kR(\phi')]}{R(\phi')}\,\mathrm{d}\phi' \tag{9.95}$$

$$A_{y1} = \frac{\mu a}{4\pi}\int_0^{2\pi} I_1(\phi')\cos\phi' \frac{\exp[-\mathrm{j}kR(\phi')]}{R(\phi')}\,\mathrm{d}\phi' \tag{9.96}$$

On using (9.56) and (9.94), (9.95) and (9.96) can be expressed as

$$A_{x1} = \frac{\mu a V B_1}{8\eta\pi^2 r}\exp(-\mathrm{j}kr)(\mathscr{I}_1 - \mathscr{I}_2) \tag{9.97}$$

$$A_{y1} = -\frac{\mathrm{j}\mu a V B_1}{8\eta\pi^2 r}\exp(-\mathrm{j}kr)(\mathscr{I}_1 + \mathscr{I}_2 + \mathscr{I}_3) \tag{9.98}$$

where

$$\mathscr{I}_1 = \int_0^{2\pi} \exp(2\mathrm{j}\phi')\exp[\mathrm{j}ka\sin\theta\cos(\phi - \phi')]\,\mathrm{d}\phi' \tag{9.99}$$

$$\mathscr{I}_2 = \int_0^{2\pi} \exp(-2\mathrm{j}\phi')\exp[\mathrm{j}ka\sin\theta\cos(\phi - \phi')]\,\mathrm{d}\phi' \tag{9.100}$$

$$\mathscr{I}_3 = 2\int_0^{2\pi} \exp[\mathrm{j}ka\sin\theta\cos(\phi - \phi')]\,\mathrm{d}\phi' \tag{9.101}$$

Consider the expression for $\mathscr{I}_1$. If we let $\psi = \phi' - \phi$, we have

$$\mathscr{I}_1 = \exp(2\mathrm{j}\phi)\int_{-\phi}^{2\pi - \phi} \exp(2\mathrm{j}\psi)\exp(\mathrm{j}ka\sin\theta\cos\psi)\,\mathrm{d}\psi \tag{9.102}$$

On using the integral representation of the Bessel function of the first kind of order n given by the formula

$$J_n(x) = \frac{1}{2\pi(\mathrm{j})^n} \int_{\alpha}^{\alpha+2\pi} \exp[\mathrm{j}(x \cos \psi + n\psi)] \, \mathrm{d}\psi \qquad (9.103)$$

where α is any real number, (9.102) can be put into the form:

$$\mathscr{I}_1(x) = (-2\pi) \exp(2\mathrm{j}\phi) J_2(ka \sin \theta) \qquad (9.104)$$

Similarly, we can express $\mathscr{I}_2$ and $\mathscr{I}_3$ as

$$\mathscr{I}_2(x) = (-2\pi) \exp(-2\mathrm{j}\phi) J_2(ka \sin \theta) \qquad (9.105)$$

$$\mathscr{I}_3(x) = 4\pi J_0(ka \sin \theta) \qquad (9.106)$$

Thus

$$A_{x1} = -\frac{\mathrm{j}\mu a B_1 V}{2\pi r \eta} \exp(-\mathrm{j}kr) J_2(ka \sin \theta) \sin(2\phi) \qquad (9.107)$$

$$A_{y1} = \frac{\mathrm{j}\mu a B_1 V}{2\pi r \eta} \exp(-\mathrm{j}kr) [\cos(2\phi) J_2(ka \sin \theta) - J_0(ka \sin \theta)] \qquad (9.108)$$

The electric field $\mathbf{E}$ is related to $\mathbf{A}$ by

$$\mathbf{E} = -\mathrm{j}\omega \mathbf{A} + \frac{\nabla(\nabla \cdot \mathbf{A})}{\mathrm{j}\omega\varepsilon\mu} \qquad (9.109)$$

In the far zone, the second term is negligible. The θ- and ϕ-components of $\mathbf{E}$ due to I_1 are then given by

$$E_{\theta 1} = -\mathrm{j}\omega A_{\theta 1} = -\mathrm{j}\omega(A_{x1} \cos \phi + A_{y1} \sin \phi) \cos \theta \qquad (9.110)$$

$$E_{\phi 1} = -\mathrm{j}\omega A_{\phi 1} = -\mathrm{j}\omega(-A_{x1} \sin \phi + A_{y1} \cos \phi) \qquad (9.111)$$

Substitution of (9.107) and (9.108) into (9.110) and (9.111) yields

$$E_{\theta 1} = \omega \cos \theta \, C(r) \{ -J_2(ka \sin \theta) \sin(2\phi) \cos \phi + [\cos(2\phi) J_2(ka \sin \theta) - J_0(ka \sin \theta)] \sin \phi \} \qquad (9.112)$$

$$E_{\phi 1} = \omega C(r) \{ J_2(ka \sin \theta) \sin(2\phi) \sin \phi + [\cos(2\phi) J_2(ka \sin \theta) - J_0(ka \sin \theta)] \cos \phi \} \qquad (9.113)$$

where

$$C(r) = \frac{\mu a V B_1}{2\pi r \eta} \exp(-\mathrm{j}kr) \qquad (9.114)$$

The total far-zone electric field is the vectorial sum of the contributions due to I_0 and I_1 and is

$$\mathbf{E} = E_{\theta 1} \hat{\boldsymbol{\theta}} + (E_{\phi 0} + E_{\phi 1}) \hat{\boldsymbol{\phi}} \qquad (9.115)$$

where $E_{\phi 0}$ is given by (9.93).

Equation (9.115) shows that, unlike the case of the small loop, the electric field has both θ- and ϕ-components and the pattern in the plane of the loop is no longer omnidirectional.

9.6 A SMALL CIRCULAR LOOP WITH AN AXIAL DIPOLE AT ITS CENTRE

It will be recalled that the far-zone electric fields of both the corner reflector with tilted dipole and the turnstile antenna, discussed in sections 8.8 and 9.2 respectively, have very complicated polarization characteristics. With proper design, the polarizations of both antennas can be made circular in the broadside directions. However, once off the broadside, they become elliptical, with ellipticities changing with the coordinates of the point of observation. Sometimes it is desirable to have an antenna system that radiates the same type of elliptically or circularly polarized wave in all directions. One way of achieving this is to combine a small horizontal circular loop and a short vertical dipole situated at its centre, fed with in-phase currents $I_\phi \hat{\boldsymbol{\phi}}$ and $I_z \hat{\mathbf{z}}$ respectively. For simplicity, let the dipole be of length $\mathrm{d}l \ll \lambda$. If a is the radius of the loop, the far-zone electric field of the combination is

$$\mathbf{E}(r) = \mathrm{j}k\eta \sin\theta \frac{\exp(-\mathrm{j}kr)}{4\pi r}(I_z \,\mathrm{d}l\, \hat{\boldsymbol{\theta}} - \mathrm{j}kI_\phi \pi a^2\, \hat{\boldsymbol{\phi}}) \tag{9.116}$$

The polarization is thus right-elliptical and is the same for all directions. If the phase of the current in the dipole is changed by 180°, the sense of the polarization is reversed.

It is clear from (9.116) that, if the magnitudes of the currents are adjusted such that

$$I_z \,\mathrm{d}l = kI_\phi \pi a^2 \tag{9.117}$$

the radiation is circularly polarized in all directions.

9.7 THE HELICAL ANTENNA

9.7.1 Qualitative Description

Another basic non-linear radiator is the helical antenna, consisting of a conductor wound in the shape of a screw thread and a flat metal plate serving as the ground plane. It is usually fed by a coaxial transmission line, with the centre conductor of the line connected to the helix via a feed wire and the outer conductor connected to the ground plane, as shown in Figure 9.13(a). The geometry of the helix proper is described by its diameter D, its turn spacing S, and the number of turns N. The total axial length of an N-turn helix is equal to NS, and the circumference $C = \pi D$. The pitch angle ψ is the angle that a line tangent to the helix wire makes with the plane perpendicular to the axis. If one turn of the helix is unrolled on a flat plane, the relation between S, C, ψ, and the length

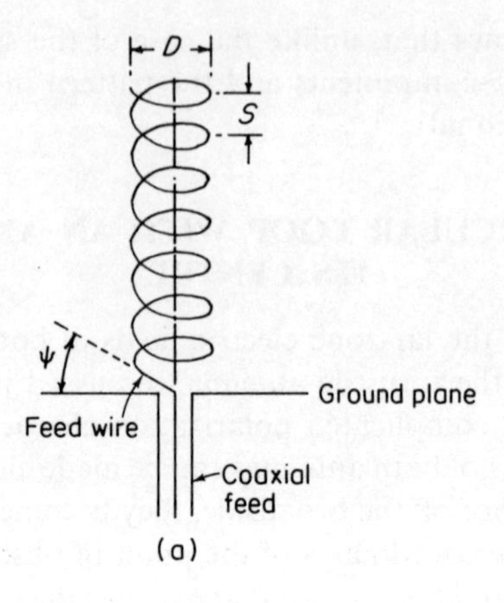

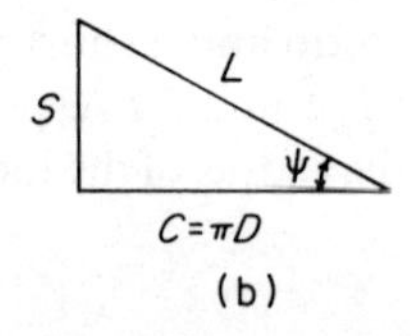

Figure 9.13 (a) A helical antenna; (b) relation between turn spacing S, length per turn L, pitch angle ψ, and circumference C

of wire per turn, L, are illustrated in Figure 9.13(b). Thus

$$\sin \psi = S/L \tag{9.118}$$

or

$$\tan \psi = S/(\pi D) = S/C \tag{9.119}$$

and

$$L = (S^2 + C^2)^{1/2} = (S^2 + \pi^2 D^2)^{1/2} \tag{9.120}$$

Note that, when the spacing is zero, $\psi = 0$ and the helix becomes a loop. On the other hand, when the diameter is zero, $\psi = 90°$ and the helix becomes a linear conductor. The feed wire shown in Figure 9. 13(a) lies in a plane through the helix axis and is inclined at an angle approximately equal to ψ with respect to the ground plane. The component of the feed wire parallel to the axis is about equal to $S/2$. The ground plane, which may be either a solid metal sheet or a wire grid, should be at least half a wavelength in diameter. The size of the conductor diameter does not appreciably affect the performance of the antenna.

Unlike the dipole or the loop antenna, the theoretical problem of determining the current distribution along the helix is extremely complicated. It has not been solved completely, although some numerical results for the axial mode range ($3/4 < C/\lambda < 4/3$) have recently been published in the literature (Nakano and Yamauchi, 1979). However, experimental data on the current distribution have long been available (e.g. Kraus, 1950). By assuming a current on the helix resembling this observed distribution, it is possible to calculate the radiation characteristics of the helical antenna. Thus the theory of the helical antenna is not a complete one, since part of it relies on experimental data. Before going into the details, we present a brief summary of the properties of the antenna.

When the dimensions of the helix are very small compared to a wavelength, the maximum radiation is in the plane perpendicular to the helix axis, as illustrated in Figure 9.14(a). This mode of radiation is referred to as the 'normal mode'. The polarization is in general elliptical and is the same in all directions. It can be made circularly polarized by choosing the dimensions properly. However, the small size of the antenna results in low efficiency and small bandwidth.

When the diameter and spacing (D and S) are appreciable fractions of a wavelength, entirely different radiation is obtained. The maximum intensity is in the direction of the axis, in the form of a directional beam with minor lobes at oblique angles, as illustrated in Figure 9.14(b). The radiation in the main lobe is nearly circularly polarized. The sense of the polarization is determined by the sense of the helix windings, i.e. a left- (right-) hand wound helix produces left- (right-) hand sensed polarization. This mode is referred to as the 'axial mode'.

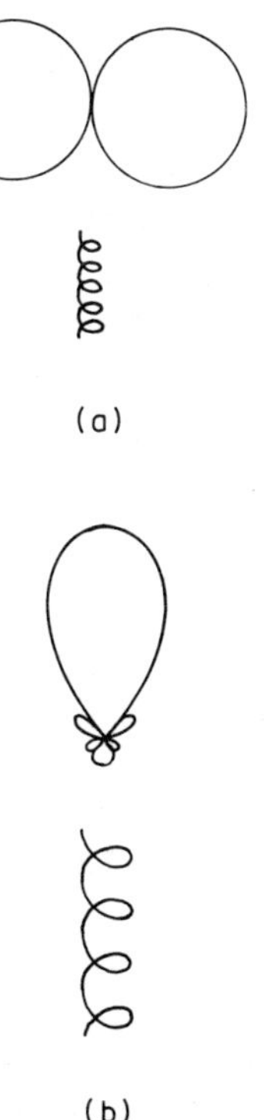

Figure 9.14 (a) Normal mode radiation pattern of the helical antenna; (b) axial mode radiation pattern of the helical antenna

It is in this mode of radiation that most helical antennas are designed to operate in practice. The axial mode is found to persist when C and ψ are in the ranges of $3/4 < C/\lambda < 4/3$ and $12° < \psi < 15°$. The ratio of the upper and lower frequencies of the axial mode is $f_U/f_L = 16/9 = 1.78$.

In the literature, the following empirical formulae are listed for helical antennas with $N > 3$ operating in the axial mode:

Half-power beamwidth

$$\mathrm{HPBW} = \frac{K_B}{C}\sqrt{\frac{\lambda^3}{NS}}\ \mathrm{deg} \qquad K_B = 52 \tag{9.121}$$

Gain

$$G = \frac{K_G N S C^2}{\lambda^3} \qquad K_G = 15 \tag{9.122}$$

Input impedance for the feed arrangement of Figure 9.13(a)

$$Z_{in} = 140\frac{C}{\lambda}\ \mathrm{ohm\ (purely\ resistive)} \tag{9.123}$$

Axial ratio

$$\mathrm{AR} = \frac{2N+1}{2N} \tag{9.124}$$

For a helical antenna with $N = 10$, $\psi = 12.5°$, $D = 4.3$ inch, and $C/\lambda = 1.1$ ($f = 962$ MHz), the above formulae yield the following results: HPBW = 30.3°, gain = 43.9, $Z_{in} = 154$ ohm, AR = 1.05.

Although the above formulae are listed in virtually all textbooks and handbooks, the following points should be kept in mind. First, they were originally deduced by Kraus and his co-workers in the late 1940s based on measurements (e.g. Kraus, 1950). Secondly, they have never been derived theoretically. Thirdly, in the recent measurements of King and Wong (1980), the factors K_B in (9.121) and K_G in (9.122) were found to be in the ranges $61 < K_B < 70$, $4.2 < K_G < 7.7$. The former is somewhat larger and the latter considerably smaller than the values obtained by Kraus. Moreover, the gain–frequency curves of King and Wong (1980) showed a peak around $C/\lambda \simeq 1.1$, while according to (9.122) the gain should increase with frequency over the entire axial mode range. Thus some unsettled questions seem to exist regarding (9.121) and (9.122) even from an experimental viewpoint.

When the circumference is much in excess of a wavelength, the radiation becomes multilobed. These modes of radiation are seldom used in practice.

9.7.2 Current Distribution and Phase Velocity

As in the case of the dipole and the circular loop, the current distribution along a helix can in principle be solved as a boundary value problem. This, however, turns out to be extremely complicated and has not been solved completely, although some numerical results for the axial mode range have recently been published (Nakano and Yamauchi, 1979). However, experimental data on the current distributions of helical antennas has along been available (e.g. Kraus, 1950). It was found that when the circumference is less than about two-thirds of a wavelength, the current distribution is nearly sinusoidal as on a straight antenna. This can be interpreted as being caused by the superposition of two oppositely travelling waves of nearly equal amplitudes, resulting in a standing wave.

When the circumference of the helix is of the order of one wavelength, the current distribution can be divided into two regions. In the input region, the current amplitude decays smoothly to a minimum in about two turns; the length of this region being independent of N. In the second region, comprising the bulk of the helix, the current amplitude is relatively uniform and is well approximated by a single travelling wave. Recent numerical results obtained by solving the integral equation for the current distribution showed the same behaviour. Kraus (1950) had therefore suggested that, as far as calculating the

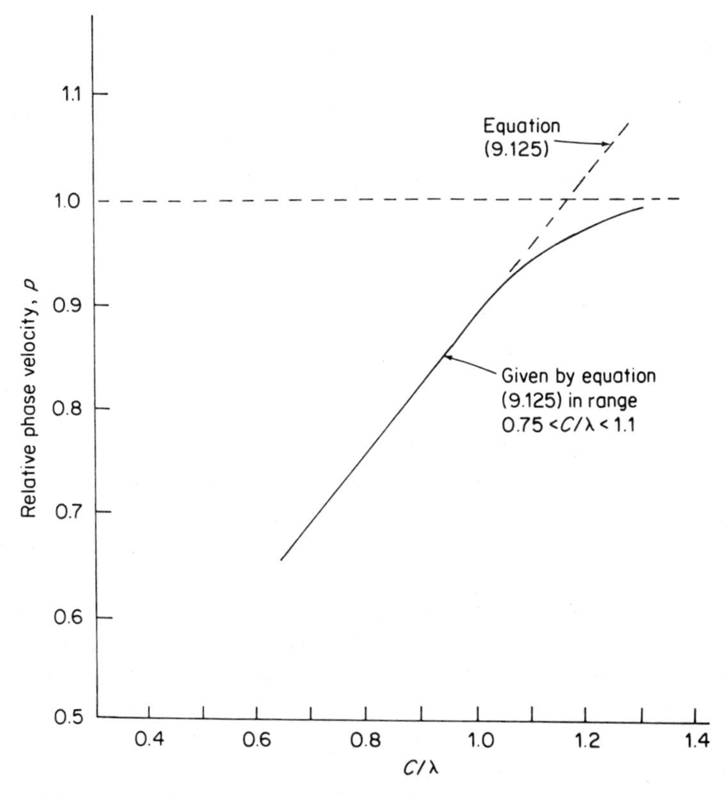

Figure 9.15 Sketch of the relative phase velocity versus circumference in the axial mode range

axial mode radiation is concerned, the variation in the input region can be neglected and the current can be taken to be a single travelling wave of constant amplitude.

The phase velocity v of the current wave propagating along the helical conductor is not the same as the velocity of light in free space, c. It depends on frequency in a complicated manner. Results of measurements on the relative phase velocity, $p = v/c$, can be summarized as follows. For $C < 0.6\lambda$, p is approximately unity while for $C > 1.33\lambda$, it is approximately 0.9. In the range $0.75\lambda < C < 1.1\lambda$, p increases almost linearly with frequency. It is well represented by the formula

$$p = \frac{1}{\sin\psi + [(2N+1)/2N](\lambda\cos\psi)/C} \tag{9.125}$$

The form of p given by (9.125) is known as the Hansen–Woodyard (HW) condition. It has a physical meaning which we shall discuss in section 9.7.4.

In the range $1.1\lambda < C < 1.33\lambda$, p does not increase linearly with frequency as indicated by (9.125) but shows a tendency to level off. A sketch of the behaviour of p as a function of C/λ in the range $0.75 < C/\lambda < 1.33$ is illustrated in Figure 9.15.

In addition to measurements, there has been some theoretical work on the phase velocity along helices. The case of an infinite helix was treated by Sensiper (1951). Recently, some numerical results on the current distribution and phase velocity on finite helices operating in the axial mode have also been published (Nakano and Yamauchi, 1979). These theoretical works appeared to support the behaviour of the phase velocity described above.

9.7.3 Theory of Normal Mode

Consider a helix with dimensions that are small compared to the wavelength ($D \ll \lambda$, $NL \ll \lambda$). The current can then be assumed to be uniform in both magnitude and phase over the entire length of the helix. Calculation of the far field is facilitated by considering the helix to be made up of a number of small loops in planes parallel to the $z = 0$ plane connected by short dipoles parallel to the z-axis, as shown in Figure 9.16. The diameter of the loops is the same as the helix diameter and the length of the dipoles is the same as the spacing between turns of the helix.

Since the helix is small, the far-field pattern is independent of the number of turns and it is sufficient to calculate the field contributed by a single small loop and a short dipole. This has been done in section 9.6. Applying (9.116) to the present situation, we have

$$\mathbf{E}(r) = \mathrm{j}\eta kI \sin\theta \frac{\exp(-\mathrm{j}kr)}{4\pi r}\left(S\hat{\boldsymbol{\theta}} - \mathrm{j}\frac{\pi^2 D^2}{2\lambda}\hat{\boldsymbol{\phi}}\right) \tag{9.126}$$

The θ- and ϕ-components are in phase quadrature. The polarization is there-

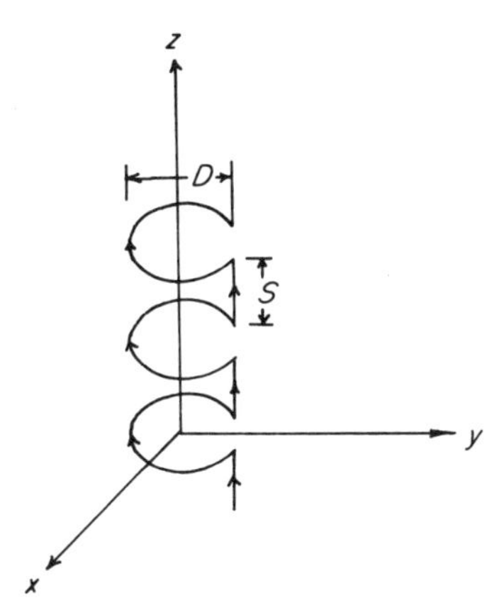

Figure 9.16 Approximating a helical turn with a loop of diameter D and a dipole of length S in the calculation of the normal mode radiation

fore elliptical, with the axial ratio given by

$$\mathrm{AR} = \frac{|E_\theta|}{|E_\phi|} = \frac{2S\lambda}{\pi^2 D^2} \tag{9.127}$$

Note that the polarization is the same in all directions except along the z-axis, where the field is zero. If the diameter and the spacing are such that

$$\pi D = \sqrt{(2S\lambda)} \tag{9.128}$$

the axial ratio is unity and circular polarization results. It follows from (9.128) that the pitch angle required for circular polarization is

$$\psi = \sin^{-1}\left(\frac{-1 + [1 + (L/\lambda)^2]^{1/2}}{L/\lambda}\right) \tag{9.129}$$

Equation (9.126) shows that radiation is maximum in the broadside direction ($\theta = 90°$). As mentioned in section 9.7.1, this mode of radiation is referred to as the 'normal' mode. For the case when the parameters are adjusted for circular polarization, the radiation resistance R_r and the directive gain $g(\theta)$ are readily shown to be

$$R_r = (\eta k^2 S^2)/3\pi \tag{9.130}$$

$$g(\theta) = 1.5 \sin^2 \theta \tag{9.131}$$

The theory of this section assumes that the current is uniform in both magnitude and phase over the entire length of the helix. This requires $D \ll \lambda$ and $NL \ll \lambda$. The radiation efficiency of such a small helix is low and its bandwidth very narrow.

If $D \ll \lambda$ but NL is an appreciable fraction of λ, the measured current shows a sinusoidal variation along the helix. The analysis of this section is no longer valid and an integration of the contributions due to the standing wave of current is required.

9.7.4 Approximate Theory of the Axial Mode for Long Helices

As mentioned in section 9.7.2, a helical antenna radiating in the axial mode may be assumed to have a single travelling wave of uniform amplitude along its conductor. A helix of N turns can be considered as an array of N elements, each spaced a distance S apart, as shown in Figure 9.17. By the principle of pattern multiplication, the far field of the helix is the product of the pattern of one turn and the array factor. When the helix is long ($NS \gg \lambda$), the array factor is much sharper than the element (single-turn) pattern and it largely determines the shape of the total pattern. Let us now study this factor in some detail.

The array factor $f(u)$ for a uniform array of N elements is given by (7.67):

$$f(u) = \sum_n \exp(\mathrm{j}nu) = \frac{\sin(\frac{1}{2}Nu)}{\sin(\frac{1}{2}u)} \tag{9.132}$$

where

$$u = kS\cos\theta + \alpha \tag{9.133}$$

In (9.133), θ is the angle that the helix axis makes with the line joining the point of observation and the origin; α is the progressive phase delay along the helix and is

$$\alpha = -k_{\mathrm{h}}L \tag{9.134}$$

where

$$k_{\mathrm{h}} = 2\pi/\lambda_{\mathrm{h}} = 2\pi/(v/f) = 2\pi/(p\lambda) = k/p \tag{9.135}$$

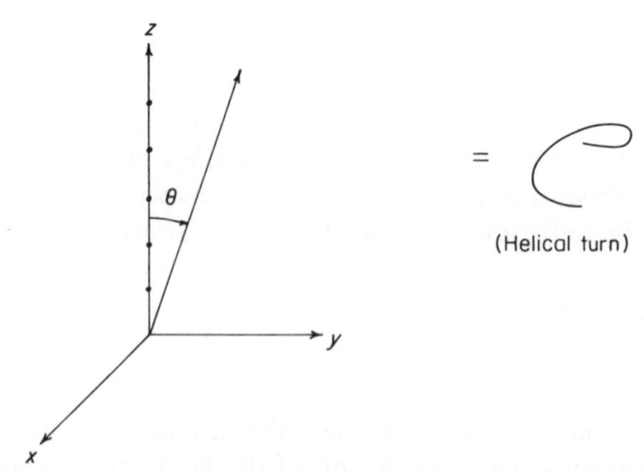

Figure 9.17 A helical antenna as a uniform array. Each dot represents a helical turn

Substitution of (9.135) into (9.133) yields

$$u = k(S \cos \theta - L/p) \tag{9.136}$$

Using the observed value of p in the axial mode range described in section 9.7.2, the array factor $f(u)$ can be calculated. For long helices, the variation of $|f(u)|$ as a function of θ obtained in this manner is in good agreement with the measured patterns. However, the array factor does not reveal any information on the polarization of the radiation, and its approximation to the far-field pattern becomes poorer as the helix becomes shorter. A more satisfactory theory can be obtained following two approaches. The first is to calculate the element (single-turn) pattern and multiply it by the array factor to obtain the total pattern. The second is to calculate the far field by directly integrating the contributions due to the current distributions from one end of the helix to another, without resorting to array theory. Since each of the turns of the helix does not lie in a plane, the geometry of a single turn is almost as complicated as that of the entire helix, with the result that the first approach is not significantly simpler than the second. Moreover, it has the disadvantage of not being applicable to helices with a non-integral number of turns. For these reasons, we shall pursue the problem using the second approach rather than the first. Before doing this, however, let us further examine the array factor given by (9.132).

Some physical insight into the observed phase velocity in the axial mode range can be obtained in the following manner. Given the experimental fact that the radiation pattern of a helical antenna operating in the axial mode is maximum on the helix axis ($\theta = 0°$), let us see what phase velocity is required to produce this result. Clearly, one way of realizing this is for the array elements to satisfy the ordinary end fire condition, namely, the fields from all the elements arrive at points on the axis in phase. Reference to (9.132) shows that this requires that

$$u = -2\pi m \qquad m = 0, 1, 2, 3, \ldots \tag{9.137}$$

The minus sign in (9.137) results from the fact that the phase of source 2 is retarded with respect to source 1, etc. Putting $\theta = 0$ and $u = -2\pi m$ in (9.136), we obtain

$$p = \frac{L/\lambda}{m + S/\lambda} \tag{9.138}$$

For $m = 0$, (9.138) yields $p = L/S$. Since this is greater than unity, it is not in accord with the observations (section 9.7.2) and must therefore be discarded. For $m = 1$, (9.138) yields

$$p = \frac{L/\lambda}{1 + (S/\lambda)} = \frac{1}{\sin \psi + (\lambda \cos \psi)/C} \tag{9.139}$$

The values of p predicted by (9.139) are slightly larger than the observed values. Moreover, when (9.139) is used in (9.132), it yields patterns that are much broader than those observed. It therefore appears that the ordinary endfire condition is not what actually exists in the helical antenna when operating in the axial mode.

Let us recall that, in a uniform array, the increased directivity condition of Hansen and Woodyard also produces maximum radiation in the axial direction. If this condition is presumed to exist in the helix, the value of α will be larger than the ordinary endfire case by the factor $-\pi/N$ (see section 7.5.2). Thus, for $m = 1$, we have

$$\alpha = -kS - 2\pi - \pi/N \tag{9.140}$$

Equating (9.140) and (9.134) and solving for p, there results

$$p = \frac{L/\lambda}{(S/\lambda) + (2N+1)/2N} = \frac{1}{\sin\psi + [(2N+1)/2N](\lambda\cos\psi)/C} \tag{9.141}$$

Equation (9.141) is the same as (9.125) and is known as the Hansen–Woodyard condition. As discussed in section 9.7.2, it yields values of p in good agreement with observations for the range $0.75\lambda < C < 1.1\lambda$. Moreover, when it is used in (9.132), it yields patterns that are in better agreement with measured patterns than when (9.139) is used. Note that for $2N \gg 1$, (9.141) reduces to (9.139).

It appears from the above discussion that the phase velocity in a helix adjusts itself to a value yielding an endfire array of increased directivity for $0.75\lambda < C < 1.1\lambda$, which spans most of the axial mode range. However, p does not seem to follow (9.141) for $1.1\lambda < C < 1.33\lambda$. Indeed, a recent calculation (Lee *et al.* 1982) showed that, if (9.141) is assumed to hold for the entire axial mode range ($0.75\lambda < C < 1.33\lambda$), it leads to the result that the magnitude of the axial electric field increases monotonically, which is in violation of the experimental fact that a peak exists around $C/\lambda \simeq 1.1$. (King and Wong, 1980). We shall return to this point in the next section.

9.7.5 Theory of the Axial Mode by Direct Integration

Far-Field Expressions

In this section, we calculate the far field of the helical antenna radiating in the axial mode by directly integrating the contributions of the current elements from one end of the helix to another. For this purpose, it was mentioned in section 9.7.2 that the current can be approximated as a single travelling wave of constant amplitude along the antenna. The geometry of the helix is shown in Figure 9.18. It begins at $z' = 0$, $\phi' = 0$ and ends with $z' = NS$ and $\phi' = \phi'_m$. Let l be the length of wire from the beginning of the helix to an arbitrary point Q on the helix and $\hat{\mathbf{l}}$ be the unit vector along the wire. Then

$$\mathbf{I}(l) = I_0 \exp(-jh_{\text{h}}l)\hat{\mathbf{l}} = I_0 \exp[(-\omega/pc)l]\hat{\mathbf{l}} \tag{9.142}$$

If L_{T} is the total length of the helix and ϕ' the azimuthal coordinate of point Q, then

$$l/L_{\text{T}} = \phi'/\phi'_{\text{m}} \tag{9.143}$$

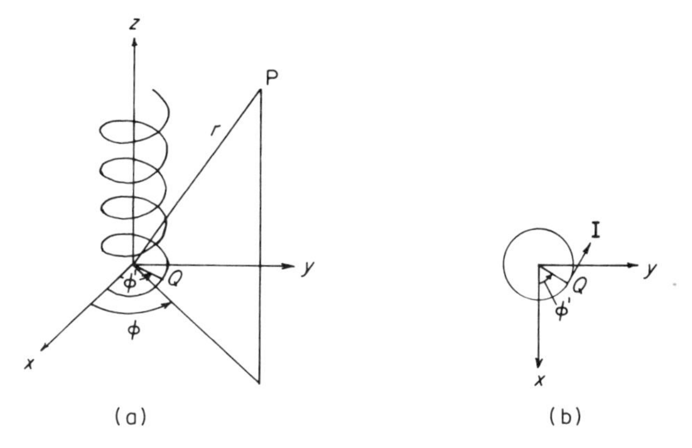

Figure 9.18 Geometry used in calculating the axial mode radiation from a helical antenna

and (9.142) can be written as

$$\mathbf{I}(l) = I_0 \exp(-jg\phi')\hat{\mathbf{I}} \tag{9.144}$$

where

$$g = \frac{\omega L_T}{pc\phi'_m} \tag{9.145}$$

For a helix of N turns (N not necessarily an integer),

$$\phi'_m = 2\pi N \tag{9.146}$$

The unit vector $\hat{\mathbf{I}}$ along the wire can be resolved into rectangular components as follows:

$$\hat{\mathbf{I}} = -\hat{\mathbf{x}} \sin\phi' + \hat{\mathbf{y}} \cos\phi' + \hat{\mathbf{z}} \sin\psi \tag{9.147}$$

From Figure 9.18 and Figure 9.13(b) it is evident that the rectangular co-ordinates of point Q are given by

$$x' = a\cos\phi' \tag{9.148}$$

$$y' = a\sin\phi' \tag{9.149}$$

$$z' = a\phi'\tan\psi \tag{9.150}$$

where $a = D/2$. Let P be the field point with rectangular coordinates (x, y, z) and spherical coordinates (r, θ, ϕ). If R is the distance between P and Q, we have

$$\begin{aligned} R^2 &= (x - x')^2 + (y - y')^2 + (z - z')^2 \\ &= r^2 + a^2 - 2ax\cos\phi' - 2ay\sin\phi' - 2az\phi'\tan\psi + a^2\phi'^2\tan^2\psi \\ &= r^2\left(1 + \frac{a^2}{r^2} - \frac{2a}{r}\sin\theta\cos(\phi - \phi') - \frac{2a}{r}\cos\theta\,\phi'\tan\psi + \frac{a^2}{r^2}\phi'^2\tan^2\psi\right) \end{aligned} \tag{9.151}$$

For P in the far zone, (9.151) simplifies to

$$R \simeq r - a \sin\theta \cos(\phi - \phi') - a\phi' \cos\theta \tan\psi \tag{9.152}$$

The magnetic vector potential at P is

$$\mathbf{A}(r) = \frac{\mu}{4\pi}\iiint \frac{\mathbf{J}\exp(-jkR)}{R}\,\mathrm{d}\tau = \frac{\mu}{4\pi}\int_0^{L_T} \frac{\exp(-jkR)}{R}\mathbf{I}\,\mathrm{d}l$$

$$= \frac{\mu}{4\pi}\int_0^{\phi'_m} \frac{\exp(-jkR)}{R}\mathbf{I}\,a\,\mathrm{d}\phi' \tag{9.153}$$

Substitution of (9.144) and (9.147) into (9.153) yields

$$\mathbf{A}(r) = \frac{\mu a I_0 \exp(-jkr)}{4\pi r} \times \int_0^{\phi'_m} \exp[ju\cos(\phi-\phi')]\exp(jd\phi')(-\hat{\mathbf{x}}\sin\phi' + \hat{\mathbf{y}}\cos\phi' + \hat{\mathbf{z}}\sin\psi)\,\mathrm{d}\phi' \tag{9.154}$$

where

$$u = ka\sin\theta \tag{9.155}$$

$$B = ka\cos\theta\tan\psi \tag{9.156}$$

$$d = B - g \tag{9.157}$$

The first term in the integrand can be expressed as a series of Bessel functions of the first kind:

$$\exp[ju\cos(\phi-\phi')] = J_0(u) + 2\sum_{b=1}^{\infty}(-1)^b J_{2b}(u)\cos[2b(\phi-\phi')] + 2j\sum_{b=0}^{\infty}(-1)^b J_{2b+1}(u)\cos[(2b+1)(\phi-\phi')] \tag{9.158}$$

Using (9.158), (9.154) becomes

$$\mathbf{A}(r) = \frac{\mu a I_0 \exp(-jkr)}{4\pi r}\int_0^{\phi'_m} \exp(jd\phi')(-\hat{\mathbf{x}}\sin\phi' + \hat{\mathbf{y}}\cos\phi' + \hat{\mathbf{z}}\sin\psi) \times \left(J_0(u) + 2\sum_{b=1}^{\infty}(-1)^b J_{2b}(u)\cos[2b(\phi-\phi')] + 2j\sum_{b=0}^{\infty}(-1)^b J_{2b+1}(u)\cos[(2b+1)(\phi-\phi')]\right)\mathrm{d}\phi' \tag{9.159}$$

Since the Bessel functions are independent of ϕ', the integration in (9.159) can be performed by considering the integral

$$\mathbf{T}(m) = \int_0^{\phi'_m} (-\hat{\mathbf{x}} \sin \phi' + \hat{\mathbf{y}} \cos \phi' + \hat{\mathbf{z}} \sin \psi) \exp(\mathrm{j}d\phi') \cos[m(\phi - \phi')] \, \mathrm{d}\phi' \tag{9.160}$$

The first integral in (9.159) corresponds to $\mathbf{T}(m = 0)$, the second to $\mathbf{T}(m = 2b)$, and the third to $\mathbf{T}(m = 2b + 1)$. The quantity $\mathbf{T}(m)$ can be written as

$$\mathbf{T}(m) = T_x(m)\hat{\mathbf{x}} + T_y(m)\hat{\mathbf{y}} + T_z(m)\hat{\mathbf{z}} \tag{9.161}$$

where

$$T_x(m) = \int_0^{\phi'_m} - \sin \phi' \exp(\mathrm{j}d\phi') \cos[m(\phi - \phi')] \, \mathrm{d}\phi' \tag{9.162}$$

$$T_y(m) = \int_0^{\phi'_m} \cos \phi' \exp(\mathrm{j}d\phi') \cos[m(\phi - \phi')] \, \mathrm{d}\phi' \tag{9.163}$$

$$T_z(m) = \int_0^{\phi'_m} \sin \psi \exp(\mathrm{j}d\phi') \cos[m(\phi - \phi')] \, \mathrm{d}\phi' \tag{9.164}$$

The integrals (9.162)–(9.164) can be evaluated in a straightforward manner. The results are given below.

$$T_x(m) = a(m) + a(-m) \tag{9.165}$$

$$T_y(m) = b(m) + b(-m) \tag{9.166}$$

$$T_z(m) = c(m) + c(-m) \tag{9.167}$$

where

$$a(m) = \frac{\exp(\mathrm{j}m\phi)}{4}\left[\exp[\mathrm{j}(d - m)\phi'_\mathrm{m}]\left(\frac{\exp(\mathrm{j}\phi'_\mathrm{m})}{d - m + 1} - \frac{\exp(-\mathrm{j}\phi'_\mathrm{m})}{d - m - 1}\right) + \frac{2}{(d - m)^2 - 1}\right] \tag{9.168}$$

$$b(m) = \frac{\exp(\mathrm{j}m\phi)}{4\mathrm{j}}\left[\exp[\mathrm{j}(d - m)\phi'_\mathrm{m}]\left(\frac{\exp(\mathrm{j}\phi'_\mathrm{m})}{d - m + 1} + \frac{\exp(-\mathrm{j}\phi'_\mathrm{m})}{d - m - 1}\right) - \frac{2(d - m)}{(d - m)^2 - 1}\right] \tag{9.169}$$

$$c(m) = \frac{\exp(\mathrm{j}m\phi)}{2\mathrm{j}}\left(\frac{\exp[\mathrm{j}(d - m)\phi'_\mathrm{m}]}{d - m} - \frac{1}{d - m}\right)\sin \psi \tag{9.170}$$

Equation (9.159) can be written as

$$\mathbf{A}(r) = \frac{\mu a I_0 \exp(-\mathrm{j}kr)}{4\pi r}\left(\mathbf{T}(0)J_0(u) + 2\sum_{b=1}^{\infty}(-1)^b \mathbf{T}(2b)J_{2b}(u) + 2\mathrm{j}\sum_{b=0}^{\infty}(-1)^b \mathbf{T}(2b + 1)J_{2b+1}(u)\right) \tag{9.171}$$

The electric field in the far zone is related to **A** by $\mathbf{E} = -\mathrm{j}\omega\mathbf{A}$. Hence

$$E_x = -\mathrm{j}\omega A_x \tag{9.172}$$

$$E_y = -\mathrm{j}\omega A_y \tag{9.173}$$

$$E_z = -\mathrm{j}\omega A_z \tag{9.174}$$

The θ- and ϕ-components of **E** can be obtained from the rectangular components in the following manner:

$$\begin{aligned} E_\theta &= E_x \cos\phi \cos\theta + E_y \sin\phi \cos\theta - E_z \sin\theta \\ &= -\mathrm{j}\omega[(A_x \cos\phi + A_y \sin\phi)\cos\theta - A_z \sin\theta] \end{aligned} \tag{9.175}$$

$$E_\phi = E_y \cos\phi - E_x \sin\phi = -\mathrm{j}\omega(A_y \cos\phi - A_x \sin\phi) \tag{9.176}$$

The magnitude of **E** is given by

$$|\mathbf{E}| = (E_\theta^2 + E_\phi^2)^{1/2} \tag{9.177}$$

With the values of the relative phase velocity p described in section 9.7.2, the far-field patterns of a helix radiating in the axial mode can be computed using (9.175)–(9.177). The relative patterns for a 10-turn helix with $D = 4.3$ inch, $\psi = 12.5^\circ$, and $C/\lambda = 1.1$ ($f = 962$ MHz) are shown in Figure 9.19. It is seen that the maximum occurs along the axis of the helix, where $|E_\theta| \simeq |E_\phi|$. The half-power beamwidth is equal to 35.6°. It is interesting that, for this set of helix parameters, the empirical formula (9.121) of Kraus predicted the half-

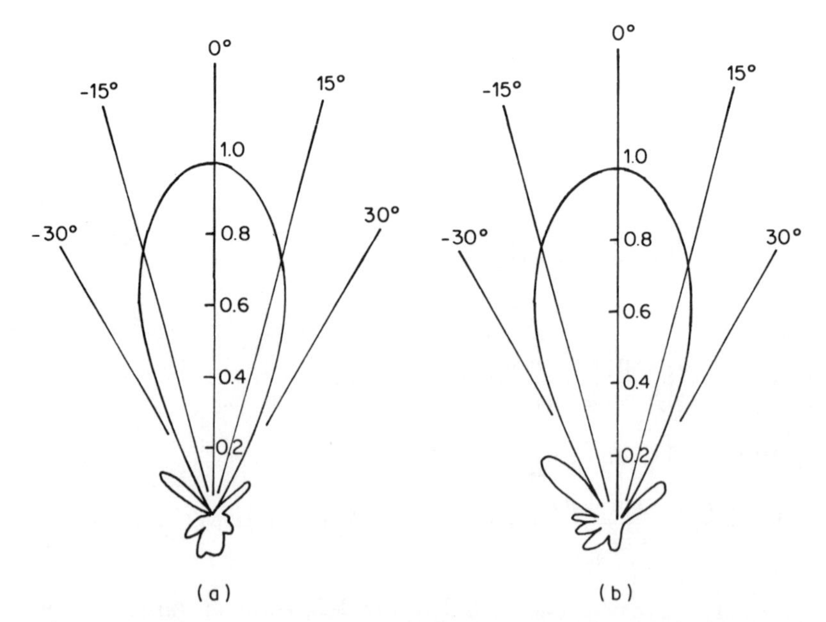

Figure 9.19 Relative E_θ (a) and E_ϕ (b) patterns of a 10-turn helix with $D = 4.3$ inch, $\psi = 12.5^\circ$, and $C/\lambda = 1.1$ ($f = 962$ MHz)

power beamwidth to be 30.3°. For agreement with the theoretical value of 35.6°, the factor K_B needs to be 62. This larger value of K_B is closer to that deduced by King and Wong (1980) in their recent measurements.

Axial Ratio

Putting $\theta = 0°$ and $\phi = 90°$ in (9.175) and (9.176), we obtain the θ- and ϕ-components of the electric field along the axis of the helix. Their ratio is given by

$$\frac{E_\theta}{E_\phi} = \frac{-A_y}{A_x} = \frac{\mathrm{j}\{\exp[\mathrm{j}(d+1)\phi'_{\mathrm{m}}](d-1) + \exp[\mathrm{j}(d-1)\phi'_{\mathrm{m}}](d+1) - 2d\}}{\exp[\mathrm{j}(d+1)\phi'_{\mathrm{m}}](d-1) - \exp[\mathrm{j}(d-1)\phi'_{\mathrm{m}}](d+1) + 2} \tag{9.178}$$

Equation (9.178) can be simplified for a helix consisting of an integral number of turns. In this case, $\phi'_{\mathrm{m}} = 2\pi N$ where N is a positive integer and (9.178) reduces to

$$E_\theta/E_\phi = -\mathrm{j}d \tag{9.179}$$

where d is obtained by putting $\theta = 0°$ in (9.157), yielding

$$d = ka \tan\psi - \frac{\omega L_{\mathrm{T}}}{pc\phi'_{\mathrm{m}}} \tag{9.180}$$

Equation (9.179) indicates that E_θ and E_ϕ are in phase quadrature and the polarization is elliptical. The axial ratio (AR) of the polarization ellipse is

$$\mathrm{AR} = |E_\theta/E_\phi| = \left|ka \tan\psi - \frac{\omega L_{\mathrm{T}}}{pc\phi'_{\mathrm{m}}}\right| \tag{9.181}$$

Since

$$\tan\psi = S/\pi D = L \sin\psi/(2\pi a)$$

and

$$\frac{\omega L_{\mathrm{T}}}{pc\phi'_{\mathrm{m}}} = \frac{2\pi c N L}{pc\lambda 2\pi N} = \frac{L}{p}$$

(9.181) can be expressed as

$$\mathrm{AR} = \left|\frac{L}{\lambda}\left(\sin\psi - \frac{1}{p}\right)\right| \tag{9.182}$$

If the value of p is taken to be given by (9.141), namely, the HW condition of increased directivity, (9.182) becomes

$$\mathrm{AR} = (2N+1)/2N \tag{9.183}$$

If $2N \gg 1$, the axial ratio approaches unity and the polarization is circular.

In summary, assuming the current to be a single travelling wave of constant amplitude along the helix, the theory predicts the axial ratio of a helix with an integral number of turns to be $(2N + 1)/2N$ if the phase velocity is taken to be the HW condition. As mentioned in section 9.7.2, this condition holds for $0.75\lambda < C < 1.1\lambda$. It does not hold for $1.1\lambda < C < 1.33\lambda$. In this range, p levels off in the manner indicated in Figure 9.15, resulting in values of AR slightly above $(2N + 1)/2N$. This feature is consistent with experiments.

Frequency Response

The fact that the HW condition does not hold for $1.1\lambda < C < 1.33\lambda$ is also evident when one computes the magnitude of the square of the axial electric field as a function of frequency. By putting $\theta = 0°$, $\phi = 90°$ in (9.177), we obtain

$$|\mathbf{E}|^2_{\text{axis}} = \omega^2(|A_y|^2 + |A_x|^2)^{1/2} \tag{9.184}$$

If the relative phase velocity p is taken to be given by the HW condition over the range $0.75\lambda < C < 1.33\lambda$, one obtains the result that $|\mathbf{E}|^2_{\text{axis}}$ increases almost linearly with frequency over the whole range (Lee *et al.*, 1982). This is in violation with the measurements of King and Wong (1980), which show that $|\mathbf{E}|^2_{\text{axis}}$ peaks around $C/\lambda \simeq 1.1$. On the other hand, if p is assumed to obey the HW condition only up to $C/\lambda \simeq 1.1$, and then levels off in the manner indicated in Figure 9.15, the frequency response curve does exhibit a peak around $C/\lambda \simeq 1.1$. The interested reader is referred to Lee *et al.* (1982) for more details.

Discussion

The direct integration method of this section describes the axial mode characteristics of the helical antenna reasonably well. The theory neglects the effect of the ground plane. This can be justified, since, if the ground plane is replaced by an image helix, the image helix would be wound in the opposite direction and have its main lobe in the direction $\theta = \pi$. Hence only a minor lobe would be added to the main lobe of the actual helix in the direction $\theta = 0°$. Experimentally, the ground plane is found to have little effect provided its dimensions are larger than about half a wavelength.

The direct integration method yields no information on the input impedance of the helical antenna. As mentioned in section 9.7.1, in the axial mode range, this quantity was found experimentally to be a resistance of about $140C/\lambda$ ohm. This value appeared to have been verified for specific cases by solving the integral equation for the current distribution numerically (Nakano and Yamauchi, 1979). However, no analytical solution of the impedance problem has been reported in the literature.

The direct integration method can also be applied to calculate the normal mode radiation patterns for helices with $D/\lambda \ll 1$ but NL comparable to λ. As discussed in section 9.7.2, the current distribution for helices radiating in the normal mode is well approximated by a sinusoidal standing wave, which can be regarded as two oppositely travelling waves of equal amplitudes. The calculations are straightforward and will not be given in detail here.

9.7.6 Applications of the Helical Antenna

The helical antenna is seldom designed for 'normal mode' operation since the small dimensions required result in low efficiency and small bandwidth. On the other hand, the high gain, large bandwidth, relative simplicity, and nearly circular polarization of the axial mode helix make it a widely used antenna for space communication applications. For example, helical antennas were installed in communication satellites and were placed on the Moon by the Apollo astronauts for transmitting telemetry data back to Earth. On the ground, they were used for transmitting command signals to rockets and space probes, and in receiving signals from weather satellites. Arrays of helical antennas have also been used as radio telescope antennas, as in the original 200–300 MHz Ohio State University radio telescope and the 300–400 MHz University of Texas radio telescope.

9.8 FREQUENCY-INDEPENDENT ANTENNAS

As mentioned in section 9.7.1, the ratio of the upper and lower frequencies of the axial mode for the helical antenna is given by $f_U/f_L = 16/9 = 1.78$. Within this range, the impedance, pattern, and polarization characteristics of the antenna do not change significantly. Referring to Figure 6.2, the ratio of frequencies at which the VSWR equals 2 is 1.17 for the thicker dipole ($a/L = 0.01$) and 1.08 for the thinner dipole ($a/L = 0.002$). In general, the axial mode helix has a wider bandwidth than a typical dipole. This is mainly due to the fact that the axial mode helix supports a travelling wave of current, the form of which is less sensitive to frequency change than the standing wave of current existing on the dipole. There are, however, a class of antennas which can achieve bandwidths of 10:1 or more. These are known as frequency-independent antennas. The structures of most of these antennas have rather unconventional appearances. With the exception of the log-periodic dipole array to be discussed in section 9.8.2, they are exceedingly difficult to analyse mathematically. Indeed, the historical development of frequency-independent antennas was based largely on intuitive reasoning and experimentation rather than on formal theory. In this section, we shall limit ourselves to a brief qualitative introduction to the subject. For more detailed description of the early development, the reader is referred to the survey articles by Isbell (1960), Dyson (1962), Elliott (1962), Mayes (1963), and Jordan *et al.* (1964), as well as the book by Rumsey (1966).

9.8.1 Log-Periodic Sheet and Wire Antennas

In the analysis of the several types of antennas we have considered up to this point, it will be noticed that the characteristics of an antenna are a function of the ratio of length to wavelength. In other words, the characteristic lengths of the structure are the features that introduce the frequency dependence of the antenna. To ensure that a given structure has the same performance at different frequencies, it is only necessary to scale the lengths in the ratio of the frequencies.

It is then reasonable to assume that, if a structure has no characteristic lengths, it will be self-scaling and therefore frequency-independent. Such a structure will be one that (a) is completely described by angles and (b) is infinite in size. Since (b) is a practical impossibility, the successful structures for frequency-independent antennas are those that, though finite, are 'effectively infinite'.

To illustrate these ideas, let us begin with the bi-triangular metal sheet shown in Figure 9.20(a). The sheet is assumed to be infinitesimally thin in the x-direction. If it begins with the vertex at the origin and extends to infinity on both sides, it will be a structure defined by angles alone, without any characteristic length. If energized at the vertex, this structure becomes an antenna, the analysis of which shows that its characteristics are indeed independent of

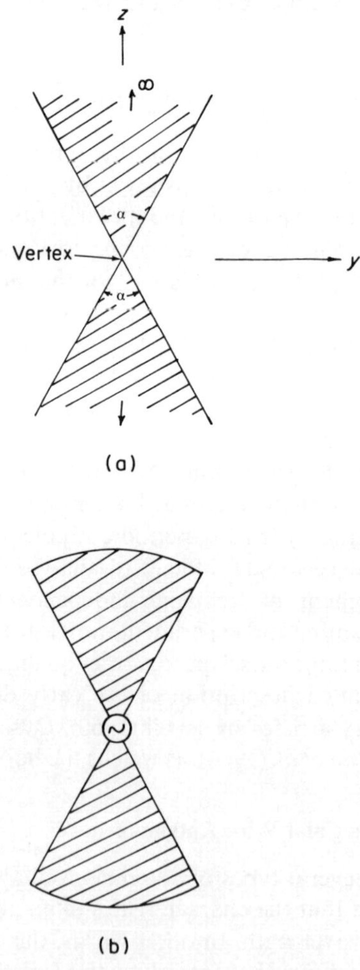

Figure 9.20 (a) A bi-triangular metal sheet of infinite extent in the y–z plane; (b) a bowtie antenna

frequency. However, to be practicable, the gap between the feed points must be finite rather than infinitesimal and the antenna size must also be finite. A practical form is shown in Figure 9.20(b). This is known as the bowtie or bifin antenna. Since the structure now has characteristic lengths, its bandwidth is limited. The radiation is linearly polarized, with a bidirectional pattern having broad main beams perpendicular to the plane of the antenna.

The bandwidth of the bowtie antenna can be expected to increase if it is modified in such a way that, although finite in size, it is 'effectively infinite'. This would be the case if the currents are negligible at the places where the truncation of the original infinite structure occurs, i.e. along the curved edges. One way of making the currents decay rapidly away from the feed point is to introduce discontinuities, for example, teeth in the fins. However, the presence of discontinuities or teeth would destroy the self-scaling nature of the structure. A clever way out of this difficulty is to arrange the positions and sizes of the teeth in the following manner. As shown in Figure 9.21, the radii of the circular arcs forming the corresponding parts of the successive teeth are in a constant ratio $R_{n+1}/R_n = r_{n+1}/r_n = \tau$. This same ratio τ also defines the widths of successive teeth. Consider a frequency at which the pth tooth is, say, one-eighth of a wavelength in size. Then, at a lower frequency, τf_1, the $(p-1)$th tooth will be one-eighth of a wavelength in size, and all the rest of the structure will be scaled accordingly. Thus, if the structure extends from zero to infinity, and is excited at the vertex, its characteristics at a frequency f_1 will be repeated at frequencies $\tau^n f_1$, where n is an integer. This would also be the case for a finite structure if the current is negligible at the places where truncation occurs. When plotted on a logarithmic scale, these frequencies are equally spaced with a period equal to the logarithm of τ. A structure having this property is known as a 'log-periodic' structure, and the one of Figure 9.21 is known as a log-periodic toothed antenna.

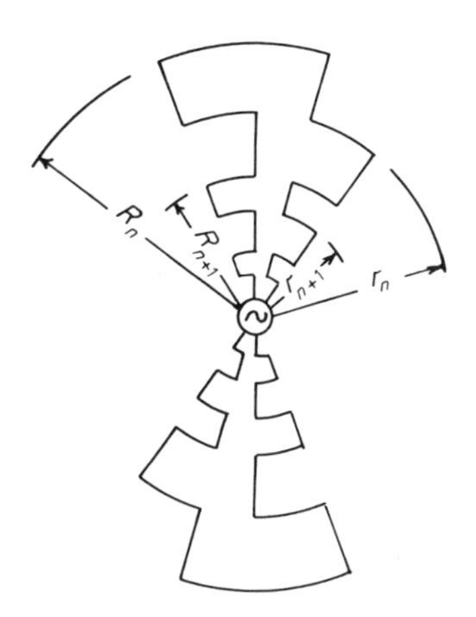

Figure 9.21 Log-periodic toothed antenna

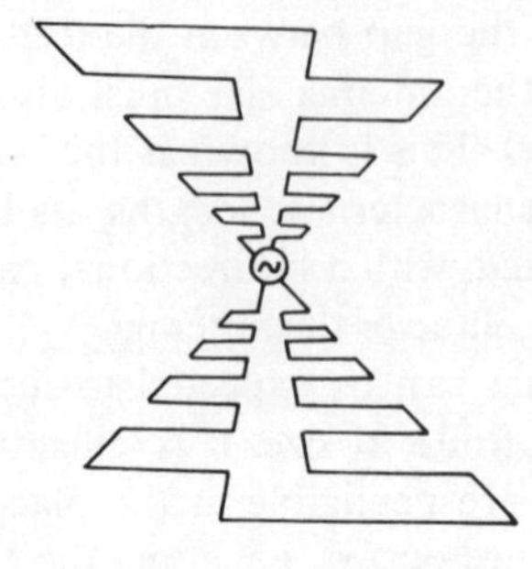

Figure 9.22 Log-periodic trapezoidal toothed antenna

Thus, while the presence of teeth destroys the ability of the structure to self-scale on a continuous basis, by arranging the teeth in the manner described, the structure does scale on a discrete basis. If the variation of its characteristics between these adjacent discrete frequencies is small, the structure will then be effectively frequency-independent. For the log-periodic toothed antenna, this was indeed found to be the case. The pattern is bidirectional, with maxima in directions normal to the plane of the antenna. The polarization is linear, with a strong component perpendicular to that normally radiated by a bowtie antenna. This indicated that significant currents were flowing along the teeth. Moreover, it was found that most of the current appeared on the teeth that were about a quarter-wavelength long. The frequency limits of operation are therefore set by the frequencies where the largest and smallest teeth arc about a quarter-wavelength long.

When the teeth was cut straight instead of curved, a log-periodic trapezoidal toothed antenna (Figure 9.22) results. It was found experimentally that the performance was comparable to the curved teeth structure. It was further found that the currents were strongest along the edges of the teeth, so that these metal sheet structures could be simulated with wires or tubes which outline the perifery of the sheets. This results in the log-periodic trapezoid wire antenna (Figure 9.23). While the sheet version is practicable for short wavelengths, the wire version is convenient for low frequencies. If either of these versions is bent into a wedge, a unidirectional pattern is obtained. The maximum of the radiation occurs off the tip (feed point) of the structure. The bandwidth of the wedge structure was found to be practically the same as the planar structure but the input impedance was lower.

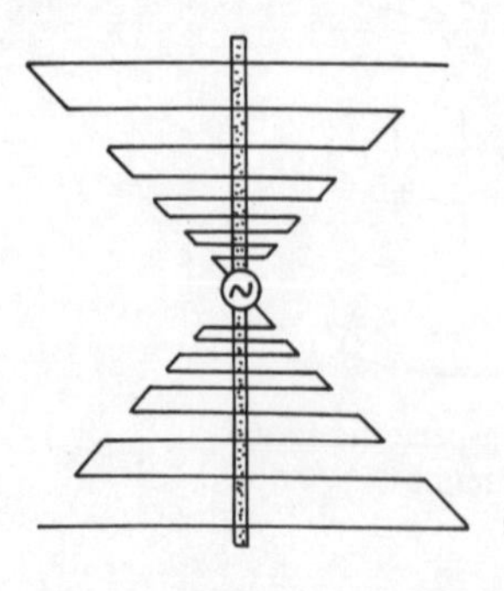

Figure 9.23 Log-periodic trapezoidal wire antenna

9.8.2 Log-Periodic Dipole Array

Imagine that the log-periodic toothed trapezoidal antenna of Figure 9.22 is folded onto itself, forming a wedge with zero included angle. The two centre fins of metal then form a parallel-wire transmission line, with the teeth coming out from them on alternate sides of the fins. When the teeth are replaced by dipoles, a log-periodic dipole array (LPDA) results. This is illustrated in Figure 9.24. The array is fed at the small end of the structure, and the maximum radiation is towards this end. The lengths of the dipoles and their positions from the apex obey the relation

$$\frac{R_{n+1}}{R_n} = \frac{L_{n+1}}{L_n} = \tau \tag{9.185}$$

where τ is called the scale factor of the array. Since the element spacings

$$d_n = R_n - R_{n-1} = (1 - \tau)R_n \tag{9.186}$$

it follows that

$$d_{n+1}/d_n = \tau \tag{9.187}$$

The spacing factor for the LPDA is defined as

$$\sigma = d_n/(2L_n) \tag{9.188}$$

Since

$$\tan(\tfrac{1}{2}\alpha) = \tfrac{1}{2}L_n/R_n \tag{9.189}$$

we have

$$d_n = (1 - \tau)\frac{L_n}{2\tan(\frac{1}{2}\alpha)} \tag{9.190}$$

and

$$\sigma = \frac{1 - \tau}{4\tan(\frac{1}{2}\alpha)} \tag{9.191}$$

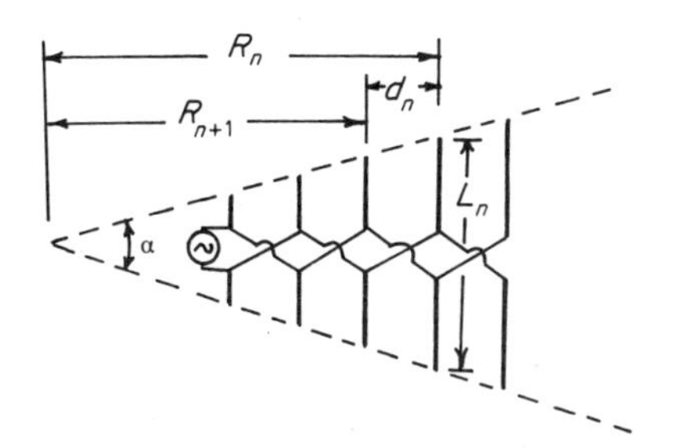

Figure 9.24 Log-periodic dipole array

Unlike the other types of frequency-independent antennas, for which mathematical analysis is difficult, the LPDA can be analysed using the theory of transmission lines and linear dipoles. This was first done by Carrel (1961). Examples of subsequent developments based on computer-aided techniques are the works of Cheong and King (1967), Wolter (1970), De Vito and Stracca (1973, 1974), and Sinnott (1974). Based on both experimental and theoretical results, the following features of the LPDA are established:

(a) The active region in the array comes from the few dipoles that were near the one that is a half-wavelength long. The currents on these dipoles are much larger than those on the rest of the elements. Indeed, the currents at the larger end of the structure should be negligible within the operating band of frequencies for the antenna to be 'effectively infinite'.

(b) Maximum radiation of the LPDA occurs in the direction of the apex. In a sense, the LPDA is similar to that of the Yagi antenna. The longer dipoles behind the active ones behave as reflectors and the shorter dipoles in front of the active ones behave as directors.

(c) As the operating frequency changes, the active region shifts to a different portion of the antenna. The operating band is determined approximately by the frequencies at which the longest and shortest dipoles are equal to about a half-wavelength.

9.8.3 Spiral Antennas

In section 9.8.1, it was noted that, when the bi-triangular sheet structure of Figure 9.20(a) was truncated, discontinuities or teeth had to be introduced so that the currents became negligible at the places of truncation, thereby rendering the finite structure 'effectively infinite'. There are structures defined by angles which, when truncated, will remain 'effectively infinite' without the necessity of introducing discontinuities. An example is the spiral antenna.

Consider a curve in the x–y plane defined by the equation

$$r = r_0 \exp[a(\phi - \phi_0)] \tag{9.192}$$

where r_0, ϕ_0, and a are constants. The curve is known as an equiangular spiral, the shape of which is illustrated in Figure 9.25(a). It is called equiangular since

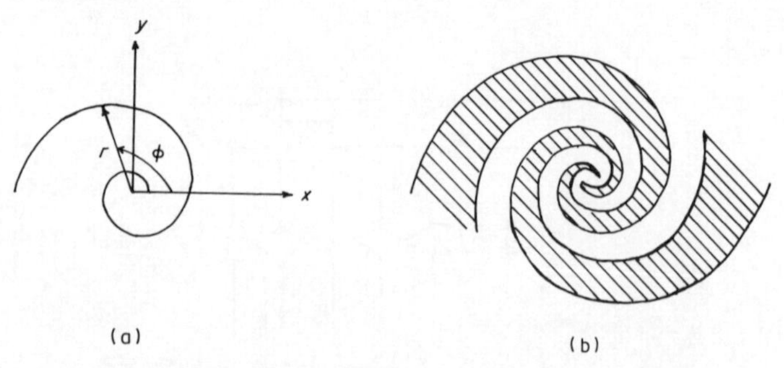

Figure 9.25 (a) Equiangular spiral; (b) equiangular spiral structure

the angle between the radius vector and the tangent to the curve is the same for all points on the spiral. Now consider the metal sheet structure shown in Figure 9.25(b). The four edges of the sheet are defined by four spirals corresponding to $\phi_0 = 0$, $\frac{1}{2}\pi$, π, and $\frac{3}{2}\pi$ in (9.192). If all the points in the spirals were to expand or contract by a factor K, the operation would merely correspond to a rotation of the structure through an angle, with the shape remaining exactly the same. This follows since

$$\frac{r}{K} = \frac{r_0}{K} \exp[a(\phi - \phi_0)] = r_0 \exp[a(\phi' - \phi_0)] \tag{9.193}$$

where

$$\phi' = \phi - \frac{\ln K}{a} \tag{9.194}$$

Thus the equiangular spiral structure, if extending to infinity, is self-scaling.

The equiangular spiral antenna is shown in Figure 9.26. It is driven by a source at the origin. The current is found to be strongest in a region around $r \simeq \frac{1}{2}\lambda$. Even without introducing any discontinuities, the current decreases rapidly beyond this region. Consequently, if the structure is truncated at points corresponding to $r \simeq \lambda$, the practical antenna, although finite, is 'effectively infinite'. Conversely, for a given size antenna, the low-frequency limit is set by the overall radius R being approximately equal to λ. The high-frequency end of the operating band is set by the feed structure, a rough criterion being that the separation of the feed points be equal to λ. Within the operating band, the structure self-scales continuously rather than discretely, as in the case of the log-periodic structures discussed in section 9.8.2.

The equiangular spiral antenna has a bidirectional pattern which is maximum broadside to the plane of the antenna and is approximately given by $\cos\theta$. The half-power beamwidth is about 90°. The polarization is circular. If the spirals are wrapped around a cone, a conical equiangular spiral antenna results. The pattern of this antenna is unidirectional with its maximum occurring off the apex of the cone.

Figure 9.26 Equiangular spiral antenna

9.9 WORKED EXAMPLES

Example 1

Consider a cross-dipole of infinitesimal current elements situated in the x–z plane, as shown in Figure 9.27. The antenna is used to receive electromagnetic energy in an incoming wave, the electric field of which is of the form $\mathbf{E} = E_0(\hat{\mathbf{x}} - 2\mathrm{j}\,\hat{\mathbf{z}})\exp(\mathrm{j}ky)$. Find the magnitude of the open-circuit voltage induced at the terminals AA′ of the cross-dipole.

Solution

To find the open-circuit voltage, we first find the complex effective length $\mathbf{h}$ when the cross-dipole is used as a transmitting antenna. Then $|V_{\mathrm{oc}}| = |\mathbf{h}\cdot\mathbf{E}|$ where $\mathbf{E}$ is the electric field of the incoming wave. From equations (9.5) and (9.34), we have

$$\mathbf{h} = \mathrm{d}l[\hat{\boldsymbol{\theta}}(\sin\theta - \mathrm{j}\cos\theta\cos\phi) + \hat{\boldsymbol{\phi}}\mathrm{j}\sin\phi]$$

In the direction of the incoming wave, $\theta = \phi = \pi/2$. Hence

$$\mathbf{h}(\pi/2, \pi/2) = \mathrm{d}l(\hat{\boldsymbol{\theta}} + \mathrm{j}\hat{\boldsymbol{\phi}})$$

Using the relationships

$$\hat{\mathbf{x}} = \sin\theta\cos\phi\hat{\mathbf{r}} + \cos\theta\cos\phi\hat{\boldsymbol{\theta}} - \sin\theta\hat{\boldsymbol{\phi}}$$

$$\hat{\mathbf{z}} = \cos\theta\,\hat{\mathbf{r}} - \sin\theta\,\hat{\boldsymbol{\theta}}$$

we have, for $\theta = \phi = \pi/2$, $\hat{\mathbf{x}} = -\hat{\boldsymbol{\phi}}$ and $\hat{\mathbf{z}} = -\hat{\boldsymbol{\theta}}$. Hence the electric field of the incoming wave can be expressed as

$$\mathbf{E} = E_0(\hat{\mathbf{x}} - \mathrm{j}2\hat{\mathbf{z}})\exp(\mathrm{j}ky) = E_0(-\hat{\boldsymbol{\phi}} + \mathrm{j}2\hat{\boldsymbol{\theta}})\exp(\mathrm{j}ky)$$

The open-circuit voltage induced at terminals AA′ is

$$|V_{\mathrm{oc}}| = |\mathbf{h}\cdot\mathbf{E}| \quad E_0\,\mathrm{d}l|(-\mathrm{j} + 2\mathrm{j})| = E_0\,\mathrm{d}l$$

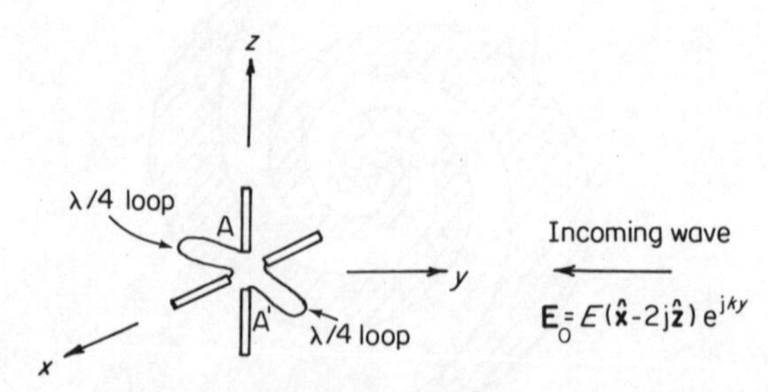

Figure 9.27 Worked example 1: a cross-dipole in the field of an incoming elliptically polarized wave

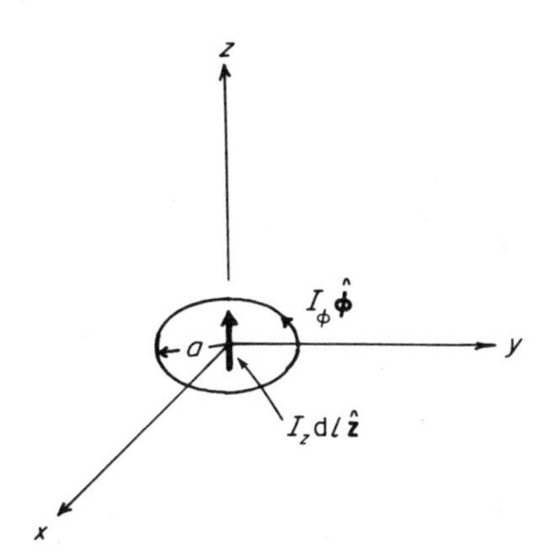

Figure 9.28 Geometry for worked example 2

Example 2

Figure 9.28 shows an antenna composed of a small circular loop of radius a lying in the x–y plane carrying a current $I_\phi \hat{\boldsymbol{\phi}}$ and a Hertzian dipole $I_z\,\mathrm{d}l\,\hat{\mathbf{z}}$ at the centre of the loop. The antenna is designed to radiate left-circularly polarized waves in all directions.

(a) Find the directive gain of the antenna.
(b) The radiation from the antenna described in (a) is being received by a cross-dipole having a complex effective length

$$\mathbf{h}\,\mathrm{d}l[\hat{\boldsymbol{\theta}}(\sin\theta - \mathrm{j}\cos\theta\cos\phi) + \hat{\boldsymbol{\phi}}\mathrm{j}\sin\phi]$$

What should be the relative orientations of the two antennas if the open-circuit voltage induced at the terminals of the receiving antenna is to be the maximum possible.

Solution

(a) The far-zone electric field of the antenna is given by (9.116):

$$\mathbf{E}(r) = \frac{\exp(-\mathrm{j}kr)}{4\pi r}\mathrm{j}k\eta\sin\theta(I_z\,\mathrm{d}l\,\hat{\boldsymbol{\theta}} - \mathrm{j}kI_\phi\pi a^2\,\hat{\boldsymbol{\phi}})$$

The radiation will be left-circularly polarized if $I_z\mathrm{d}l = -kI_\phi\pi a^2$. Hence

$$\mathbf{E}(r) = \frac{\exp(-\mathrm{j}kr)}{4\pi r}\mathrm{j}k\eta\sin\theta\, I_z\mathrm{d}l(\mathrm{j}\hat{\boldsymbol{\theta}} + \hat{\boldsymbol{\phi}})$$

$$\mathbf{H}(r) = \frac{\exp(-\mathrm{j}kr)}{4\pi r}\mathrm{j}k\sin\theta\, I_z\mathrm{d}l(-\hat{\boldsymbol{\theta}} + \mathrm{j}\hat{\boldsymbol{\phi}})$$

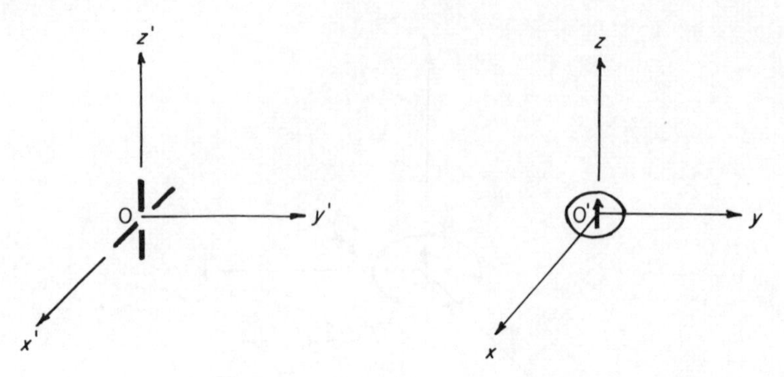

Figure 9.29 Worked example 2: a cross-dipole receiving the radiation from an antenna consisting of a small loop and a dipole

Radiation intensity

$$U = r^2\hat{\mathbf{r}}\cdot\tfrac{1}{2}\mathrm{Re}(\mathbf{E}\times\mathbf{H}^*) = \frac{k^2\eta\sin\theta}{16\pi^2}(I_z\mathrm{d}l)^2$$

Power radiated

$$W = \int_0^{2\pi}\int_0^{\pi}\frac{(I_z\mathrm{d}l)^2k^2\eta}{16\pi^2}\sin^3\theta\,\mathrm{d}\theta\,\mathrm{d}\phi = \frac{(I_z\mathrm{d}l)^2\eta k^2}{6\pi}$$

Directive gain

$$g = 4\pi U/W = 1.5\sin^2\theta$$

(b) The receiving antenna should (i) be placed in the direction where the incoming field has maximum intensity and (ii) have a complex effective length corresponding to left-circular polarization in the direction of the incoming wave. These requirements will be met if the two antenna are oriented as shown in Figure 9.29. The line OO′ joining the two centres lies in the plane of the loop and is perpendicular to the plane containing the axes of the dipoles.

Example 3

Figure 9.30 shows an x-directed Hertzian dipole $\mathrm{d}l$ and a circular loop of radius $a \ll \lambda$ lying in the x–y plane. The vector $\mathbf{r}$ makes an angle ϕ' with respect to $-\hat{\mathbf{x}}$ and the length of r is comparable to the wavelength.

(a) For what value of ϕ' will the mutual coupling be maximum?
(b) Obtain an equation for determining the distance r in units of wavelength such that the mutual impedance between the antennas is purely imaginary.

Solution

The geometry in the x–y plane is shown in Figure 9.30(b). Let the loop be driven by a uniform current I. The electric field at the Hertzian dipole is along the

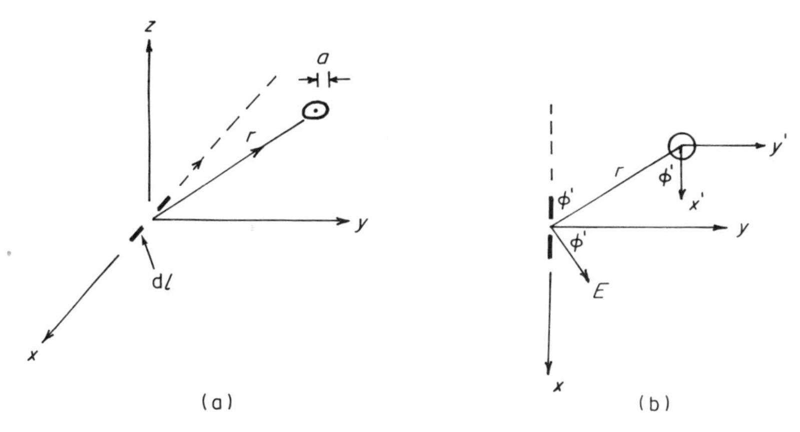

Figure 9.30 Worked example 3: mutual coupling between a small circular loop and a Hertzian dipole

direction shown. Its value is, by putting $\theta = 90^\circ$ in (9.76), given by

$$\mathbf{E} = \omega\mu I\pi a^2 \frac{\exp(-\mathrm{j}kr)}{4\pi r} k\left(1 + \frac{1}{\mathrm{j}kr}\right)\hat{\boldsymbol{\phi}}'$$

(a) Maximum mutual coupling occurs when **E** is parallel to the dipole, i.e. $\phi' = 90^\circ$.

(b) The electric field parallel to the axis of the dipole is

$$E_{\parallel} = \frac{\omega\mu I\pi a^2 k}{4\pi r}[\cos(kr) - \mathrm{j}\sin(kr)]\left(1 - \frac{\mathrm{j}}{kr}\right)\sin\phi'$$

If this is purely imaginary, the mutual impedance will be purely imaginary. The condition is therefore

$$\cos(kr) - \frac{\sin(kr)}{kr} = 0$$

or

$$\tan(kr) = kr$$

The above equation can be solved by plotting the left- and right-hand sides as a function of kr and locating the points of intersection.

9.10 PROBLEMS

1. Consider a cross-dipole of infinitesimal current elements in which $I_z = I_0$ and $I_x = I_0 \exp(\mathrm{j}\delta)$. Show that the far-zone electric field is given by equation (9.25).

2. For the cross-dipole of problem 1, obtain an expression for the radiation intensity and show that the power radiated is independent of δ, the phase angle between the two current elements.

3. For a cross-dipole of half-wave elements, show that the far-zone electric field and the radiation intensity are given by (9.29) and (9.30) respectively.

4. In the corner reflector shown in Figure 8.5, if $\beta = 90°$, $S = 0.5\lambda$, and $L = 0.5\lambda$, determine its effective length in the direction $\theta = 90°$, $\phi = 0$ and compare it with the effective length in the same direction of the half-wave dipole when the reflector is removed.

5. Consider the turnstile antenna of Figure 9.4 being used as the transmitting antenna and the corner reflector with tilted dipole of Figure 8.20 being used as the receiving antenna. The dipole-to-apex spacing S of the latter is equal to 0.5λ. If the centre of the transmitting antenna is at $(0, 0, 0)$, what should be the orientations of the receiving antenna for maximum signal pickup if:

   (a) its centre is located at $(d, 0, 0)$,
   (b) its centre is located at $(0, d, 0)$.

   In both cases, $d \gg \lambda$.

6. Consider a small rectangular loop of sides L_1 and L_2 lying in the x–y plane with its centre at the origin, situated so that the currents are directed along x and y. Assume that the current is the same in all parts of the loop. Calculate the electromagnetic fields in the far zone and the radiation resistance. Compare the results with those of the small circular loop.

7. A small circular loop antenna of N turns, each of radius $a = 0.01\lambda$, is to have a radiation resistance of 5 ohm. Determine N and the effective length of the antenna.

8. Compare the radiation resistance and effective length of a single-turn circular loop antenna of radius $a = 0.01\lambda$ with the corresponding values of a dipole of length $L = 2\pi a$.

9. At $f = 500$ kHz, compare the radiation efficiencies of two circular loop antennas of radius $a = 15$ cm, one made of AWG 12 ($b = 1$ *mm*) copper wire and the other of AWG 8 ($b = 1.6$ *mm*) copper wire. Take the conductivity of copper to be 5.7×10^7 ohm^{-1} m^{-1}.

10. A ferrite-core circular loop antenna is 20 cm long and 1 cm in diameter. It has 22 turns and an effective permeability of 38. It is used as a receiving antenna. The wave normal and polarization of the incident wave are in the plane of the loop. If the magnitude of its electric field vector is 10^{-4} V m^{-1}, find the open-circuit voltage induced at the terminals of the loop.

11. For a circular loop with $ka = 0.5$, plot the relative patterns in the three coordinate planes. Comment on their differences with the relative patterns of the small circular loop.

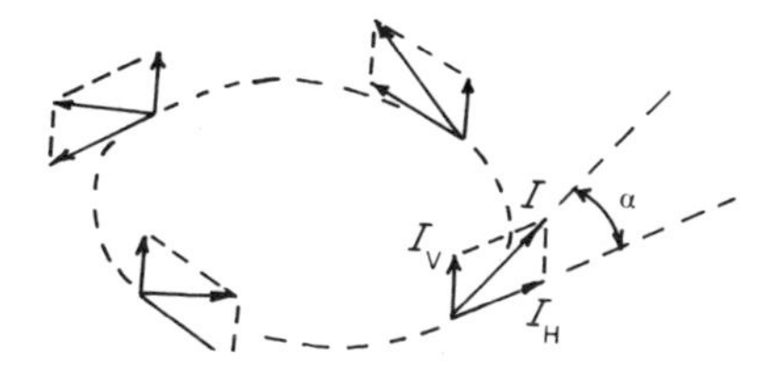

Figure 9.31 Geometry for problem 13: four slanted Hertzian dipoles I dl disposed uniformly around the periphery of a circle of radius S

12. Compare the radiation resistance of a circularly polarized normal mode helical antenna consisting of three turns with that of (a) a dipole of length equal to the turn spacing S, (b) a circular loop of diameter equal to the helix diameter.

13. Figure 9.31 shows a system of four slanted Hertzian dipoles disposed uniformly around the periphery of a circle of radius S lying in the horizontal plane. For each dipole, the plane formed by the vertical and horizontal components of the current is a tangent to the circle. The slant angle α and the current I are the same for all dipoles.

    (a) Obtain an expression for the far-zone electric field.
    (b) If $S \ll \lambda$, show that the polarization is circular at all points in space if the condition $\tan \alpha = \frac{1}{2}kS$ is satisfied.

14. Use equation (9.121) to construct a graph of the HPBW of an axial mode helical antenna as a function of axial length NS in units of λ in the range $0.7 < NS/\lambda < 10$ for $C/\lambda = 0.9$, 1.0, and 1.1. Plot the curves on log–log paper.

15. Verify equations (9.165)–(9.170).

16. Verify equation (9.178).

17. Describe how linearly polarized radiation can be obtained in a helical structure.

BIBLIOGRAPHY

Elliott, R. S. (1962). 'A view of frequency independent antennas', *Microwave J.*, November, 61–68.

Jasik, H. (ed.) (1961). *Antenna Engineering Handbook*, McGraw-Hill, New York, Chapter 28.

Jordan, E. C. and Balmain, K. G. (1968). *Electromagnetic Waves and Radiating Systems*, 2nd edn, Prentice-Hall, Englewood Cliffs, NJ, Chapter 15.

Jordan, E. C., Deschamps D. A., Dyson, J. D., and Mayes, P. E. (1964). 'Developments in broadband antennas', *IEEE Spectrum*, **1**, 58–71.

Kraus, J. (1950). *Antennas*, McGraw-Hill, New York, Chapter 7.

Sinclair, G. (1950). 'The transmission and reception of elliptically polarized waves', *Proc. IRE*, 148–151.

CHAPTER 10
Aperture Antennas

10.1 INTRODUCTION

In all of the antennas considered up to this point, the currents on the structures are assumed to be known to a reasonable degree of accuracy. The analyses of these antennas are concerned basically with calculating the radiation fields set up by these currents. For example, the current distributions on the dipole, the cross-dipole, and the loop are described approximately by sinusodial standing waves; that of the helix radiating in the axial mode by an outward travelling wave with a certain frequency-dependent phase velocity. Even the current distributions on complicated frequency-independent antennas can be estimated by a combination of intelligent guesses and experimentation. These antennas, as well as others in which the currents on them are known to good approximations, are referred to as wire antennas.

There is another class of antennas in which the source current distribution is difficult to obtain, but for which the electromagnetic field configuration over a specific surface enclosing the source is known to a reasonable degree of accuracy. For such antennas, it is simpler to calculate their radiation fields from knowledge of the fields that exist in the aforementioned closed surface. This is possible because Huygen's principle states that any wavefront can be considered to be the source of secondary waves that add to produce distant wavefronts. Such antennas are referred to as aperture antennas. The most prominent of these are the horn, the slot, and the parabolic reflector. These are illustrated in Figure 10.1. In all of them, the current distributions that actually exist in the conductors

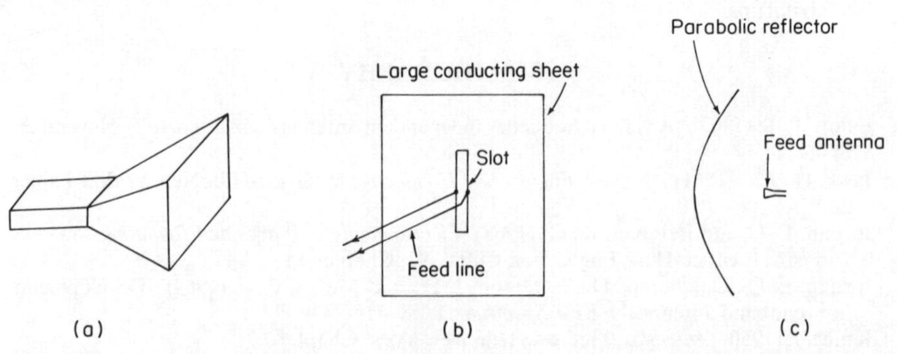

Figure 10.1 Examples of aperture antennas: (a) the horn antenna; (b) the slot antenna; (c) the parabolic reflector antenna

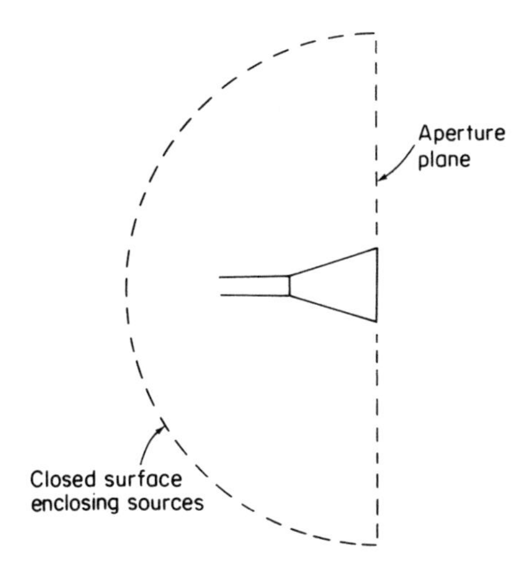

Figure 10.2 Dividing space into two regions by a closed surface consisting of an infinite plane and a hemisphere of infinite radius enclosing all the sources

which make up the structures are difficult to obtain. Even if they are known, their complexities would render the calculation of the radiation field very difficult. On the other hand, the region of space may be divided into two regions by a closed surface consisting of an infinite plane and a hemisphere of infinite radius enclosing all the sources. Since the infinite plane is usually chosen to contain the aperture of the antenna, it is referred to as the aperture plane. This is illustrated in Figure 10.2 for the horn antenna. The fields over the hemisphere are zero since it is infinitely far away. Although exact knowledge of the fields over the entire aperture plane is rarely available, the fields over that portion containing the antenna aperture can frequently be estimated with reasonable accuracy. The radiation to the right of the antenna is usually calculated from the aperture fields only, assuming that the fields elsewhere in the aperture plane are negligible.

In section 10.2, the fields as sources of radiation (Huygen's principle) are formulated. The results are applied to the horn antenna, the slot antenna, and the parabolic reflector. Our discussion is mostly concerned with an individual element, keeping in mind that the results of Chapter 7 concerning array factors can be used to deal with an array of aperture antennas.

10.2 FIELDS AS SOURCES OF RADIATION

10.2.1 General Formulation for the Far Field

Consider a source ($\mathbf{J}$, ρ) localized in a certain region of space. At any point P outside the source, the electromagnetic fields can be calculated in terms of

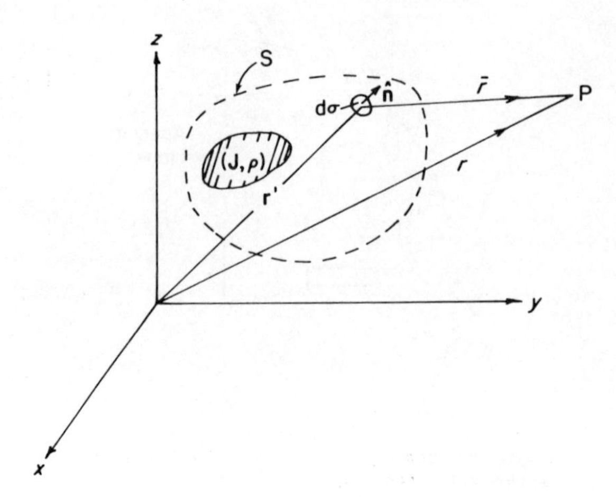

Figure 10.3 The electromagnetic field at a point P can be calculated either directly in terms of (**J**, ρ) or in terms of the tangential electric and magnetic fields on a closed surface enclosing the sources. The point P is outside S

(**J**, ρ) in the manner described in Chapter 2. Let us now imagine a closed surface S containing the source distribution, as shown in Figure 10.3. The field over this surface is of course determined by the source (**J**, ρ). It then turns out that the fields at any point P outside the surface S can be expressed in terms of the tangential components of the electric and magnetic fields existing on the closed surface S, with the actual source (**J**, ρ) within S removed. To keep the complexity of the algebra to a minimum, we shall show this in detail for the case when the distance from P to any point in the surface, denoted by $\bar{r}$, satisfies the far-field criterion, namely, $k\bar{r} \gg 1$. For this case, the electric field at P is given by

$$\mathbf{E} = \frac{jk}{4\pi}\iint_S \frac{\exp(-jk\bar{r})}{\bar{\mathbf{r}}}\{(\mathbf{n} \times \mathbf{E}_s) \times \hat{\bar{\mathbf{r}}} + \eta[(\hat{\mathbf{n}} \times \mathbf{H}_s) \times \hat{\bar{\mathbf{r}}}] \times \hat{\bar{\mathbf{r}}}\}\, d\sigma \qquad (10.1)$$

where $\mathbf{E}_s$ and $\mathbf{H}_s$ are the electric and magnetic fields on an element of area $d\sigma$, the outward normal of which is $\hat{\mathbf{n}}$.

It follows from (10.1) that the contribution to **E** from the element $d\sigma$ is

$$d\mathbf{E} = \frac{jk}{4\pi\bar{r}}\exp(-jk\bar{r})\{(\hat{\mathbf{n}} \times \mathbf{E}_s) \times \hat{\bar{\mathbf{r}}} + \eta[(\hat{\mathbf{n}} \times \mathbf{H}_s) \times \hat{\bar{\mathbf{r}}}] \times \hat{\bar{\mathbf{r}}}\}\, d\sigma \qquad (10.2)$$

To prove (10.1) and (10.2), consider two sets of sources and fields. The first set corresponds to the situation shown in Figure 10.3. The second set corresponds to the same geometry but with the source within S removed and a Hertzian dipole I **dl** placed at P. For convenience, let the magnitude of I be unity. Then, from (4.17), we have

$$\nabla\cdot(\mathbf{E}_1 \times \mathbf{H}_2) - \nabla\cdot(\mathbf{E}_2 \times \mathbf{H}_1) = \mathbf{E}_2\cdot\mathbf{J}_1 - \mathbf{E}_1\cdot\mathbf{J}_2 \qquad (10.3)$$

where subscripts 1 and 2 refer to the two situations mentioned above.

Let us integrate (10.3) over a region lying between S and a sphere of infinite radius. Since the only source in the region is the dipole at P, we have

$$\iiint \nabla\cdot(\mathbf{E}_1 \times \mathbf{H}_2 - \mathbf{E}_2 \times \mathbf{H}_1)\mathrm{d}\tau = -\iiint \mathbf{E}_1\cdot\mathbf{J}_2\mathrm{d}\tau \tag{10.4}$$

The right-hand side of (10.4) evaluates to $-\mathbf{E}\cdot\mathrm{dl}$. On using the divergence theorem to convert the left-hand side to a surface integral, and noting that the integral over the surface of the infinite sphere vanishes, we can write (10.4) as

$$\iint_{\mathrm{S}} (\mathbf{E}_\mathrm{s} \times \mathbf{H}_2 - \mathbf{E}_2 \times \mathbf{H}_\mathrm{s})\cdot(-\hat{\mathbf{n}}\,\mathrm{d}\sigma) = -\mathbf{E}\cdot\mathrm{dl} \tag{10.5}$$

The quantities $\mathbf{E}_2$ and $\mathbf{H}_2$ are the fields existing on points in the surface S due to the dipole dl at P. If the distance $\bar{r}$ satisfies the far-field criterion $k\bar{r} \gg 1$, we have, from (3.15) and (3.16),

$$\mathbf{H}_2 = -\frac{jk}{4\pi\bar{r}}\exp(-jk\bar{r})\,\mathrm{dl} \times \hat{\bar{\mathbf{r}}} \tag{10.6}$$

$$\mathbf{E}_2 = -\eta\mathbf{H}_2 \times \hat{\bar{\mathbf{r}}} \tag{10.7}$$

where the minus signs arise from the fact that $\hat{\bar{\mathbf{r}}}$ is directed towards the dipole and not away from it. Equation (10.7) can be written as

$$\mathbf{E}_2 = \frac{jk\eta}{4\pi\bar{r}}\exp(-jk\bar{r})(\mathrm{dl} \times \hat{\bar{\mathbf{r}}}) \times \hat{\bar{\mathbf{r}}} = -\frac{jk\eta}{4\pi\bar{r}}\exp(-jk\bar{r})\,\mathrm{dl}^\mathrm{T} \tag{10.8}$$

where

$$\mathrm{dl}^\mathrm{T} = \hat{\bar{\mathbf{r}}} \times (\mathrm{dl} \times \hat{\bar{\mathbf{r}}}) \tag{10.9}$$

and is the component of dl transverse to $\hat{\bar{\mathbf{r}}}$.

On using (10.6) and (10.8), we can write the contribution to $\mathbf{E}\cdot\mathrm{dl}$ from an element $\mathrm{d}\sigma$ as

$$\mathrm{d}\mathbf{E}\cdot\mathrm{dl} = \frac{jk}{4\pi\bar{r}}\exp(-jk\bar{r})[-\mathbf{E}_\mathrm{s} \times (\mathrm{dl} \times \hat{\bar{\mathbf{r}}}) + \eta(\mathrm{dl}^\mathrm{T} \times \mathbf{H}_\mathrm{s})]\cdot\hat{\mathbf{n}}\,\mathrm{d}\sigma \tag{10.10}$$

From the vector identity $\mathbf{A}\cdot(\mathbf{B} \times \mathbf{C}) = \mathbf{B}\cdot(\mathbf{C} \times \mathbf{A}) = \mathbf{C}\cdot(\mathbf{A} \times \mathbf{B})$, the terms inside the square bracket in (10.10) can be manipulated as follows:

$$(\mathrm{dl}^\mathrm{T} \times \mathbf{H}_\mathrm{s})\cdot\hat{\mathbf{n}} = \hat{\mathbf{n}}\cdot\mathrm{dl}^\mathrm{T} \times \mathbf{H}_\mathrm{s} = \mathrm{dl}^\mathrm{T}\cdot\mathbf{H}_\mathrm{s} \times \hat{\mathbf{n}}$$

Since dl^T is perpendicular to $\hat{\bar{\mathbf{r}}}$, only the same component of $\mathbf{H}_\mathrm{s} \times \hat{\mathbf{n}}$ is relevant, so that we can write this term as

$$\mathrm{dl}^\mathrm{T}\cdot(\mathbf{H}_\mathrm{s} \times \hat{\mathbf{n}})^\mathrm{T} = \mathrm{dl}\cdot(\mathbf{H}_\mathrm{s} \times \hat{\mathbf{r}})^\mathrm{T} \tag{10.11}$$

Similarly,

$$\begin{aligned}\mathbf{n}\cdot\mathbf{E}_\mathrm{s} \times (\mathrm{dl} \times \hat{\bar{\mathbf{r}}}) &= \mathbf{E}_\mathrm{s}\cdot[(\mathrm{dl} \times \hat{\bar{\mathbf{r}}}) \times \hat{\mathbf{n}}] = (\mathrm{dl} \times \hat{\bar{\mathbf{r}}})\cdot\hat{\mathbf{n}} \times \mathbf{E}_\mathrm{s} \\ &= -(\hat{\mathbf{n}} \times \mathbf{E}_\mathrm{s})\cdot(\hat{\bar{\mathbf{r}}} \times \mathrm{dl}) = -\mathrm{dl}\cdot(\hat{\mathbf{n}} \times \mathbf{E}_\mathrm{s}) \times \hat{\bar{\mathbf{r}}} \\ &= -(\hat{\mathbf{n}} \times \mathbf{E}_\mathrm{s}) \times \hat{\bar{\mathbf{r}}}\cdot\mathrm{dl}\end{aligned} \tag{10.12}$$

Substitution of (10.11) and (10.12) into (10.10) yields the result

$$\mathrm{d}\mathbf{E}\cdot\mathrm{d}\mathbf{l} = \frac{\mathrm{j}k}{4\pi\bar{r}}\exp(-\mathrm{j}k\bar{r})[(\hat{\mathbf{n}}\times\mathbf{E}_{\mathrm{s}})\times\hat{\bar{\mathbf{r}}} + \eta(\mathbf{H}_{\mathrm{s}}\times\hat{\mathbf{n}})^{\mathrm{T}}]\cdot\mathrm{d}\mathbf{l}\,\mathrm{d}\sigma$$

Since the orientation of **dl** is entirely arbitrary, it follows that

$$\mathrm{d}\mathbf{E} = \frac{\mathrm{j}k}{4\pi\bar{r}}\exp(-\mathrm{j}k\bar{r})[(\hat{\mathbf{n}}\times\mathbf{E}_{\mathrm{s}})\times\hat{\bar{\mathbf{r}}} + \eta(\mathbf{H}_{\mathrm{s}}\times\hat{\mathbf{n}})^{\mathrm{T}}]\,\mathrm{d}\sigma \tag{10.13}$$

Since

$$(\mathbf{H}_{\mathrm{s}}\times\hat{\mathbf{n}})^{\mathrm{T}} = \hat{\bar{\mathbf{r}}}\times[(\mathbf{H}_{\mathrm{s}}\times\hat{\mathbf{n}})\times\hat{\bar{\mathbf{r}}}] = [(\hat{\mathbf{n}}\times\mathbf{H}_{\mathrm{s}})\times\hat{\bar{\mathbf{r}}}]\times\hat{\bar{\mathbf{r}}},$$

it follows that (10.1) and (10.2) are established.

If $\bar{r}$ is large compared to the linear dimension of S, then $\bar{r}\simeq r-\mathbf{r}'\cdot\hat{\mathbf{r}}$ and $\hat{\bar{\mathbf{r}}}\simeq\hat{\mathbf{r}}$. Equation (10.1) then simplifies to

$$\mathbf{E}(\mathbf{r}) = -\frac{\mathrm{j}k\exp(-\mathrm{j}kr)}{4\pi r}\hat{\mathbf{r}}\times\iint_{\mathrm{S}}[\hat{\mathbf{n}}\times\mathbf{E}_{\mathrm{s}} - \eta\hat{\mathbf{r}}\times(\hat{\mathbf{n}}\times\mathbf{H}_{\mathrm{s}})]\exp(\mathrm{j}k\mathbf{r}'\cdot\hat{\mathbf{r}})\,\mathrm{d}\sigma \tag{10.14}$$

As in the case of current sources, the presence of a flat infinite conducting

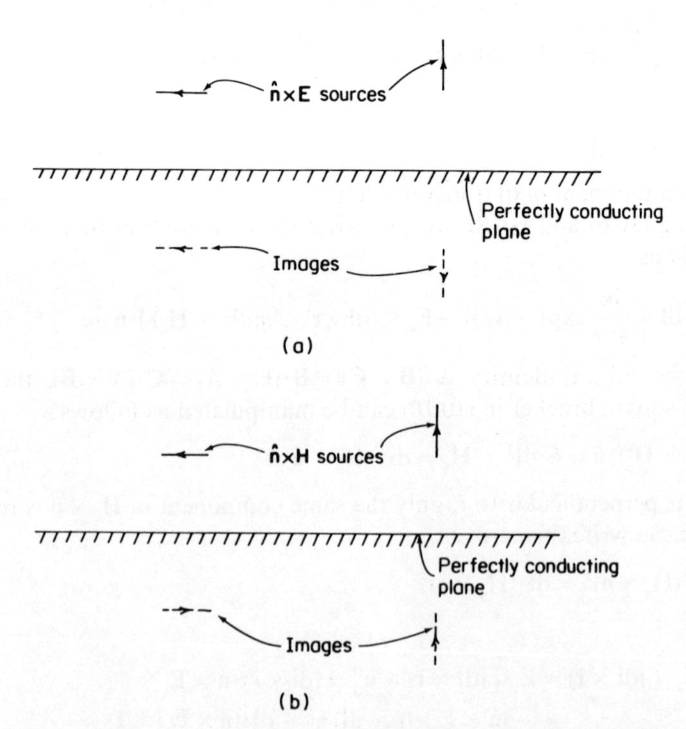

Figure 10.4 (a) Images of a tangential electric field source; (b) image of a tangential magnetic field source

plane can be accounted for by using image sources. For a tangential electric field source ($\hat{\mathbf{n}} \times \mathbf{E}$) parallel (perpendicular) to the conducting plane, the image is in the same (opposite) direction. For a tangential magnetic field source ($\hat{\mathbf{n}} \times \mathbf{H}$) parallel (perpendicular) to the conducting plane, the image is in the opposite (same) direction. These results, which are illustrated in Figure 10.4, can be readily deduced from equation (10.2) (problems 3 and 4 in section 10.7). Although the validity of (10.2) is based on the assumption $k\bar{r} \gg 1$, it turns out that the orientations of the images described above are valid for the general case when the source is placed at any arbitrary distance from the conducting plane.

10.2.2 Uniform Rectangular Aperture

Figure 10.5 shows a rectangular aperture of sides a and b lying in the x–y plane. A distant localized source with centre at the $-z$ axis produces a plane electromagnetic wave across the aperture of the form

$$\mathbf{E}_s = E_0\hat{\mathbf{y}} \tag{10.15}$$

$$\mathbf{H}_s = -(E_0/\eta)\hat{\mathbf{x}} \tag{10.16}$$

For simplicity, let us assume that the rest of the x–y plane is made up of perfectly absorbing material so that the electromagnetic field on it is zero. Then, if we take the closed surface enclosing the source to consist of the x–y plane and a hemisphere of infinite radius to the left of the x–y plane, the electric field in the far zone to the right of the x–y plane is, from (10.14), given by

$$\mathbf{E}(\mathbf{r}) = -\frac{jk}{4\pi r}\exp(-jkr)\hat{\mathbf{r}} \times \int_{-b/2}^{b/2}\int_{-a/2}^{a/2} [\hat{\mathbf{n}} \times \mathbf{E}_s - \eta\hat{\mathbf{r}} \times (\hat{\mathbf{n}} \times \mathbf{H}_s)] \times \exp(jk\mathbf{r}'\cdot\hat{\mathbf{r}})\,\mathrm{d}x'\,\mathrm{d}y'$$

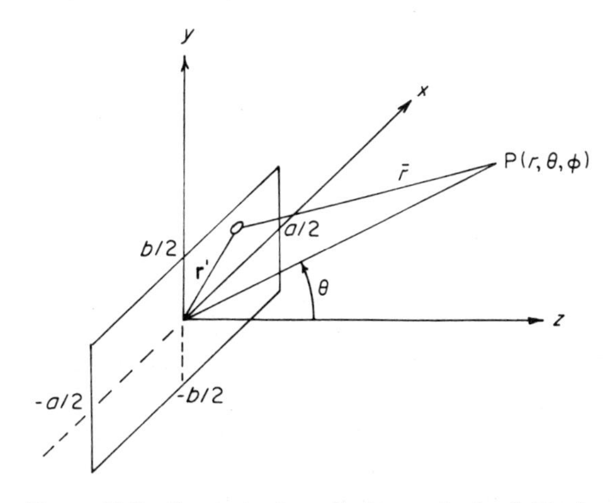

Figure 10.5 Geometry for calculating the far field of a rectangular aperture

$$= -\frac{jk}{4\pi r}\exp(-jkr)\hat{\mathbf{r}} \times [\hat{\mathbf{z}} \times \mathbf{E}_s - \eta\hat{\mathbf{r}} \times (\hat{\mathbf{z}} \times \mathbf{H}_s)] \int_{-b/2}^{b/2}\int_{-a/2}^{a/2} \times \exp(jk\mathbf{r}' \cdot \hat{\mathbf{r}})\, dx'\, dy' \quad (10.17)$$

Since

$$\mathbf{r}' = x'\hat{\mathbf{x}} + y'\hat{\mathbf{y}} \quad (10.18)$$

$$\hat{\mathbf{r}} = \hat{\mathbf{x}} \sin\theta\cos\phi + \hat{\mathbf{y}} \sin\theta\sin\phi + \hat{\mathbf{z}}\cos\theta \quad (10.19)$$

we have

$$\mathbf{r}' \cdot \hat{\mathbf{r}} = x'\sin\theta\cos\phi + y'\sin\theta\sin\phi = x'\cos\chi + y'\cos\chi' \quad (10.20)$$

where

$$\cos\chi = \hat{\mathbf{x}} \cdot \hat{\mathbf{r}} = \sin\theta\cos\phi \quad (10.21)$$

$$\cos\chi' = \hat{\mathbf{y}} \cdot \hat{\mathbf{r}} = \sin\theta\sin\phi \quad (10.22)$$

Using (10.15), (10.16), and (10.18)–(10.22), we obtain

$$\int_{-b/2}^{b/2}\int_{-a/2}^{a/2} \exp(jk\mathbf{r}' \cdot \hat{\mathbf{r}})\, dx'\, dy' = ab\frac{\sin(\frac{1}{2}ka\cos\chi)}{(\frac{1}{2}ka\cos\chi)}\frac{\sin(\frac{1}{2}kb\cos\chi')}{(\frac{1}{2}kb\cos\chi')} \quad (10.23)$$

and

$$-\hat{\mathbf{r}} \times [\hat{\mathbf{z}} \times \mathbf{E}_s - \eta\hat{\mathbf{r}} \times (\hat{\mathbf{z}} \times \mathbf{H}_s)] = E_0(\cos\theta\cos\phi\,\hat{\boldsymbol{\phi}} + \sin\phi\,\hat{\boldsymbol{\theta}}) + E_0(\cos\theta\sin\phi\,\hat{\boldsymbol{\theta}} + \cos\phi\,\hat{\boldsymbol{\phi}}) \quad (10.24)$$

Hence (10.17) becomes

$$\mathbf{E}(\mathbf{r}) = \frac{jk\exp(-jkr)}{4\pi r}E_0 ab\frac{\sin(\frac{1}{2}ka\cos\chi)}{\frac{1}{2}ka\cos\chi}\frac{\sin(\frac{1}{2}kb\cos\chi')}{\frac{1}{2}kb\cos\chi'} \times [\hat{\boldsymbol{\theta}}\sin\phi(1+\cos\theta) + \hat{\boldsymbol{\phi}}\cos\phi(1+\cos\theta)] \quad (10.25)$$

In the y–z ($\phi = 90°$) and x–z ($\phi = 0°$) planes, (10.25) reduces to

$$\mathbf{E}_{yz} = \frac{jk\exp(-jkr)}{4\pi r}E_0 ab(1+\cos\theta)\frac{\sin(\frac{1}{2}kb\sin\theta)}{\frac{1}{2}kb\sin\theta}\hat{\boldsymbol{\theta}} \quad (10.26)$$

$$\mathbf{E}_{xz} = \frac{jk\exp(-jkr)}{4\pi r}E_0 ab(1+\cos\theta)\frac{\sin(\frac{1}{2}ka\sin\theta)}{\frac{1}{2}ka\sin\theta}\hat{\boldsymbol{\phi}} \quad (10.27)$$

It is clear from (10.25)–(10.27) that maximum radiation occurs in the broadside direction ($\theta = 0°$). For large apertures ($ka \gg 1$ and $kb \gg 1$), the width of the main beam will be narrow. To illustrate, in Figure 10.6, $\sin(\frac{1}{2}ka\sin\theta)/(\frac{1}{2}ka\sin\theta)$ is plotted against θ for $ka = 20\pi$, corresponding to a equal to 10 wavelengths across. For this case, the width of the main beam in the x–z plane is about 11°. Putting $\cos\theta \simeq 1$ in (10.26) and (10.27), it can be shown (problem 5 in section 10.7) that the half-power beamwidths are given by

$$(\text{HPBW})_{yz} = 0.886\lambda/b \quad (10.28)$$

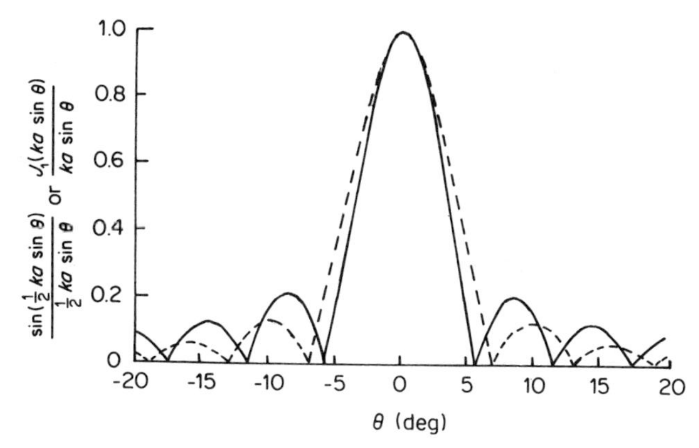

Figure 10.6 The functions $\sin(\frac{1}{2}ka\sin\theta)/(\frac{1}{2}ka\sin\theta)$ (full curve) and $J_1(ka\sin\theta)/(ka\sin\theta)$ (broken curve) plotted against θ for $ka = 20$

$$(\text{HPBW})_{xz} = 0.886\lambda/a \tag{10.29}$$

For $\cos\theta \simeq 1$, (10.25) shows that the polarization of the radiation field is given by $\hat{\boldsymbol{\theta}}\sin\phi + \hat{\boldsymbol{\phi}}\cos\phi$. Since

$$\hat{\mathbf{y}} = \hat{\mathbf{r}}\sin\theta\sin\phi + \hat{\boldsymbol{\theta}}\cos\theta\sin\phi + \hat{\boldsymbol{\phi}}\cos\phi \simeq \hat{\boldsymbol{\theta}}\sin\phi + \hat{\boldsymbol{\phi}}\cos\phi$$

it follows that the polarization is along $\hat{\mathbf{y}}$, namely, the same as that of the aperture electric field.

10.2.3 Uniform Circular Aperture

Consider a circular aperture of radius a in the x–y plane, as shown in Figure 10.7. The aperture is illuminated by a distant source such that the electromagnetic fields in the aperture are uniform in both amplitude and phase and are given by (10.15) and (10.16). The far-zone electric field to the right of the x–y plane is then

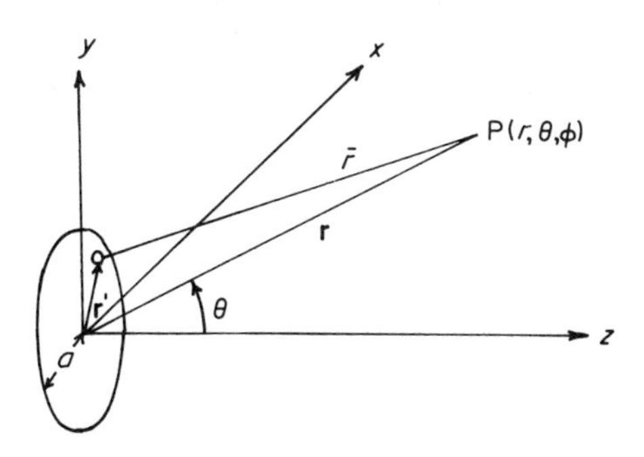

Figure 10.7 Geometry for the calculation of the far field of a circular aperture

given by

$$\mathbf{E}(\mathbf{r}) = -\frac{jk}{4\pi r}\exp(-jkr)[\hat{\boldsymbol{\theta}}\sin\phi(1+\cos\theta) + \hat{\boldsymbol{\phi}}\cos\phi(1+\cos\theta)]$$

$$\times \int_0^{2\pi}\int_0^a \exp(jk\mathbf{r}'\cdot\hat{\mathbf{r}})r'\,dr'\,d\phi' \tag{10.30}$$

The exponent in (10.30) can be written as

$$\begin{aligned} jk\mathbf{r}'\cdot\hat{\mathbf{r}} &= jk(x'\sin\theta\cos\phi + y'\sin\theta\sin\phi) \\ &= jk(r'\cos\phi'\sin\theta\cos\phi + r'\sin\phi'\sin\theta\sin\phi) \\ &= jkr'\sin\theta\cos(\phi-\phi') \end{aligned} \tag{10.31}$$

On making use of the Bessel function identities

$$J_0(x) = \frac{1}{2\pi}\int_0^{2\pi}\exp(jx\cos\alpha)\,d\alpha \tag{10.32}$$

$$\int xJ_0(x)\,dx = xJ_1(x) \tag{10.33}$$

the integral in (10.30) can be evaluated and the following result is obtained:

$$\mathbf{E}(\mathbf{r}) = \frac{jk}{2\pi r}\exp(-jkr)E_0(\pi a^2)\frac{J_1(ka\sin\theta)}{ka\sin\theta}[\hat{\boldsymbol{\theta}}\sin\phi(1+\cos\theta)$$

$$+\hat{\boldsymbol{\phi}}\cos\phi(1+\cos\theta)] \tag{10.34}$$

In the y–z and x–z planes, (10.34) reduces to

$$\mathbf{E}_{yz} = \frac{jk\exp(-jkr)}{2\pi r}E_0(\pi a^2)\frac{J_1(ka\sin\theta)}{ka\sin\theta}(1+\cos\theta)\,\hat{\boldsymbol{\theta}} \tag{10.35}$$

$$\mathbf{E}_{xz} = \frac{jk\exp(-jkr)}{2\pi r}E_0(\pi a^2)\frac{J_1(ka\sin\theta)}{ka\sin\theta}(1+\cos\theta)\,\hat{\boldsymbol{\phi}} \tag{10.36}$$

Equations (10.34)–(10.36) show that maximum radiation occurs in the broadside direction. For large apertures ($ka \gg 1$), the width of the main lobe will be narrow. To illustrate this, the function $J_1(ka\sin\theta)/(ka\sin\theta)$ is plotted against θ for $ka = 20\pi$ in Figure 10.6. For this case in which the aperture diameter is equal to 10 wavelengths, the width of the main beam in both the y–z and x–z planes is about 14°.

In the general case, let the first null occur at $\theta = \theta_0$. Then

$$k(\tfrac{1}{2}D)\sin\theta_0 = 3.83 \qquad (D = 2a)$$

or

$$\theta_0 = \sin^{-1}(1.22\lambda/D) \tag{10.37}$$

For large apertures,

$$\theta_0 \simeq 70/(D/\lambda)\ \text{deg} \tag{10.38}$$

The width of the main beam is

$$2\theta_0 = 140/(D/\lambda)\ \text{deg} \tag{10.39}$$

For $\cos\theta \simeq 1$, the half-power point occurs at $k(\frac{1}{2}D)\sin\theta_{\text{HP}} = 1.6$, so that the half-power beamwidth is

$$\begin{aligned}\text{HPBW} = 2\theta_{\text{HP}} &= 2\sin^{-1}(3.2/kD)\\ &\simeq 58.44(\lambda/D)\ \text{deg}\end{aligned} \tag{10.40}$$

The ratio of the first side lobe to the main lobe is readily shown to be -17.6 dB. As in the case of the large rectangular aperture, the polarization of the far-zone electric field for the large uniformly illuminated circular aperture is the same as that of the aperture electric field.

10.2.4 Directivity of Uniform Apertures

The directivity of an antenna is given in terms of the maximum radiation intensity U_{m} and the power radiated W by

$$D = 4\pi U_{\text{m}}/W \tag{10.41}$$

where

$$U_{\text{m}} = \frac{r^2}{2\eta}(|E_\theta|^2 + |E_\phi|^2)^2_{\text{m}} \tag{10.42}$$

$$W = \iint U \sin\theta \,\mathrm{d}\theta\,\mathrm{d}\phi \tag{10.43}$$

For the rectangular and circular apertures with fields that are uniform in both amplitude and phase, (10.25) and (10.34) show that U_{m} occur in the broadside direction and are given by

$$(U_{\text{m}})_{\text{rec}} = \frac{k^2E_0^2a^2b^2}{8\eta\pi} \tag{10.44}$$

$$(U_{\text{m}})_{\text{cir}} = \frac{k^2E_0^2a^4}{8\eta} \tag{10.45}$$

In obtaining (10.45), we have made use of the formula

$$J_1(x) \to x/2 \qquad \text{as } x \to 0 \tag{10.46}$$

The power radiated can be obtained by performing the integral indicated in (10.43). However, a much similar way is to note that the total power reaching the far field must have passed through the aperture. Since the power density in the aperture is $E_0^2/2\eta$, we have

$$W = \frac{1}{2\eta}\iint E_0^2\,\mathrm{d}\sigma = \frac{E_0^2S_a}{2\eta} \tag{10.47}$$

where S_a is the aperture area, being equal to ab for the rectangular aperture and πa^2 for the circular aperture. Substitution of (10.44), (10.45), and (10.47) in (10.41) yields

$$D_{\text{rec}} = \frac{4\pi}{\lambda^2}(ab) \tag{10.48}$$

$$D_{\text{cir}} = \frac{4\pi}{\lambda^2}(\pi a^2) \tag{10.49}$$

Comparing (10.48) and (10.49) with the general relation between the directivity and the effective area of an antenna $D = (4\pi/\lambda)A_e$, we see that the effective areas for the uniform rectangular and circular apertures are equal to their respective physical areas. This result turns out to hold for an aperture of any shape in which the fields are uniform in both amplitude and phase. If the fields are not uniform in either amplitude or phase, it can be shown that the effective area is less than the physical area A_p. It is convenient to relate the two areas by

$$A_e = \varepsilon_{ap} A_p \qquad 0 \leqslant \varepsilon_{ap} \leqslant 1 \tag{10.50}$$

where ε_{ap} is called the aperture efficiency and is a measure of how efficiently the antenna physical area A_p is utilized. The range of ε_{ap} for aperture antennas is about 30–90%. Horn antennas designed for optimum gain have an aperture efficiency of about 50%. For circular parabolic reflector antennas, it is typically 55%.

10.3 HORN ANTENNA

10.3.1 Types of Horns and Their Uses

A practical class of antennas for which the radiation field can be calculated from the distribution of electromagnetic fields over an aperture is the horn antenna. There are many types of horns for general and specific purposes. We shall limit our discussions to the most common ones, namely, the rectangular sectoral and pyramidal horns.

Consider a rectangular waveguide designed to operate in the dominant TE_{10} mode. If one end of the waveguide is excited by a source and the other is left open, radiation will occur from the open end. However, such a simple structure is seldom used for two reasons. First, since the intrinsic impedance of free space is not the same as the wave impedance of the dominant mode, the abrupt ending of the guide causes some of the incident energy to be reflected back towards the source. Secondly, as the result of section 10.2.2 shows, a large cross-section of the guide is required to obtain a narrow beam. However, if the cross-section is made too large, the waveguide can support higher-order modes. This makes the determination of the aperture field and the calculation of the radiation pattern difficult. Both of these problems can be overcome by flaring the walls of the guide. This allows for a gradual transition from the guided wave to the free-space wave, with the result that reflection is minimized. For a gradual flare, the higher-order modes generated at the throat will be attenuated before they arrive at a position where the dimension is large enough to propagate such modes. Since the flare results in a larger radiating aperture, the beamwidth will be narrower and the directivity higher compared to the case of the open-ended waveguide. The term horn antenna usually refers to this flared structure.

A sectoral horn is a horn where the walls are flared in one dimension but not in the other. If the rectangular waveguide designed for the TE_{10} mode is flared on the broad walls, the result is called an *E*-plane sectoral horn since the flare is in the direction of the electric field. When the narrow walls are flared, the result is called an *H*-plane sectoral horn since the flare is in the plane containing the magnetic field. Flaring both walls results in a pyramidal horn. These are illustrated in Figure 10.8.

The horn antennas are of interest both as directive radiators in themselves and also as primary feed antennas for parabolic reflectors. They are extensively used in the microwave region above about 1 GHz. They are relatively high in gain, wide in bandwidth, rugged, and simple to construct. The sectoral horns are generally used to obtain fan-shaped beams of specified sharpness in the plane containing the flare. In the other plane, the pattern is broad and is essentially the same as that of an open-ended waveguide. The pyramidal horn can be used to obtain specified beamwidths independently in the two principal planes.

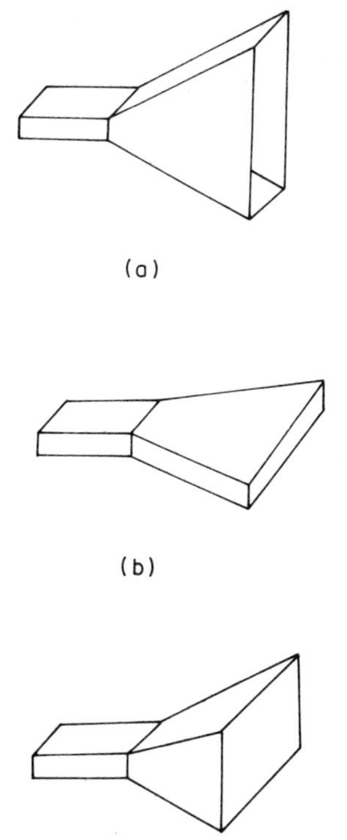

Figure 10.8 (a) An *E*-plane sectoral horn; (b) an *H*-plane sectoral horn; (c) a pyramidal horn

It is also used as a primary gain standard, since its gain can be calculated to within a tenth of a decibel if it is accurately constructed.

In the next three sections, the open-ended rectangular waveguide, the sectoral horns, and the pyramidal horn will be analysed in some detail.

10.3.2 The Open-Ended Rectangular Waveguide

Although in practice the open-ended waveguide is seldom used as a radiator, for the reasons mentioned in section 10.3.1, its analysis serves as a useful preliminary to the analyses of horn antennas. Consider the system shown in Figure 10.9. It will be assumed that the dimensions a and b are such that only the TE_{10} mode can propagate. The current distribution in this structure consists of the current in the probe and the current on the waveguide walls. While calculating the electromagnetic fields to the right of the waveguide mouth directly in terms of these current systems is possible in principle, it is difficult to carry out in practice. A much simpler method based on fields as sources of radiation is as follows.

Let us enclose all the current sources by a closed surface S, consisting of an infinite plane (aperture plane) containing the waveguide mouth and a hemisphere of infinite radius, as shown in Figure 10.9. The fields over the hemisphere are zero since it is infinitely far away. As for the fields over the aperture plane, it will be assumed as a first approximation that they are negligible except in the waveguide mouth. If $\mathbf{E}_t$ and $\mathbf{H}_t$ are the tangential components in the mouth, the radiation field in the far zone to the right of the surface S is, on using (10.14), given by

$$\mathbf{E}(r) = -\frac{jk\exp(-jkr)}{4\pi r}\hat{\mathbf{r}} \times \int_{-b/2}^{b/2}\int_{-a/2}^{a/2} [\hat{\mathbf{z}} \times \mathbf{E}_t - \eta\hat{\mathbf{r}} \times (\hat{\mathbf{z}} \times \mathbf{H}_t)]\exp(jkr'\cdot\hat{\mathbf{r}})d\sigma \tag{10.51}$$

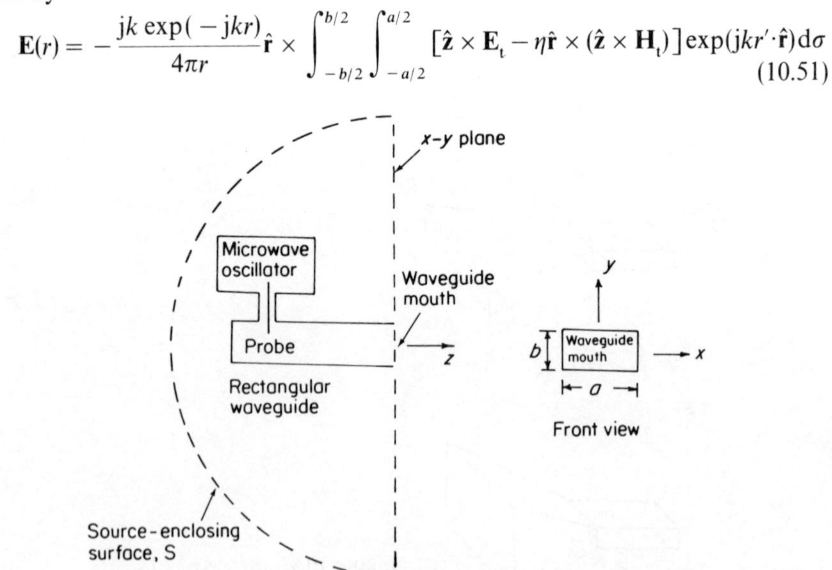

Figure 10.9 Geometry of an open-ended rectangular waveguide

Each of the quantities $\mathbf{E}_t$ and $\mathbf{H}_t$ consists of an incident and a reflected wave of the TE_{10} mode as well as higher-order modes excited by the discontinuity. For the TE_{10} mode, we have

$$\mathbf{E}_{ti} = E_i \cos(\pi x'/a) \exp(-jk_{10}z)\hat{\mathbf{y}} \tag{10.52}$$

$$\mathbf{H}_{ti} = -\frac{E_i}{Z_{10}} \cos(\pi x'/a) \exp(-jk_{10}z)\hat{\mathbf{x}} \tag{10.53}$$

$$\mathbf{E}_{tr} = \Gamma E_i \cos(\pi x'/a) \exp(jk_{10}z)\hat{\mathbf{y}} \tag{10.54}$$

$$\mathbf{H}_{tr} = \frac{\Gamma E_i}{Z_{10}} \cos(\pi x'/a) \exp(jk_{10}z)\hat{\mathbf{x}} \tag{10.55}$$

where Z_{10}, k_{10}, and Γ are the wave impedance, propagation constant, and reflection coefficient of the TE_{10} mode. The quantities Z_{10} and k_{10} are given by (e.g. Jordan and Balmain, 1968)

$$Z_{10} = \eta[1 - (\lambda/2a)^2]^{-1/2} \tag{10.56}$$

$$k_{10} = k[1 - (\lambda/2a)^2]^{1/2} \tag{10.57}$$

The value of Γ is typically 0.3, corresponding to a VSWR of about 1.5.

Assuming that the incident and reflected TE_{10} modes comprise the bulk of the fields in the mouth, we have, on setting $z = 0$,

$$\mathbf{E}_t = E_i(1 + \Gamma) \cos(\pi x'/a)\hat{\mathbf{y}} = E_0 \cos(\pi x'/a)\hat{\mathbf{y}} \tag{10.58}$$

$$\mathbf{H}_t = -\frac{E_0}{Z_1} \cos(\pi x'/a)\hat{\mathbf{x}} \tag{10.59}$$

where

$$Z_1 = Z_{10}\left(\frac{1 + \Gamma}{1 - \Gamma}\right) \tag{10.60}$$

On using (10.58) and (10.59) and noting that

$$\begin{aligned}\int_{-a/2}^{a/2} &\cos(\pi x'/a) \exp(jkx' \cos\chi)\,dx' \\ &= \frac{1}{2}\int_{-a/2}^{a/2} [\exp(j\pi x'/a) + \exp(-j\pi x'/a)] \exp(jkx' \cos\chi)\,dx' \\ &= \frac{(2a/\pi)\cos(\frac{1}{2}ka\cos\chi)}{1 - (k^2a^2/\pi^2)\cos^2\chi}\end{aligned} \tag{10.61}$$

$$\int_{-b/2}^{b/2} \exp(jky' \cos\chi')\,dy' = b\frac{\sin(\frac{1}{2}kb\cos\chi')}{\frac{1}{2}kb\cos\chi'} \tag{10.62}$$

$$\hat{\mathbf{x}} = \hat{\mathbf{r}} \sin\theta \cos\phi + \hat{\boldsymbol{\theta}} \cos\theta \cos\phi - \hat{\boldsymbol{\phi}} \sin\phi \tag{10.63}$$

we have

$$\hat{\mathbf{r}} \times \int_{-b/2}^{b/2} \int_{-a/2}^{a/2} (\hat{\mathbf{z}} \times \mathbf{E}_t) \exp(jk\mathbf{r}' \cdot \hat{\mathbf{r}})\, d\sigma = -E_0 (\cos\theta \cos\phi\, \hat{\boldsymbol{\phi}} + \sin\phi\, \hat{\boldsymbol{\theta}}) F \tag{10.64}$$

$$\hat{\mathbf{r}} \times \int_{-b/2}^{b/2} \int_{-a/2}^{a/2} [-\eta \hat{\mathbf{r}} \times (\hat{\mathbf{z}} \times \mathbf{H}_t)] \exp(jkr' \cdot \hat{\mathbf{r}})\, d\sigma = -\frac{\eta E_0 F}{Z_1} [\cos\theta \sin\phi\, \hat{\theta} + \cos\phi\, \hat{\boldsymbol{\phi}}] \tag{10.65}$$

where

$$F = 2ab\pi \frac{\cos(\frac{1}{2}ka\cos\chi)}{(\pi^2 - k^2a^2\cos^2\chi)} \frac{\sin(\frac{1}{2}kb\cos\chi')}{(\frac{1}{2}kb\cos\chi')} \tag{10.66}$$

Substitution of (10.64) and (10.65) into (10.51) yields

$$\mathbf{E}(\mathbf{r}) = \frac{jk\exp(-jkr)}{4\pi r} E_0 F \left\{ \hat{\boldsymbol{\theta}} \sin\phi \left(1 + \cos\theta \frac{\eta}{Z_1} \right) + \hat{\boldsymbol{\phi}} \left[\cos\theta\cos\phi + \frac{\eta}{Z_1}\cos\phi \right] \right\} \tag{10.67}$$

The E- and H-planes correspond to the y–z ($\phi = 90°$) and x–z ($\phi = 0°$) planes respectively, in which (10.67) reduces to

$$\mathbf{E}_{yz} = \frac{jk\exp(-jkr)}{2\pi^2 r} E_0 ab \frac{\sin[(\pi b/\lambda)\sin\theta]}{(\pi b/\lambda)\sin\theta} \left(1 + \cos\theta \frac{\eta}{Z_1} \right) \hat{\boldsymbol{\theta}} \tag{10.68}$$

$$\mathbf{E}_{xz} = \frac{jk\exp(-jkr)}{2r} E_0 ab \frac{\cos[(a\pi/\lambda)\sin\theta]}{\pi^2 - (4\pi^2a^2/\lambda^2)\sin^2\theta} \left[\cos\theta + \frac{\eta}{Z_1} \right] \hat{\boldsymbol{\phi}} \tag{10.69}$$

The patterns in the H- and E-planes are governed mainly by the factors

$$\frac{\cos[(\pi a/\lambda)\sin\theta]}{\pi^2 - (4a^2\pi^2/\lambda^2)\sin^2\theta}$$

and

$$\frac{\sin[(\pi b/\lambda)\sin\theta]}{(\pi b/\lambda)\sin\theta}$$

respectively. These functions are plotted in Figure 10.10. For a rectangular waveguide with $a = 2b$, the criterion for single-mode operation is $a < \lambda < 2a$. Thus a and b are fractions of wavelength and the pattern is quite broad.

The directivity can be obtained in the manner similar to section 10.2.4. The result is

$$D = \frac{8Z_1}{\eta\pi\lambda^2} \left(1 + \frac{\eta}{Z_1} \right)^2 ab \tag{10.70}$$

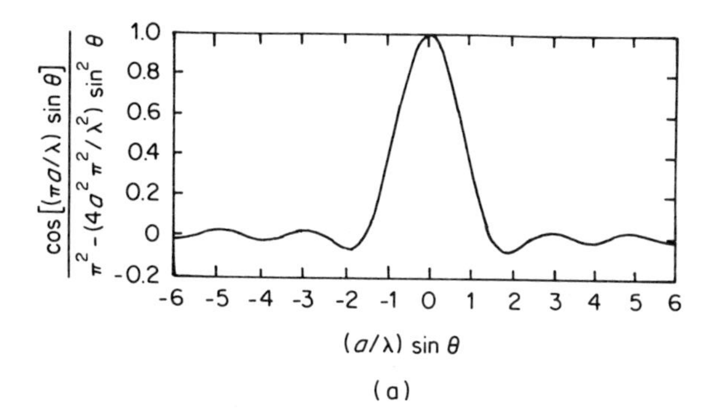

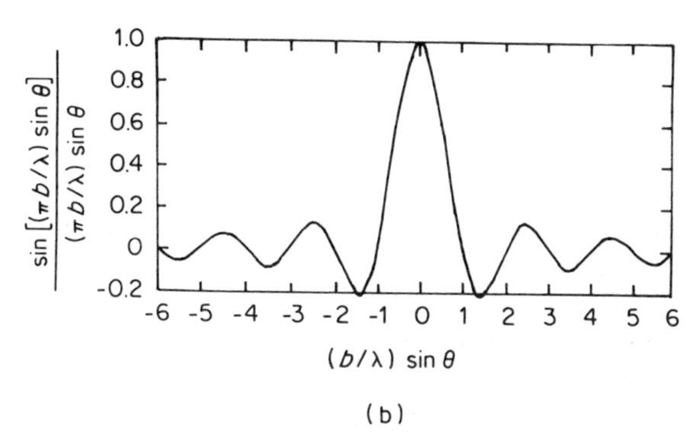

Figure 10.10 The main factors for the principal plane patterns of an open-ended rectangular waveguide

10.3.3 The *E*-Plane Sectoral Horn

If a rectangular waveguide designed for dominant TE_{10} mode operation is flared on the broad walls, an *E*-plane sectoral horn is obtained. The geometry of such a horn is shown in Figure 10.11. The mouth of the horn has dimensions $a \times B$, where B may be several wavelengths across. This results in a fan-shaped beam which is narrow in the plane containing the flare but broad in the plane perpendicular to the flare.

As in the case of the open-ended waveguide, the radiation from the horn can be calculated from a knowledge of the fields in the x–y plane containing the mouth of the horn. As a first approximation, the fields outside the mouth are assumed to be negligible. To find the fields in the mouth, it is necessary to solve Maxwell's equations in the flared section, subject to the boundary conditions of zero tangential **E** and zero normal **H** at the walls. The details are somewhat involved and will not be given here. It is found that, as in the case of the rectangular guide, the flared section can in general support an infinite number of modes, each with a cutoff frequency. The number of propagating modes is determined by the minimum dimension of the flared section, namely b. For the

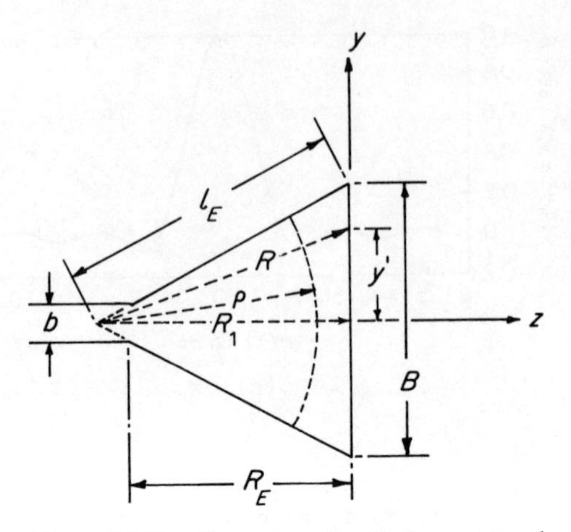

Figure 10.11 Geometry of an E-plane sectoral horn

case in which the horn is driven by a rectangular waveguide with a single TE_{10} mode, it turns out that the size of b is such that only the lowest mode can propagate. The surfaces of constant phase of this mode are the surfaces for which ρ = constant, where ρ is the distance measured from the point of intersection of the projection of the walls, as indicated in Figure 10.11. The transverse fields at the mouth of the horn are approximately given by

$$\mathbf{E}_t = E_0 \cos(\pi x'/a) \exp[-jf(y')]\hat{\mathbf{y}} \tag{10.71}$$

$$\mathbf{H}_t = -(E_0/\eta) \cos(\pi x'/a) \exp[-jf(y')]\hat{\mathbf{x}} \tag{10.72}$$

where $f(y')$ represents the phase variation due to the curvature of the wavefront. Taking the zero of phase at the centre of the mouth, we have

$$\exp[-jf(y')] = \exp[-jk(R - R_1)] = \exp[-jkR_1(R/R_1 - 1)] \tag{10.73}$$

From the geometry,

$$R = (R_1^2 + y'^2)^{1/2} = R_1[1 + (y'/R_1)^2]^{1/2}$$
$$\simeq R_1[1 + \tfrac{1}{2}(y'/R_1)^2] \qquad \text{for} \quad B/2 \ll R_1$$

Hence

$$\exp[-jf(y')] \simeq \exp[-j(ky'^2/2R_1)] \qquad \text{for} \quad B/2 \ll R_1 \tag{10.74}$$

and

$$\mathbf{E}_t = E_0 \cos(\pi x'/a) \exp[-j(ky'^2/2R_1)]\hat{y} \tag{10.75}$$

$$\mathbf{H}_t = -(E_0/\eta) \cos(\pi x'/a) \exp[-j(ky'^2/2R_1)]\hat{x} \tag{10.76}$$

Thus, under the assumption $B/2 \ll R_1$, the mouth of the sectoral horn can be regarded as having approximately a plane wave field distribution with a quadratic phase error.

The far-field pattern resulting from this aperture field is found by using (10.75) and (10.76) in (10.51), where b is replaced by B. The electric field on the z-axis, where $\mathbf{r}'\cdot\hat{\mathbf{r}} = 0$, is particularly easy to obtain. For this case

$$\begin{aligned}\int_{-B/2}^{B/2}\int_{-a/2}^{a/2} (\hat{\mathbf{z}}\times\mathbf{E}_t)\exp(jk\mathbf{r}'\cdot\hat{\mathbf{r}})\,d\sigma &= \int_{-B/2}^{B/2}\int_{-a/2}^{a/2}(\hat{\mathbf{z}}\times\mathbf{E}_t)\,d\sigma \\ &= -\hat{\mathbf{x}}E_0\int_{-B/2}^{B/2}\exp(-jky'^2/2R_1)\,dy' \\ &\quad\times\int_{-a/2}^{a/2}\cos(\pi x'/a)\,dx' \\ &= -\hat{\mathbf{x}}\frac{4aE_0}{\pi}\int_0^{B/2}\exp(-jky'^2/2R_1)\,dy' \end{aligned} \tag{10.77}$$

On letting

$$u^2 = ky'^2/R_1\pi \tag{10.78}$$

we have

$$dy' = \sqrt{(\lambda R_1/2)}\,du \tag{10.79}$$

and

$$\int_{-B/2}^{B/2}\int_{-a/2}^{a/2}(\hat{\mathbf{z}}\times\mathbf{E}_t)\,d\sigma = -\hat{\mathbf{x}}\frac{4aE_0}{\pi}\sqrt{\left(\frac{\lambda R_1}{2}\right)}\int_0^{(B/2)\sqrt{(2/\lambda R_1)}}\exp(-j\pi u^2/2)\,du \tag{10.80}$$

Similarly, for points on the z-axis,

$$\begin{aligned}&-\int_{-B/2}^{B/2}\int_{-a/2}^{a/2}\eta\hat{\mathbf{r}}\times(\hat{\mathbf{z}}\times\mathbf{H}_t)\exp(jk\mathbf{r}'\cdot\hat{\mathbf{r}})\,d\sigma \\ &= -\int_{-B/2}^{B/2}\int_{-a/2}^{a/2}\eta\hat{\mathbf{z}}\times(\hat{\mathbf{z}}\times\mathbf{H}_t)\,d\sigma \\ &= -\hat{\mathbf{x}}E_0\frac{4a}{\pi}\sqrt{\left(\frac{\lambda R_1}{2}\right)}\int_0^{(B/2)\sqrt{(2/\lambda R_1)}}\exp(-j\tfrac{1}{2}\pi u^2)\,du\end{aligned} \tag{10.81}$$

On using (10.80) and (10.81) in (10.51), we arrive at the following expression for the electric field on the axis:

$$\mathbf{E}_{\text{axis}} = \frac{2jkE_0a}{\pi^2 r}\exp(-jkr)\sqrt{\left(\frac{\lambda R_1}{2}\right)}[C(v)-jS(v)]_{v=(B/2)\sqrt{(2/\lambda R_1)}}\hat{\mathbf{y}} \tag{10.82}$$

where $C(v)$ and $S(v)$ are the Fresnel integrals defined by

$$C(v) = \int_0^v \cos(\tfrac{1}{2}\pi u^2)\, du \tag{10.83}$$

$$S(v) = \int_0^v \sin(\tfrac{1}{2}\pi u^2)\, du \tag{10.84}$$

The directivity of the E-plane sectoral horn can be obtained in the manner similar to section 10.2.4. The result is

$$D_E = \frac{64aR_1}{\pi\lambda B}\left[C^2(v) + S^2(v)\right]_{v=(B/2)\sqrt{(2/\lambda R_1)}} \tag{10.85}$$

At set of universal directivity curves for the E-plane sectoral horn can be obtained by plotting $(D_E\lambda/a)$ versus B/λ, with R_1/λ as parameter. These are shown in Figure 10.12. It is seen that, for a given axial length R_1, there is an optimum value for B, corresponding to the peak of these curves. If the values of B/λ corresponding to the peaks are plotted against R_1/λ, it is found that the curve passing these points can be described by the equation

$$B = \sqrt{(2\lambda R_1)} \qquad \text{(optimum)} \tag{10.86}$$

The fact that an optimum condition exists is due to the quadratic phase deviation. If there is no phase deviation, the directivity would increase as B, and hence the aperture area, increases. However, for a given R_1, the phase deviation

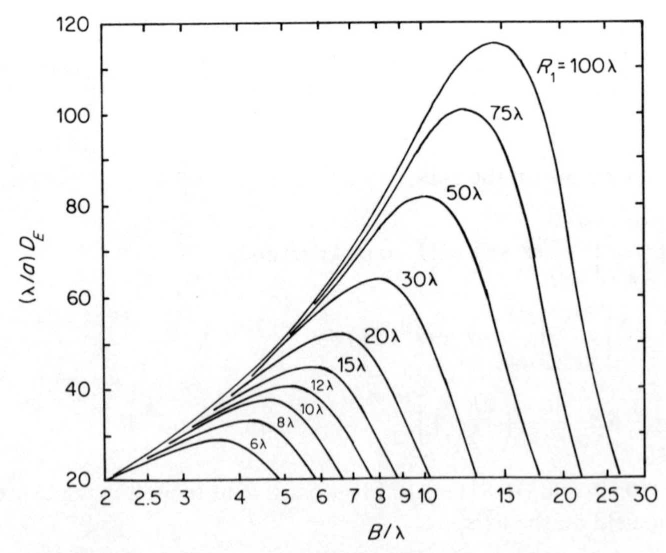

Figure 10.12 Universal directivity curves of an E-plane sectoral horn (Source: Schelkuneff and Friis (1952), *Antennas, Theory and Practice*. Reproduced by permission of John Wiley and Sons Inc. and Bell Laboratories, Copyright 1952)

also increases as B increases. This departure from the in-phase condition leads to cancellations in the far field, resulting in a decrease in directivity. The optimum value of B is reached when these two opposing factors balance each other.

For points not on the z-axis, the integrals in (10.51) can be similarly expressed in terms of the Fresnel integrals. The result is

$$\mathbf{E}(\mathbf{r}) = jkE_0\sqrt{\left(\frac{\lambda R_1}{2}\right)}\frac{a\exp(-jkr)}{\pi^2 r}\exp[j(\tfrac{1}{2}kR_1)(\cos\chi')^2]$$

$$\times(\hat{\boldsymbol{\theta}}\sin\phi + \hat{\boldsymbol{\phi}}\cos\phi)\left(\frac{1+\cos\theta}{2}\right)\frac{\cos(\frac{1}{2}ka\cos\chi)}{1-[(ka/\pi)^2\cos\chi]^2}$$

$$\times[C(v_2) - jS(v_2) - C(v_1) + jS(v_1)] \qquad (10.87)$$

where

$$v_1 = \sqrt{\left(\frac{k}{\pi R_1}\right)}(-\tfrac{1}{2}B - R_1\cos\chi') \qquad v_2 = \sqrt{\left(\frac{k}{\pi R_1}\right)}(+\tfrac{1}{2}B - R_1\cos\chi) \qquad (10.88)$$

The normalized patterns in the H ($\phi = 0°$) and E ($\phi = 90°$) planes are given by

$$F_H(\theta) = \left(\frac{1+\cos\theta}{2}\right)\frac{\cos(\frac{1}{2}ka\sin\theta)}{1-(\frac{1}{2}ka\sin\theta)^2} \qquad (10.89)$$

$$|F_E(\theta)| = \left(\frac{1+\cos\theta}{2}\right)\left\{\frac{[C(v_4)-C(v_3)]^2+[S(v_4)-S(v_3)]^2}{4[C^2(2\sqrt{s})+S^2(2\sqrt{s})]}\right\}^{1/2} \qquad (10.90)$$

where

$$v_3 = 2\sqrt{s}\left[-1-\frac{1}{4s}\left(\frac{B\sin\theta}{\lambda}\right)\right] \qquad v_4 = 2\sqrt{s}\left[+1-\frac{1}{4s}\left(\frac{B\sin\theta}{\lambda}\right)\right] \qquad (10.91)$$

$$s = B^2/(8\lambda R_1) = \tfrac{1}{8}(B/\lambda)^2(\lambda/R_1) \qquad (10.92)$$

The quantity s defined by (10.92) is equal to the maximum phase error across the aperture divided by 2π. This follows since the phase distribution in (10.74) is $\delta = (k/2R_1)y'^2$. The maximum phase error occurs at $y' = \pm B/2$, where it is equal to $\delta_{max} = 2\pi B^2/(8\lambda R_1)$. Using (10.86) in (10.92), we obtain the value of s for optimum conditions as

$$s = 1/4 \qquad \text{(optimum)} \qquad (10.93)$$

A set of universal curves for the patterns of the E-plane horn in the two principal planes are shown in Figure 10.13. The factor $\frac{1}{2}(1+\cos\theta)$ is not included in these plots. Its effect can be included by adding $20\log[\frac{1}{2}(1+\cos\theta)]$ to the

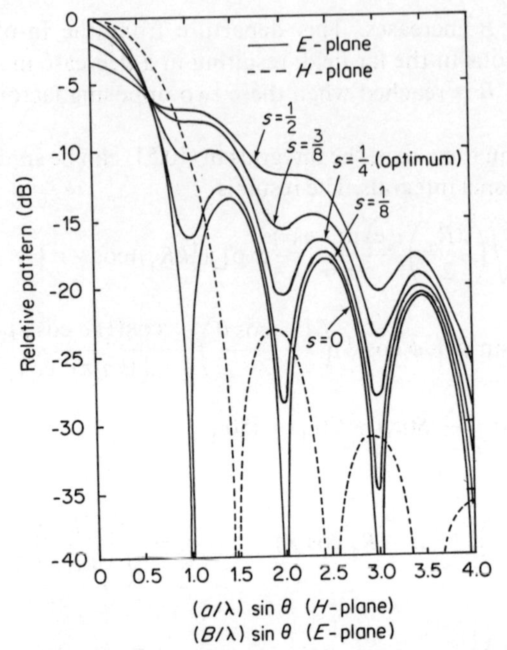

Figure 10.13 Principal plane radiation patterns of the E-plane sectoral horn. The factor $\frac{1}{2}(1 + \cos \theta)$ is not included (Source: Wolff (1966), *Antenna Analysis*. Reproduced by permission of E. A. Wolff)

values obtained from the curves. Usually this is small and can be neglected.

In Figure 10.13, the E-plane curves (full curves) for various values of s are not normalized to 0 dB at the maximum point, but rather are given relative to the zero phase error case ($s = 0$). For a horn designed for optimum performance, the half-power beamwidth in the E-plane can be obtained from the $s = \frac{1}{4}$ curve in Figure 10.13. The point which is 3 dB down from the maximum occurs for $(B/\lambda) \sin \theta = 0.47$, so that

$$\begin{aligned}(\mathrm{HPBW})_E &= 2 \sin^{-1}(0.47\lambda/B) \\ &\simeq 0.94\lambda/B = 54\lambda/B \text{ deg} \qquad \text{for } B \gg \lambda \end{aligned} \tag{10.94}$$

10.3.4 The *H*-Plane Sectoral Horn

If a rectangular waveguide designed for dominant mode operation is flared on the narrow walls, an H-plane sectoral horn is obtained. The geometry of such a horn is shown in Figure 10.14. The mouth of the horn has dimensions $A \times b$, where A may be several wavelengths across. This results in a fan-shaped beam which is narrow in the plane containing the flare but broad in the plane perpendicular to the flare.

As in the case of the E-plane sectoral horn, the fields across the mouth can be obtained by solving Maxwell's equations in the flared portion, subject to the

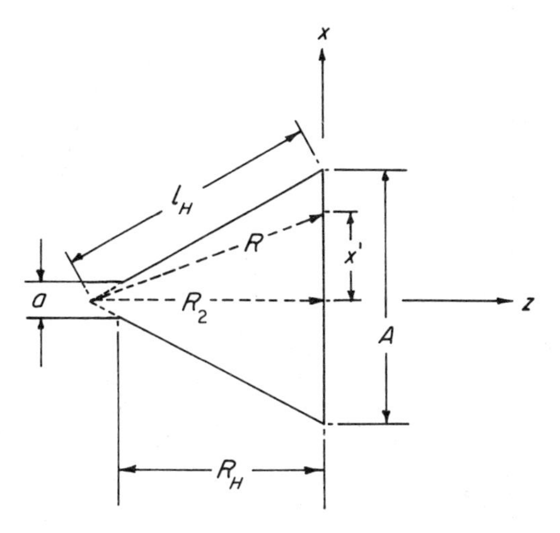

Figure 10.14 Geometry of an H-plane sectoral horn

boundary conditions of zero tangential **E** and zero normal **H** at the walls. For the case in which horn is driven by a rectangular waveguide with a single TE_{10} mode, only the lowest mode can propagate. The transverse fields of this mode at the mouth are approximately given by

$$\mathbf{E}_t = E_0 \cos(\pi x'/A)\exp[-jk(R - R_2)]\hat{\mathbf{y}} \tag{10.95}$$

$$\mathbf{H}_t = -(E_0/\eta)\cos(\pi x'/A)\exp[-jk(R - R_2)]\hat{\mathbf{x}} \tag{10.96}$$

For $A/2 \ll R_2$, $(R - R_2) \simeq \frac{1}{2}x'^2 R_1$ and

$$\mathbf{E}_t = E_0 \cos(\pi x'/A)\exp[-jkx'^2/(2R_2)]\hat{\mathbf{y}} \tag{10.97}$$

$$\mathbf{H}_t = -(E_0/\eta)\cos(\pi x'a)\exp[-jkx'^2/(2R_2)]\hat{\mathbf{x}} \tag{10.98}$$

The far-field pattern resulting from this aperture field is found by using (10.97) and (10.98) in (10.51), where a is replaced by A. To handle the integrals, the cosine function in (10.97) and (10.98) is changed into exponential, and the results can again be expressed in terms of the Fresnel integrals. On account of the change of variables required, the formulae are more complicated than those of the E-plane sectoral horn and will not be given here. However, the universal curves for directivity and principal plane patterns are shown in Figure 10.15 and 10.16 respectively. Reference to Figure 10.15 shows that, for a given axial length R_2, there is an optimum aperture width A corresponding to the peak of the appropriate curve. If values of A/λ corresponding to these peaks are plotted against R_2/λ, it is found that the curve passing through these points can be described by the equation

$$A\sqrt{(3\lambda R_2)} \qquad \text{(optimum)} \tag{10.99}$$

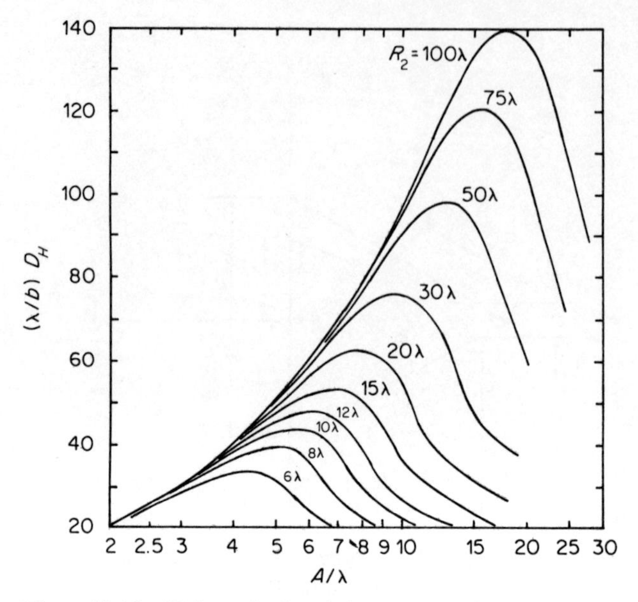

Figure 10.15 Universal directivity curves of an H-plane sectoral horn (Source: Schelkunoff and H. Friis (1952), *Antennas, Theory and Practice.* Reproduced by permission of John Wiley and Sons Inc. and Bell Laboratories, Copyright 1952)

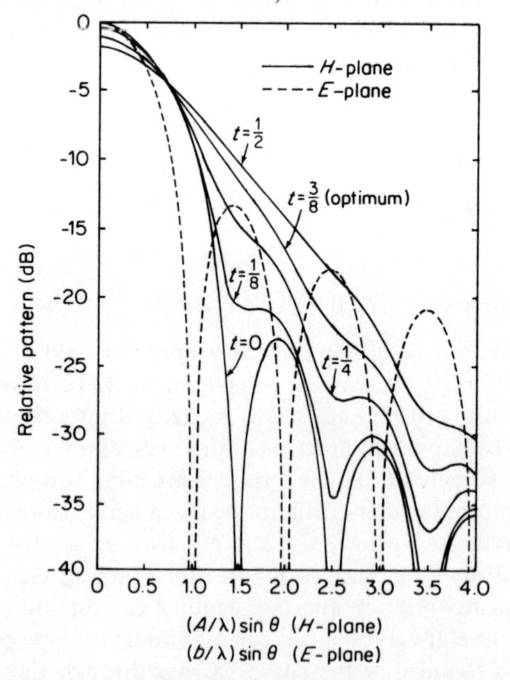

Figure 10.16 Principal plane radiation patterns of the H-plane sectoral horn. The factor $\frac{1}{2}(1+\cos\theta)$ is not included (Source: Wolff (1966), *Antenna Analysis.* Reproduced by permission of E. A. Wolff)

As in the case of the E-plane sectoral horn, the patterns for the principal planes of an H-plane sectoral horn are found to contain the factor $\frac{1}{2}(1 + \cos\theta)$. This factor is not included in the curves for the relative patterns shown in Figure 10.16. The parameter t is the maximum phase error divided by 2π and is given by

$$t = A^2/(8\lambda R_2) = \tfrac{1}{8}(A/\lambda)^2(\lambda/R_2) \tag{10.100}$$

For an optimum horn, A is given by (10.99) and hence

$$t = 3/8 \tag{10.101}$$

The half-power beamwidth in the H-plane for the optimum horn can be determined from the $t = \frac{3}{8}$ curve in Figure 10.1. For $A \gg \lambda$, it is given by

$$(\text{HPBW})_H \simeq 78(\lambda/A)\ \text{deg} \tag{10.102}$$

10.3.5 The Pyramidal Horn

If a rectangular waveguide designed for dominant mode operation is flared on both walls, a pyramidal horn is obtained. The geometry of such a horn is shown in Figure 10.17. The mouth of the horn is of dimensions $A \times B$, where A and B may be several wavelengths across. This leads to a pencil beam which is narrow in both principal planes.

Since the walls of the pyramidal horn do not fit the coordinate surfaces of any separable coordinate system, it is not possible to solve rigorously for the modes associated with this geometry. The usual practice is to assume that the fields in the mouth are the same as the E-plane sectoral horn in the y-direction and the same as the H-plane sectoral horn in the x-direction.

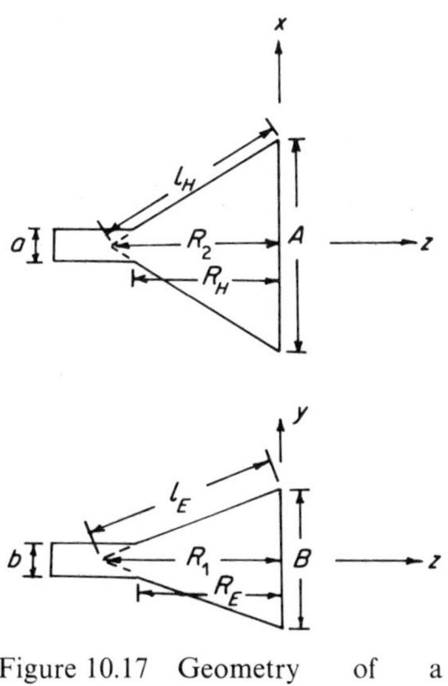

Figure 10.17 Geometry of a pyramidal horn

Thus

$$\mathbf{E}_t = E_0 \cos\left(\frac{\pi x'}{A}\right) \exp\left[-j\tfrac{1}{2}k\left(\frac{x'^2}{R_2} + \frac{y'^2}{R_1}\right)\right]\hat{\mathbf{y}} \tag{10.103}$$

$$\mathbf{H}_t = -\frac{E_0}{\eta} \cos\left(\frac{\pi x'}{A}\right) \exp\left[-j\tfrac{1}{2}k\left(\frac{x'^2}{R_2} + \frac{y'^2}{R_1}\right)\right]\hat{\mathbf{x}} \tag{10.104}$$

When the above aperture field is used to calculate the pattern, one obtains the result that the E-plane pattern is the same as that of an E-plane sectoral horn and the H-plane pattern is the same as that of an H-plane sectoral horn. Thus the full curves of Figures 10.13 and 10.16 can be used to obtain the principal plane patterns of the pyramidal horn.

The directivity of the pyramidal horn is found to be given by

$$D_P = \tfrac{1}{32}\pi(\lambda D_E/A)(\lambda D_H/B) \tag{10.105}$$

The terms in parentheses in equation (10.105) can be obtained from the curves of Figures 10.12 and 10.15, where the ordinates are to be interpreted as $\lambda D_E/A$ and $\lambda D_H/B$ respectively.

10.4 SLOT ANTENNA

10.4.1 Slots in Large Ground Plane

Slot antennas are used extensively in fast-moving vehicles such as aircraft since they do not protrude from the surface of the vehicle. The simplest form is a narrow slot cut in a large conducting plane and fed by a generator via a transmission line connected to two opposite points at the centre of the slot, as shown in Figure 10.18(a). The width w is much smaller than a wavelength and the length L is usually about a half-wavelength. The fields across the slot can be estimated as follows. Consider first a transmission line in the form of a pair of conducting half-planes lying in the x–z plane separated by the width w, as shown in Figure 10.18(b). When driven by a source at $z = 0$, the voltage across the line will be a travelling transverse electromagnetic wave propagating in the positive and negative z-directions, since the line is infinitely long. Suppose we short-circuit the line at $z = L/2$ and $z = -L/2$ and fill the space in the x–z plane beyond the shorts with metal, then the slot antenna of Figure 10.18(a) is obtained. The voltage distribution and hence the electromagnetic fields in the z-direction will be that of a shorted transmission line, namely, a standing wave of the form

$$\mathbf{E} = \hat{\mathbf{x}}\frac{V_m}{w}\sin\left[k(\tfrac{1}{2}L - |z'|)\right] \tag{10.106}$$

$$\mathbf{H} = \hat{\mathbf{y}}\frac{V_m}{w\eta}\sin\left[k(\tfrac{1}{2}L - |z'|)\right] \tag{10.107}$$

where V_m is the peak voltage.

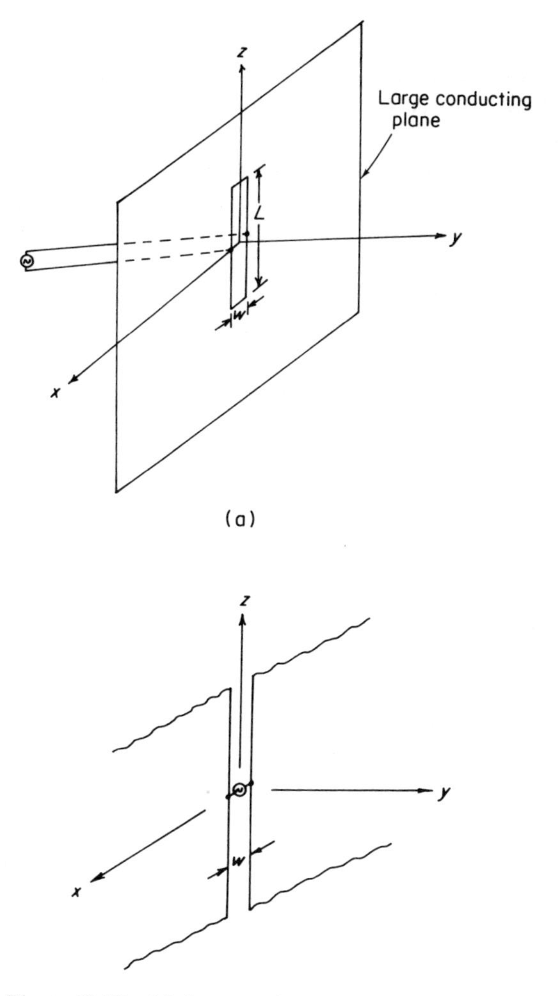

Figure 10.18 (a) A centre-fed slot antenna in a large ground plane; (b) a transmission line consisting of two half-plane conducting sheets

If the generator and the feed line are in $y < 0$, the fields in $y > 0$ can be determined as follows. First, the large ground plane is modelled by an infinite ground plane. Next consider a closed surface S consisting of a rectangular boss of size $w \times L$ just in front of the slot (S_1), an infinite plane inside the ground plane (S_2), and an infinite hemisphere in $y < 0$. The only non-zero tangential fields on this surface will be the electric field on S_1 given by

$$\hat{\mathbf{n}} \times \mathbf{E}_s = \hat{\mathbf{y}} \times \mathbf{E}_s = \frac{V_m}{w} \sin\left[k(\tfrac{1}{2}L - |z'|)\right]\hat{\mathbf{y}} \times \hat{\mathbf{x}} = -\frac{V_m}{w} \sin\left[k(\tfrac{1}{2}L - |z'|)\right]\hat{\mathbf{z}} \tag{10.108}$$

The fields in $y > 0$ can be computed from the $(\hat{\mathbf{n}} \times \mathbf{E}_s)$ source on S_1 and the actual currents induced on the ground plane. Alternatively, they can be calculated using an image source to account for the ground plane. This is possible because, as the fields in the interior of S ($y < 0$) are zero, the slot can be closed off with conductor, causing the ground plane to become an infinite plane sheet with no hole. Hence the fields in $y > 0$ can be computed from the $(\hat{\mathbf{n}} \times \mathbf{E}_s)$ source on S_1 plus its image. As discussed in Section 10.2.1, the image of a tangential electric field source is co-directed. Thus, as the boss is lowered so that S_1 approaches infinitesimally close to the surface of the ground plane, $(\hat{\mathbf{n}} \times \mathbf{E}_s)$ and its image add. The far-zone electric field in $y > 0$ is therefore given by

$$\begin{aligned}
\mathbf{E}(\mathbf{r}) &= -\frac{jk\exp(-jkr)}{4\pi r}\hat{\mathbf{r}} \times \iint (2\hat{\mathbf{n}} \times \mathbf{E}_s)\exp(jk\mathbf{r}'\cdot\hat{\mathbf{r}})\,d\sigma \\
&= \frac{jk\exp(-jkr)}{4\pi r}\frac{V_m}{w}\hat{\mathbf{r}} \times \hat{\mathbf{z}} \\
&\quad \times \int_{-L/2}^{L/2}\int_{-w/2}^{w/2} 2\sin[(\tfrac{1}{2}L - |z'|]\exp(jk\mathbf{r}'\cdot\hat{\mathbf{r}})\,dx'\,dz' \\
&= \frac{jk\exp(-jkr)}{2\pi r}\frac{V_m}{w}\hat{\mathbf{r}} \times \hat{\mathbf{z}}\int_{-w/2}^{w/2}\exp(jkx'\cos\chi)\,dx' \\
&\quad \times \int_{-L/2}^{L/2}\exp(jkz'\cos\theta)\sin[k(\tfrac{1}{2}L - |z'|)]\,dz'
\end{aligned} \tag{10.109}$$

Since

$$\int_{-w/2}^{w/2}\exp(jkx'\cos\chi)\,dx' = \frac{2\sin(\tfrac{1}{2}kw\cos\chi)}{k\cos\chi} \simeq w \qquad \text{for } w/\lambda \ll 1$$

$$\hat{\mathbf{r}} \times \hat{\mathbf{z}} = \hat{\mathbf{r}} \times (\hat{\mathbf{r}}\cos\theta - \hat{\boldsymbol{\theta}}\sin\theta) = -\sin\theta\,\hat{\boldsymbol{\phi}}$$

$$\int_{-L/2}^{L/2}\exp(jkz'\cos\theta)\sin[k(\tfrac{1}{2}L - |z'|)]\,dz' = \frac{2}{k}\left(\frac{\cos(\tfrac{1}{2}kL\cos\theta) - \cos(\tfrac{1}{2}kL)}{\sin\theta}\right)$$

we have

$$\mathbf{E}(\mathbf{r}) = -\frac{jV_m}{\pi r}\exp(-jkr)\left(\frac{\cos(\tfrac{1}{2}kL\cos\theta) - \cos(\tfrac{1}{2}kL)}{\sin\theta}\right)\hat{\boldsymbol{\phi}} \tag{10.110}$$

In the far zone, the magnetic field is given by

$$\mathbf{H} = H_\theta\hat{\boldsymbol{\theta}}$$

where

$$H_\theta = -\frac{E_\phi}{\eta} = j\frac{V_m}{\pi r\eta}\exp(-jkr)\left(\frac{\cos(\tfrac{1}{2}kL\cos\theta) - \cos(\tfrac{1}{2}kL)}{\sin\theta}\right)\hat{\boldsymbol{\theta}} \tag{10.111}$$

Equations (10.110) and (10.111) show that the radiation pattern of the slot of length L is exactly the same as that of a dipole of length L, except that the electric and magnetic field vectors are interchanged.

The dipole and the slot are an example of a pair of complementary antennas. In general, for a metal antenna, the complementary antenna is the one that results when the metal is replaced by air and air replaced by metal. It can be shown that the impedances of complementary antennas obey the relation (Booker, 1946)

$$Z_{\text{air}}Z_{\text{metal}} = \eta/4 = 35\,476 \text{ ohm}^2 \tag{10.112}$$

If Z_s and Z_d are the impedances of the slot and its complementary dipole, respectively, then

$$Z_s Z_d = 35\,476 \text{ ohm}^2 \tag{10.113}$$

For example, for a half-wave dipole, $Z_d = 73 + j42.5$ ohm. Using (10.113), the impedance of a half-wave slot is found to be $Z_s(\frac{1}{2}\lambda) = 363 - j211$ ohm. As another example, for a resonant dipole with a length-to-diameter ratio of 100, the impedance is a pure resistance of 67 ohm. The complementary slot of length 0.475λ will then have an impedance of 529 ohm.

A slot in a large ground plane can also be fed from a waveguide or a cavity, as shown in Figure 10.19. If the waveguide propagates only the dominant TE_{10} mode and the cavity is excited primarily in its fundamental mode, the electric field distributions across the slot are again described by (10.106), with maxima at the centre and tapering to zero at the edges. The arrangement of Figure 10.19(a) is sometimes called an endwall slot. In the arrangement of Figure 10.19(b), the cavity itself is fed by a probe and a coaxial line. The short-circuit is one-quarter of a TE_{10} mode wavelength from the slot. so that to a first approximation, it transforms into an open-circuit at the slot. Since the radiation is confined to one side of the plane, the radiation resistance is one-half of that of the open-plane structure.

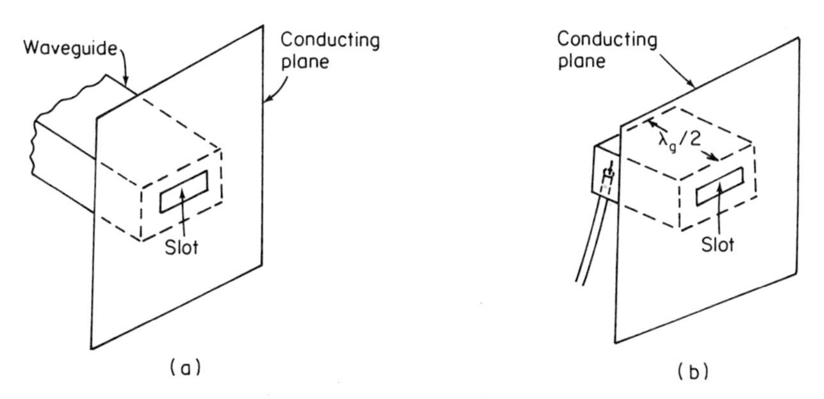

Figure 10.19 (a) An endwall slot; (b) a cavity-fed slot

10.4.2 Slots on Waveguide Walls

If a slot is cut in a proper position on the wall of a waveguide, a unidirectionally radiating slot is obtained. A slot on a waveguide wall will radiate efficiently if an appreciable tangential electric field exists in the slot. Such a field generally develops when the current flow path in the waveguide wall is interrupted. The electric field developed has the character of a displacement current which replaces the interrupted conduction current. The current patterns on the inner walls of waveguides are given in textbooks on electromagnetic theory (e.g. Jordan and Balmain, 1968). The pattern corresponding to the TE_{10} mode in a rectangular waveguide is shown in Figure 10.20. From this pattern, it is easy to determine whether a slot is an efficient radiator or not.

Figure 10.21 shows some waveguide slots that radiate efficiently and some that do not. The radiating slots are those which are so positioned that the flow of current is interrupted. The bigger the interruption, the more the radiation. The non-radiating slots are parallel to the current lines so that their presence

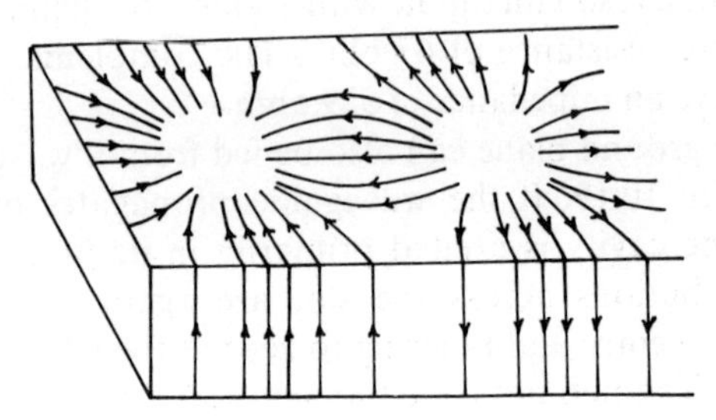

Figure 10.20 Current flow pattern of the TE_{10} mode

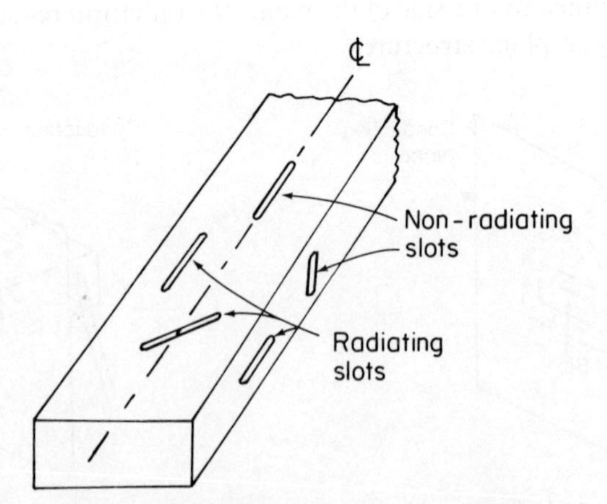

Figure 10.21 Examples of radiating and non-radiating slots cut on the walls of a rectangular waveguide operating in the TE_{10} mode

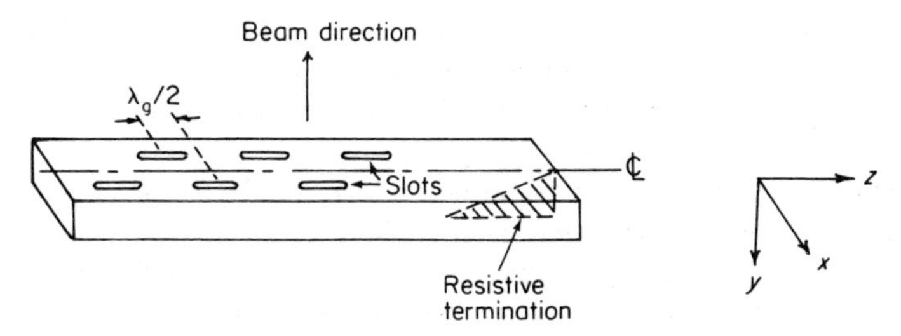

Figure 10.22 A broadside array of waveguide slots

does not perturb the flow appreciably, resulting in negligible tangential electric field across the slots.

A directive antenna can be constructed by cutting an array of slots on a waveguide wall. A typical arrangement is shown in Figure 10.22. A rectangular waveguide operating in the TE_{10} mode is terminated in a resistive load to minimize reflection and standing waves. Slots of half-wavelength long, parallel to the waveguide axis and displaced from the centre, are cut along the broad wall. The tangential electric field developed in each slot depends on the amount of interruption it presents to the current path, which in turn depends on its lateral displacement from the centre. Thus the variation of this displacement provides a means of controlling the amplitude distribution of the array.

Assuming that reflection of the TE_{10} mode is negligibe as a result of the resistive termination, the wall currents reverse every $\lambda_g/2$, where λ_g is the guide wavelength. Consequently, if a broadside pattern is desired, the slots are spaced $\frac{1}{2}\lambda_g$ apart, with adjacent slots on the opposite sides of the centre of the guide. This will result in in-phase operation. The polarization is perpendicular to the axis of the slot.

10.5 PARABOLIC REFLECTOR ANTENNA

10.5.1 Introduction

In many applications in communications, radar, and in particular radio astronomy, very high-gain antennas operating at UHF and microwave frequencies are required. As an example, suppose an antenna with a gain of 1000 (30 dB) is to be designed. Let us consider whether this can be achieved with the types of directive antennas that we have discussed so far, namely, the Yagi, arrays of active elements, the helix, the corner reflector, and the horn antenna. A careful design of the Yagi will produce an economical unit yielding about 15 dB gain. The same can be said of the helix. Broadside arrays of active elements can in principle provide the gain required. As a rule of thumb, about four array elements are needed for each square wavelength of effective area. For a gain of 1000, the number of elements is about 300. This large number of elements makes the feed system very complex.

Consider now a pyramidal horn. The directivity is given by (10.105). Reference to Figure 10.12 and 10.15 shows that, if $B/\lambda = 10$ and $A/\lambda = 10$, the axial lengths R_1 and R_2 would have to be about 100λ in order to obtain a gain of 1000. This would result in an excessively long antenna.

The remaining antenna to consider is the corner reflector. From the formula $D = 4\pi A/\lambda^2$, it would appear that, at first sight, one can realize an arbitrarily large directivity by increasing the aperture of the reflector, which can be obtained by increasing the size of the flat surfaces forming the corner reflector. Unfortunately, this does not work since the formula holds only if the fields across the aperture are the same in direction, phase, and amplitude. That this is not so for the corner reflector is apparent from the following consideration. Let us assume the reflecting surfaces are in the far field of the feed and their dimensions are large compared to the wavelength. The geometrical optics approximation can then be used to examine the manner in which rays from the feed are reflected from the conducting surfaces. This is illustrated in Figure 10.23(a). Since the angle of incidence is equal to the angle of reflection (Snell's law), it is clear that, across the aperture of the reflector, the reflected rays are not parallel. In addition, since the path lengths for the rays are different, the aperture fields vary in both amplitude and phase. This is the reason why the analysis of Chapter 8 shows that the gain can hardly exceed about 15 dB even if we assume the

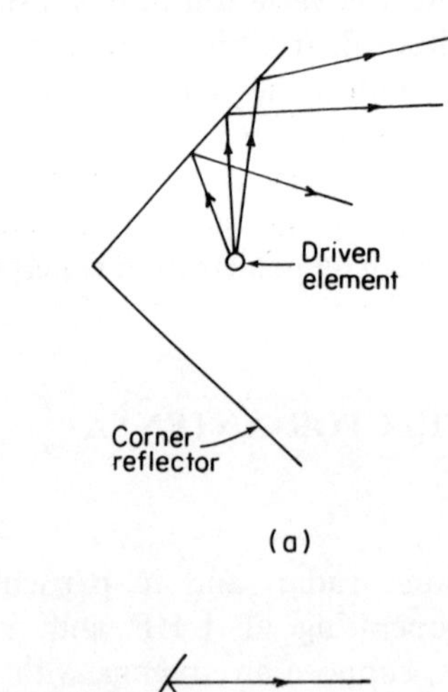

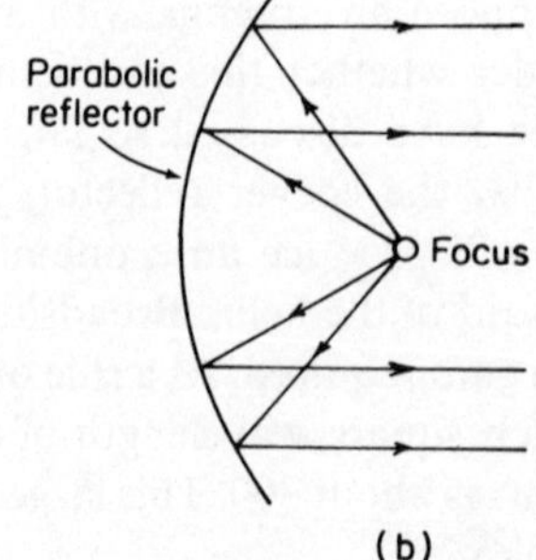

Figure 10.23 (a) The reflected rays of a corner reflector; (b) the reflected rays of a parabolic reflector

conducting sheets to be infinite so that the method of images can be applied.

Suppose that, instead of flat surfaces, the reflector shape is in the form of a paraboloid, which is formed by rotating the arc of a parabola about the line joining the vertex and the focal point. If the feed is a point source at the focus, the spherical wave produced by the point source will, upon reflection, be converted to a plane wave with wave normal parallel to the axis, as shown in Figure 10.23(b). Moreover, the fields across the mouth of the reflector will be in phase. If the amplitudes are also assumed to be the same, then the formula $D = 4\pi A/\lambda^2$ applies. This yields, for $D = 1000$, an aperture diameter of 10λ. If the amplitude distribution contains a taper, the size of the aperture will be larger. However, unless the taper is severe, it will not differ from the uniform case by more than 50%.

It appears from the above discussion that the practical solution to the problem of achieving a high-gain antenna is to use either a parabolic reflector antenna or an array of active elements. For the latter choice, if the radiation pattern is to be of the pencil-beam type, the array needs to be two-dimensional. On account of the large number of elements required in the array, with its associated complicated feed arrangement, the parabolic reflector is usually the superior system from the standpoint of design simplicity and weight. Its bandwidth potential is also better, since different frequencies can be used by simply readjusting or replacing the single feed element. The exceptions are applications that require the tramsmission or reception in two or more directions separated by several beamwidths, at scan rates exceeding those mechanically feasible by moving a single large reflector.

We now discuss the parabolic reflector in some detail.

10.5.2 Geometrical Relations, Aperture Field, and Radiation Pattern

Geometrical Relations

The geometry of a parabolic reflector (also called a 'dish') is shown in Figure 10.24(a). The intersection with the reflector of any plane containing the reflector axis (z-axis) forms a parabola. The cross-section is shown in Figure 10.24(b). For a point P on the reflector surface, let r' be the displacement from the z-axis, ρ be the distance to the focal point F, and θ' be the angle that the line FP makes with the z-axis. Then the equation describing the parabolic surface is

$$(r')^2 = 4f(f - z') \tag{10.114}$$

where $(r')^2 = x'^2 + y'^2$ and the focal length f is the distance from the focal point to the apex. The plane perpendicular to the axis containing the focal point is called the focal plane (x–y plane in Figure 10.24a).

Since $r' = \rho \sin \theta'$ and $z' = \rho \cos \theta'$, (10.114) can also be written as

$$\rho = \frac{2f}{1 + \cos \theta'} = f \sec^2(\theta'/2) \tag{10.115}$$

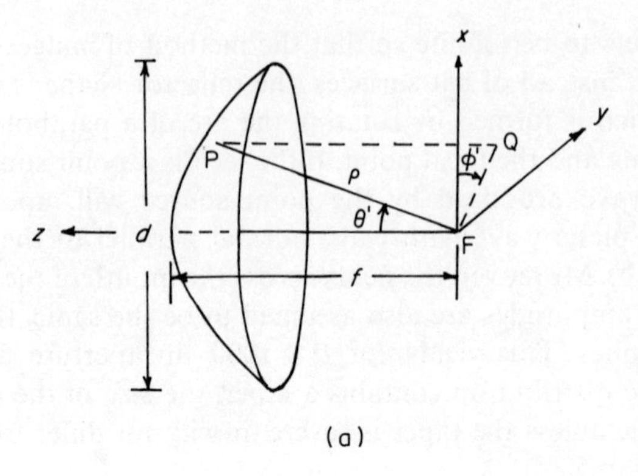

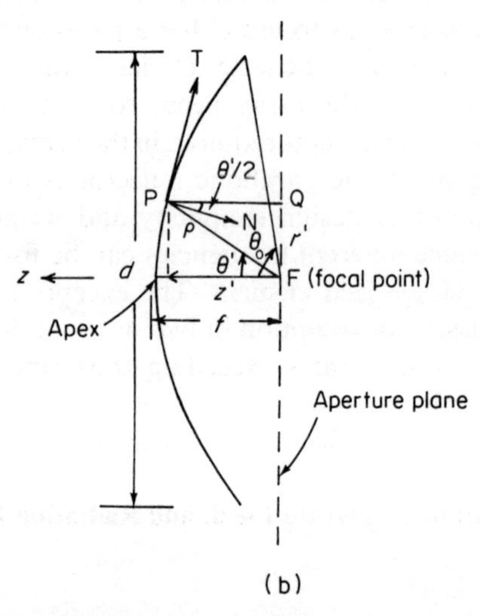

Figure 10.24 (a) Geometry of a parabolic reflector; (b) typical cross-section of a parabolic reflector

The parabolic reflector has two important properties when a point source is placed at its focus. If Snell's law of geometrical optics is assumed to govern the reflection process, then (a) for a ray incident on the surface at any point P, the reflected ray is parallel to the z-axis; (b) all path lengths from the focal point to the reflector and on to the focal plane are the same. To prove (a), it is sufficient to show that, if PQ is a straight line parallel to the z-axis and PN is the normal to the tangent PT at point P (see Figure 10.24b), then $\angle \mathrm{QPN} = \theta'/2$:

$$\tan(\angle \mathrm{TPQ}) = -\frac{\mathrm{d}r'}{\mathrm{d}z'} = \frac{2f}{r'} = \frac{2f}{\rho \sin \theta'} = \frac{2}{\sin \theta' \sec^2(\theta'/2)}$$

$$= \cot(\theta'/2) = \tan(90^\circ - \theta'/2)$$

Hence

$$\angle \mathrm{TPQ} = 90^\circ - \theta'/2$$

and

$$\angle \mathrm{QPN} = 90^\circ - \angle \mathrm{TPQ} = \theta'/2$$

If the point Q is in the focal plane, the total path length

$$\mathrm{FP} + \mathrm{PQ} = \rho + \rho \cos \theta' = 2f \tag{10.116}$$

is therefore constant for all rays emanating from F. Hence (b) is proved.

The half-angle subtended by the reflector aperture at the focal point indicates whether the reflector is deep or shallow. This half-angle is designated θ_0 in Figure 10.24(a). It is related to the focal length-to-diameter ratio (f/d) by

$$\cot(\tfrac{1}{2}\theta_0) = 4f/d \tag{10.117}$$

The value $2\theta_0$ is the total angle subtended by the reflector. Using (10.117), the following table is obtained:

f/d	$2\theta_0$ (deg)
0.25	180
0.35	142
0.50	106
0.75	74
1.00	56

Projected Aperture Field

A parabolic reflector is usually fed by an antenna placed at the focus. The radiation from the feed or primary antenna induces currents on the reflecting surface, which serves as another source of radiation. The radiation in the far field is the sum of that from the reflector and the direct radiation from the feed antenna. The pattern of the latter is known as the primary pattern while that of the entire antenna is known as the secondary pattern. Since the feed is usually designed to illuminate the reflector only, the direct radiation from it may be neglected insofar as its effect on the radiation pattern is concerned. Under such circumstances, the secondary pattern can be taken to be due to the induced currents on the reflector surface If it is calculated directly in terms of these currents, an integration over a curved surface is required. A simpler method is to estimate the electromagnetic fields on some aperture plane and then calculate the pattern in terms of these fields. To simplify matters further, the aperture plane should be chosen such that there is negligible direct radiation from the feed antenna. For this purpose, the focal plane, rather than the plane containing the reflector aperture, is usually chosen. The fields on the focal plane are then estimated from the following considerations. First, the reflector surface is assumed to be in the far field of the feed. If L is the characteristic dimension of the feed, the criterion for this is taken to be $\rho > 2L^2/\lambda$. (see p. 38) The incident field can then be described as a spherical wave. When a ray reaches a point P on the reflector surface, it is assumed to be reflected as though the reflection had taken place at an infinite tangent plane at P. This assumption is valid if the radii of curvature at P are large compared to the wavelength. The ray reflected at P is assumed to travel to the point Q in the focal plane in the manner of geometrical optics. In other words, no diffraction takes place between points P and Q so that the field at Q has the same amplitude and polarization as it has just after reflection at P, but its phase is retarded by an amount corresponding to PQ. However, from the focal plane onwards, the field is allowed to diffract. Since the path lengths from the focal point to the surface and on to the focal plane are the same, the aperture plane is illuminated by a set of in-phase fields which exist within a circle of diameter d and are zero outside. This circle can be considered as the projection of the reflector aperture on the focal plane.

Based on the above assumptions, the aperture field can be calculated quantitatively as follows. For an isotropic source at the focal point, the field intensity at the surface varies as $1/\rho$, since the wave is spherical. After reflection, the wave is planar and there is no spreading loss. Hence the intensity at the aperture plane also varies as $1/\rho$. For a feed antenna with a normalized radiation pattern $F_f(\theta', \phi')$, the aperture electric field can be written in the form

$$\mathbf{E}_a(\theta', \phi') = \frac{E_0 F_f(\theta', \phi')}{\rho} \hat{\mathbf{e}}_r \tag{10.118}$$

where E_0 is a complex constant and $\hat{\mathbf{e}}_r$ is a unit vector to be determined. Let $\mathbf{E}_i$ and $\mathbf{E}_r$ be the incident and reflected electric fields at the surface of the reflector. For large reflectors, Snell's law for planar reflection surfaces holds so that the incident and reflected rays each makes an angle $\theta'/2$ relative to the reflector surface unit normal $\hat{\mathbf{n}}$. For a perfect conductor, the boundary condition is that the tangential component of the total field $\mathbf{E}_i + \mathbf{E}_r$ be zero. On account of the symmetry about $\hat{\mathbf{n}}$, the normal components double. Thus

$$\mathbf{E}_i + \mathbf{E}_r = 2(\hat{\mathbf{n}} \cdot \mathbf{E}_i)\hat{\mathbf{n}}$$

or

$$\mathbf{E}_r = 2(\hat{\mathbf{n}} \cdot \mathbf{E}_i)\hat{\mathbf{n}} - \mathbf{E}_i \tag{10.119}$$

Since the amplitudes of the incident and reflected waves are equal, the above equation can be divided by this amplitude to yield

$$\hat{\mathbf{e}}_r = 2(\hat{\mathbf{n}} \cdot \hat{\mathbf{e}}_i)\hat{\mathbf{n}} - \hat{\mathbf{e}}_i \tag{10.120}$$

where

$$\hat{\mathbf{e}}_r = \mathbf{E}_r/|\mathbf{E}_r| \quad \text{and} \quad \hat{\mathbf{e}}_i = \mathbf{E}_i/|\mathbf{E}_i| \tag{10.121}$$

The aperture magnetic field $\mathbf{H}_a$ is related to $\mathbf{E}_a$ by

$$\mathbf{H}_a = \frac{|\mathbf{E}_a|}{\eta}\hat{\mathbf{h}}_r \qquad \text{where} \quad \hat{\mathbf{e}}_r \times \hat{\mathbf{h}}_r = -\hat{\mathbf{z}} \tag{10.122}$$

As an example, suppose the parabolic reflector is fed by an x-directed Hertzian dipole at the focus. From worked example 2 in section 3.16, we have

$$\hat{\mathbf{e}}_i = \frac{-\hat{\boldsymbol{\theta}}' \cos\theta' \cos\phi' + \hat{\boldsymbol{\phi}}' \sin\phi'}{\sqrt{(1-\sin^2\theta' \cos^2\phi')}} \tag{10.123}$$

From Figure 10.24(b), the reflector surface unit normal is seen to be given by

$$\hat{\mathbf{n}} = -\hat{\rho}\cos(\tfrac{1}{2}\theta') + \hat{\boldsymbol{\theta}}' \sin(\tfrac{1}{2}\theta') \tag{10.124a}$$

or

$$\hat{\mathbf{n}} = -\hat{\mathbf{x}}\sin(\tfrac{1}{2}\theta')\cos\phi' - \hat{\mathbf{y}}\sin(\tfrac{1}{2}\theta')\sin\phi' - \hat{\mathbf{z}}\cos(\tfrac{1}{2}\theta') \tag{10.124b}$$

Substituting (10.123) and (10.124) into (10.120), we obtain, after some algebra, the result

$$\hat{\mathbf{e}}_r = \frac{\hat{\mathbf{x}}(1-\cos^2\phi' + \cos\theta' \cos^2\phi') - \hat{\mathbf{y}}\cos\phi' \sin\phi'(1-\cos\theta')}{\sqrt{(1-\sin^2\theta' \cos^2\phi')}} \tag{10.125}$$

The normalized pattern of an x-directed Hertzian dipole, is, from worked example 2 in section 3.16, given by

$$F_f(\theta', \phi') = \sqrt{(1-\sin^2\theta' \cos^2\phi')} \tag{10.126}$$

The aperture electric field is therefore

$$\mathbf{E}_a = \frac{E_0}{\rho}[\hat{\mathbf{x}}(1-\cos^2\phi' + \cos\theta' \cos^2\phi') - \hat{\mathbf{y}}\cos\phi' \sin\phi'(1-\cos\theta')] \tag{10.127}$$

Using (10.115), we can express the parameters ρ and θ' in terms of r' and ϕ':

$$\theta' = 2\tan^{-1}\left(\frac{r'}{2f}\right) \tag{10.128}$$

$$\rho = \frac{4f^2 + r'^2}{4f} \tag{10.129}$$

Figure 10.25 shows the electric field lines given by equation (10.127). It is seem that the vector $\mathbf{E}_a$ is along x for $\phi' = 0°$ and $\phi' = 90°$. At all other values of ϕ', it has both an x-component and a y-component. These are known as the principal and cross-polarization components, respectively. There is no z-component, which is consistent with the assumption that the fields between the reflector and the aperture plane are those of a plane wave travelling parallel to the axis. In Figure 10.25, the outer circle is for the case when the rim of the paraboloid coincides with the focal plane, for which $f/d = 0.25$. The inner circle

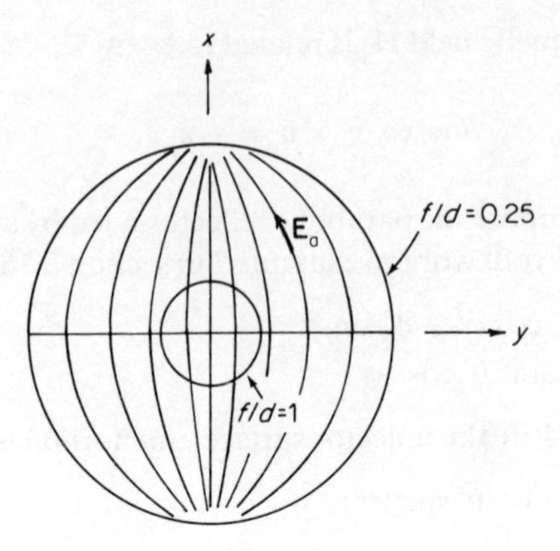

Figure 10.25 The electric field lines in the projected aperture plane due to an x-directed Hertzian dipole. The outer circle is for a focal length-to-diameter ratio $f/d = 0.25$ and the inner circle is for $f/d = 1.0$

is for the case when $f/d = 1$, which corresponds to a smaller-diameter dish. It is seen that the cross-polarized component decreases as the focal length-to-diameter ratio increases.

Analysis similar to the above can be carried out for other feed antennas. In particular, the following results are noted:

(a) For a small loop in the x–z plane with a magnetic moment along y, the aperture electric field has a principal component along x and a cross-polarized component along y. The direction of the latter is opposite to that of the x-directed dipole.
(b) By combining a dipole and a loop of the appropriate strengths, it is possible that the y-components of the two cancel, leaving a purely linear aperture electric field.
(c) For an open waveguide or horn fed with electric field polarized in the x- or y-direction, the aperture electric field again has both x- and y-components.

Secondary Radiation Pattern

Once the aperture field has been determined, the secondary radiation pattern can be found by using (10.14). The integral can usually be performed only numerically. In what follows, we discuss qualitatively the various factors which affect the radiation pattern.

(a) It has been pointed out that the fields on the aperture plane are in phase. If the amplitude distribution is uniform, then the result for a uniform circular aperture would apply, namely $D_{cir} = (4\pi/\lambda^2)(\pi a^2)$ and HPBW = $1.02\lambda/d$ radians. However, (10.118) shows that even for a feed with an isotropic pattern, there is a natural amplitude taper of the form $1/\rho$. There is also a taper introduced by the pattern of the source. The effect of amplitude taper on the pattern is that the directivity is lower and the HPBW is larger, with an accompanying decrease in the side-lobe level. Introducing the aperture efficiency ε_{ap} where $0 < \varepsilon_{ap} < 1$, we can write

$$D = \varepsilon_{ap} \frac{4\pi}{\lambda^2} (\pi a^2)$$

(b) If the beamwidth of the feed antenna is larger than θ_0, a portion of the feed radiation is not intercepted by the reflector. This is known as 'spillover', and leads to a decrease in directivity since this power is not directed in the main beam maximum.

(c) For a given feed pattern, there will be less taper but more spillover as the ratio f/d increases (θ_0 decreases). Since the former leads to an increase and the latter to a decrease in directivity, there is an optimum value for the aperture angle θ_0 that yields maximum directivity. Conversely, for a given aperture angle, there is an optimum feed pattern. Based on these two factors alone, the aperture efficiency attainable is about 80%.

(d) Aside from amplitude taper and spillover, there are other factors that will affect the aperture efficiency. If the surface deviates randomly from the true parabolic shape, the aperture fields will not be exactly in phase, leading to far-field cancellations and decrease in directivity. The presence of the feed antenna and its support in the path of the reflected rays will block the illumination of portions of the aperture. This is known as aperture blocking, which will be discussed further in section 10.5.4. As a result, the aperture efficiency usually has a value between 0.5 and 0.7.

(e) As shown in the previous paragraph, except for $\phi = 0°$ and $\phi = 90°$, the aperture electric field contains a principal component and a cross-polarized component. Since the latter is antisymmetric with respect to the lines $\phi = 0°$ and $\phi = 90°$, its contributions to the principal plane patterns cancel. Consequently, the far fields in these planes do not contain any cross-polarized component. This, however, does not hold for other directions in space. Detailed analysis shows that the cross-polarization component is maximum in the 45° plane between the principal planes.

Cross-polarization characteristics of an antenna are important in applications where, in order to save frequency bandwidth, two orthogonal polarizations are used either within the same beam or as a means of providing isolation between beams operating in the same frequency band. In such systems, the isolation between channels depends on the suppression of the cross-polarization. This depends on the feed and the f/d ratio. For a given feed, the cross-polarization decreases as f/d increases, as noted in the preceeding paragraph.

10.5.3 Feeding Arrangements

Feed Radiatiors

For metre wavelengths, a dipole can be used as the feed antenna. Since it radiates both in the forward and backward directions, a reflector in the form of a parasitic element or a plane sheet is used to eliminate the backward radiation. If circular polarization is desired, a cross-dipole or a helix can be used. At higher frequencies, the parabolic reflector is usually fed by an open waveguide or horn.

Feed Support for Prime-Focus System

When the feed antenna is placed at the focus of the parabola, it is known as a prime-focus system. If the dish is to be steerable, the feed support must be such that the feed remains in the focus as the dish moves. Methods for meeting this requirement are shown in Figure 10.26. In Figure 10.26(a) and (b), the feed antennas are supported by the waveguide or transmission line. This is adequate when the reflector is small and of short focal length. In large reflectors, a common method is that of Figure 10.26(c), in which the feed is supported by struts rising from firmly held points on the reflector, and the waveguide from the horn runs down one of the struts.

The problem of feed support is simplified if the dish need not be moved. In this case, the support can be attached to the ground rather than to the dish, and light weight is not important. This situation occurs, for example, with large reflector antennas used for tropospheric-scatter point-to-point communication at VHF or UHF frequencies.

Cassegrain Feed

For a prime-focus system, the waveguide or transmission line connecting the feed and the transmitting and/or receiving equipment located behind the dish can be quite long. In low-noise receiving systems, the losses in such lengths may be more than can be tolerated. One solution is to place the equipment close to the feed. This, however, may not be practicable since the equipment may be too large and too heavy to be placed near the focus, especially for steerable systems.

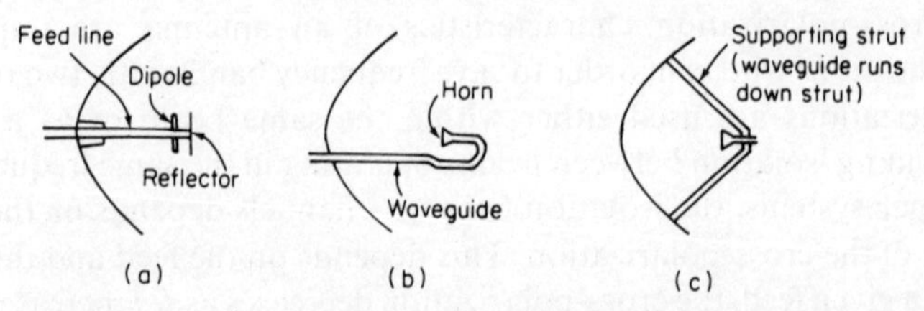

Figure 10.26 Some methods of supporting the feed antenna of a prime-focus system

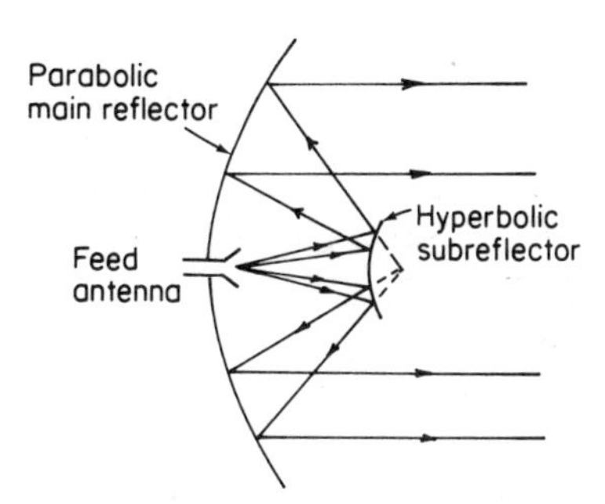

Figure 10.27 A Cassegrain system

Moreover, this location is inconvenient as far as servicing and adjustments are concerned.

An alternative solution is to use a Cassegrain feed, in which the primary antenna is located in a small opening at the vertex of the paraboloid, rather than at the focus. It is aimed at a small subreflector located between the vertex and the focus. This is illustrated in Figure 10.27.

As shown, the rays from the primary radiator are first reflected from the subreflector. If the surface of the subreflector is hyperbolic, and if the feed is at one of the foci of the hyperboloid, the reflected rays will appear to originate from the focus of the paraboloid. On reaching the main reflector, they are converted to parallel beams in the usual way. This arrangement has several advantages.

(a) Long transmission lines running to the focus of the paraboloid are avoided.
(b) The transmitter and/or receiver are relatively accessible for servicing and adjustment. They may be as large and heavy as required since they are not located in a position where mechanical support is a problem and they are not in the path of the rays.
(c) In Earth terminal receiving antennas, the feed of the Cassegrain system is directed towards the relatively low-noise sky, whereas in the prime-focus system, it is directed towards the more noisy ground. Consequently, the noise pickup from the environment surrounding the antenna is less for the Cassegrain system.
(d) There is an equivalent prime-focus parabolic reflector which produces the same aperture field as the Cassegrain system. The f/d ratio of this equivalent reflector is considerably larger than the actual reflector. This is illustrated in Figure 10.28. The ray from the feed at F_1 that strikes the subreflector at P is reflected from the main reflector at P′. It then travels parallel to the main axis. Instead, it can be imagined that the ray F_1P continues beyond the subreflector in a straight line until it strikes an imaginary paraboloid reflector at P″ such that the ray is reflected parallel to, and at the same distance from, the axis. This imaginary paraboloid is known as the equivalent paraboloid, and has a focal length which is considerably larger than that of the actual reflector. Thus the

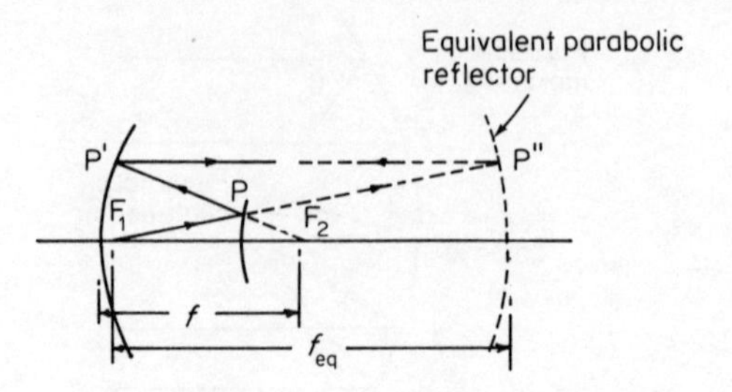

Figure 10.28 The equivalent parabolic reflector of a Cassegrain system

Cassegrain system permits the use of a parabola of conventional f/d ratio to obtain the same effect as a parabola with a larger f/d. As indicated in section 10.5.2, one of the advantages of a larger f/d is that the cross-polarization component is reduced.

The disadvantage of the Cassegrain system is that the subreflector tends to be much larger than the feed for a prime-focus system. This results in a larger aperture blockage. The effect of aperture blockage will be discussed in the next section.

10.5.4 Aperture Blockage

The presence of the feed, supporting structures, or a subreflector in the path of the rays has two effects on the performance of the antenna. First, some of the energy reflected enters the feed and acts as a travelling wave in the reverse direction in the transmission line. Standing waves are therefore produced, causing an impedance mismatch and a degradation of the transmitter performance. The mismatch can be corrected by an impedance-matching device, but this remedy is effective only over a relatively narrow frequency band.

The second effect is that the effective size of the aperture is reduced. The influence on the radiation pattern is difficult to calculate quantitatively, but can best be taken into account empirically. However, the qualitative effects of aperture blocking can be revealed by the following procedure. Cancel the aperture distribution over the geometrical shadow cast by the feed and the structures in front of the reflector. Then subtract from the radiation pattern the pattern due to an aperture the same shape as the shadow. To illustrate this, let us for simplicity regard the shadow region as a small circular aperture centred at the axis. The procedure then leads to a reduced main lobe and an increase of the first side lobe, as illustrated in Figure 10.29. The latter arises because, at the place of the first side lobe, the radiation from the aperture with radius d and that from the small aperture of radius d' are 180° out of phase.

Both the aperture blocking and the mismatch at the feed due to the intercepted radiation are eliminated with the offset-feed parabolic antenna shown in

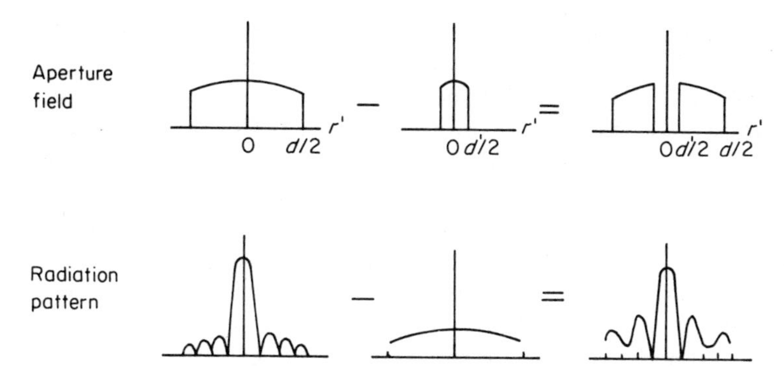

Figure 10.29 Illustrating the effect of aperture blockage on the far-field pattern

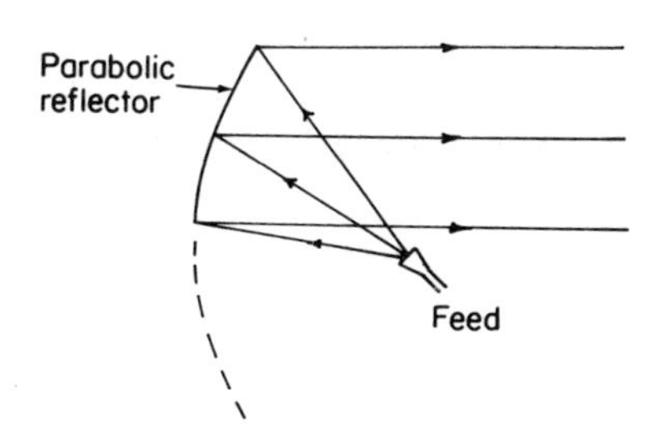

Figure 10.30 Parabolic reflector with offset feed

Figure 10.30. In this scheme, the centre of the feed is placed at the focus of the parabola, but it is tilted with respect to the axis. The major portion of the lower half of the parabola is removed, leaving that portion shown by the full curve in Figure 10.30. Since the feed is out of the path of the reflected rays, there is no pattern deterioration due to aperture blocking, nor is there any significant amount of energy intercepted by the feed to produce an impedance mismatch. However, this scheme introduces problems of its own. First, detailed analysis shows that the cross-polarization component is appreciable. Secondly, its asymmetrical geometry complicates the problem of providing the desired primary pattern taper. Thirdly, it is more difficult to support the antenna properly and to steer it.

10.5.5 Beam Steering by Feed Offset

Without moving the dish mechanically, the beam produced by a parabolic reflector antenna can be steered over a limited angle by moving the feed off the reflector axis but keeping it in the focal plane. When the feed is not at the focus, the fields in the aperture plane are no longer in phase. Consequently, as the steering angle is increased, the beam shape deteriorates, resulting in increased beamwidth, decreased directivity, and increasing side-lobe level. At large values of offset of the feed, the deterioration can become so excessive that it can hardly be referred to as a beam. The permissible steering angle is roughly proportional to the on-axis beamwidth and is generally of the order of a few beamwidths. It is greater with large than with small f/d ratios. For example, the gain of a paraboloid with $f/d = 0.25$ is reduced to 80% of its maximum value when the beam is scanned ± 3 beamwidths off-axis. A paraboloid with $f/d = 0.5$ can be scanned ± 6.5 beamwidths off-axis before the gain is reduced to 80% of maximum.

If the parabolic reflector is replaced by a spherical reflector, it is possible to achieve a wider scanning angle. However, as a spherical reflector does not produce in-phase plane waves across the antenna aperture, some means is necessary to minimize the effects produced by this 'spherical aberration'. One approach is to compensate for the spherical aberration with special feeds. These feeds are designed so that the rays leaving the different portions of the feed arrive at the spherical reflector with different path lengths so that, upon reflection, they produce an in-phase plane wave field across the aperture.

10.5.6 Reflector Surface Accuracy Requirements

In order to provide the desired plane wavefront of the reflected radiation, the surface of a reflector must in theory be a true paraboloid. Perfect accuracy is, of course, not attainable in practice. Moreover, since construction of a precise surface is very expensive, it is useful to know what degree of inaccuracy can be tolerated without serious sacrifice of performance. If the deviations are of a random nature, the generally accepted rule-of-thumb figure for the permissible root mean square deviation is $\frac{1}{32}$ to $\frac{1}{16}$ wavelength. Thus the maximum frequency at which the antenna can be used successfully is in effect determined by the irregularity of the reflector surface.

The accuracy of the surface is more difficult to maintain in a steerable than in a fixed reflector. This is because, in the former, the gravitational forces acting on different portions of the reflector are different for various elevation angles. Thus even if the shape is paraboloidal at one of the angles, the loading at other angles may be such that it will deform. This problem is very serious for large reflectors, and is one of the principal factors in limiting the size of the diameter of the dish that can be constructed.

10.5.7 Some of the World's Largest Reflector Antennas

The world's largest reflector antennas are to be found in the various radio astronomy observatories around the globe. The largest fully steerable dish is the Effelsberg radio telescope of the Max Planck Institute in West Germany. It has a diameter of 100 m and a weight of 3200 tonnes. Even under worst conditions, the deviation from the ideal paraboloid does not exceed 0.6 mm r.m.s. It picks up cosmic radiation with wavelengths of between 75 cm and 7 mm.

The measured gain at 2.8 cm wavelength is about 78 dB and does not vary by more than 1% as elevation angle changes from 30° to 90°.

The largest fixed reflector is the 1000 ft (305 m) diameter reflector of the Arecibo Observatory in Arecibo, Puerto Rico. It was conceived and designed by Cornell University engineers and built under their supervision in the early 1960s. The reflector is spherical in shape and lies fixed in a natural depression in the ground, with its aperture plane horizontal. The deviation from the ideal sphere has been held to less than 6 mm r.m.s. The feed used at 430 MHz is a tapered section of slotted waveguide, 96 ft in length and weighing almost 10 000 lb. A similar but shorter feed has also been in service at 1415 MHz. The feeds are fastened to a carriage suspended by cables some 500 ft above the reflector. The carriage allows the feed to be moved about the centre of the circular aperture. The 1415 MHz feed realizes a gain of 67.8 dB and can scan 11° away from the zenith in any direction.

At Jodrell Bank, in England, there is a 250 ft fully steerable parabolic dish. At the National Radio Astronomy Observatory (NRAO) in Green Bank, West Virginia, USA, there is a 140 ft fully steerable dish and a 300 ft dish that is movable in elevation only. Finally, it should be mentioned that NRAO also operates a Very Large Array (VLA) in the Plains of San Augustin in New Mexico, USA. The VLA consists of 28 parabolic dishes, each of 85 ft in diameter. They are arranged in the form of a Y, each of the three arms being 21 km long. It has been in operation since 1981.

10.6 WORKED EXAMPLES

Example 1

(a) Estimate the aperture efficiencies of the optimum E- and H-plane sectoral horns with axial lengths of 30λ, 10λ, and 6λ.

(b) Estimate the aperture efficiencies of the optimum pyramidal horns with axial lengths $R_1 = R_2 = 30\lambda$, 10λ, and 6λ.

Solution

(a) For a given axial length R_1 of the E-plane sectoral horn, the values of B/λ and $(\lambda D_E/a)$ under optimum conditions can be obtained from Figure 10.12. If ε_E is the aperture efficiency, then

$$D_E = \varepsilon_E \frac{4\pi Ba}{\lambda^2} \qquad \text{or} \qquad \varepsilon_E = \frac{\lambda D_E/a}{4\pi B/\lambda}$$

Similarly, for a given axial length R_2 of the H-plane sectoral horn, the values of A/λ and $(\lambda D_H/b)$ under optimum conditions can be obtained from Figure 10.15. If ε_H is the aperture efficiency, then

$$D_H = \varepsilon_H \frac{4\pi Ab}{\lambda^2} \qquad \text{or} \qquad \varepsilon_H = \frac{\lambda D_H/b}{4\pi A/\lambda}$$

The results for axial lengths of 30λ, 10λ, and 6λ are tabulated below.

R_1	B/λ	$\lambda D_E/a$	ε_E	R_2	A/λ	$\lambda D_H/b$	ε_H
30	7.7	65	0.67	30	9.5	77	0.64
10	4.5	38	0.67	10	5.5	44	0.64
6	3.5	28	0.64	6	4.2	34	0.64

(b) For the pyramidal horn, the directivity is given by (10.105):

$$D_\mathrm{p} = \frac{\pi}{32}\left(\frac{\lambda}{A}D_E\right)\left(\frac{\lambda}{B}D_H\right)$$

For a given R_1 and R_2, the dimensions of A and B for optimum performance can be found from (10.86) and (10.99):

$$A = \sqrt{(3\lambda R_2)}$$
$$B = \sqrt{(2\lambda R_1)}$$

The values $(\lambda D_E/A)$ and $(\lambda D_H/B)$ can then be read off Figures 10.12 and 10.15. If ε_p is the aperture efficiency, then

$$\varepsilon_\mathrm{p} = \frac{\lambda^2 D_\mathrm{p}}{4\pi AB} = \frac{(\lambda D_E/A)(\lambda D_H/B)}{128(A/\lambda)(B/\lambda)}$$

The results for $R_1 = R_2 = 30\lambda$, 10λ, and 6λ are tabulated below:

R_1	B/λ	A/λ	$\lambda D_E/A$	$\lambda D_H/B$	ε_p
30	7.7	9.5	65	77	0.53
10	4.5	5.5	38	44	0.53
6	3.5	4.2	28	34	0.51

Example 2

A pyramidal horn is to be fed from a waveguide of dimensions $a \times b$. Its directivity at the wavelength λ is specified. Indicate how you would determine the dimensions A, B, R_E, and R_H (see Figure 10.17) if the performance of the horn is to be optimal.

Solution

The geometry for the pyramidal horn is shown in Figure 10.17. The following geometrical relations are obtained:

$$l_H^2 = R_2^2 + (A/2)^2 \tag{10.130}$$

$$R_H = (A - a)\sqrt{[(l_H/A)^2 - \tfrac{1}{4}]} \tag{10.131}$$

$$l_E^2 = R_1^2 + (B/2)^2 \tag{10.132}$$

$$R_E = (B - b)\sqrt{[(l_E/B)^2 - \tfrac{1}{4}]} \tag{10.133}$$

If the horn is to fit together with the feed waveguide, we must have

$$R_E = R_H = R_\mathrm{p} \tag{10.134}$$

Hence

$$(A - a)\sqrt{[(l_H/A)^2 - \tfrac{1}{4}]} = (B - b)\sqrt{[(l_E/B)^2 - \tfrac{1}{4}]} \tag{10.135}$$

For large horns, $R_1 \simeq l_E$ and $R_2 \simeq l_H$. For optimum performance,

$$A = \sqrt{(3\lambda R_2)} \simeq 3\lambda l_H \tag{10.136}$$

$$B = \sqrt{(2\lambda R_1)} \simeq 2\lambda l_E \tag{10.137}$$

From worked example 1(b), the aperture efficiency of an optimum pyramidal horn is approximately 0.5. Using this information, we have

$$D_\mathrm{p} \simeq \frac{1}{2}\frac{4\pi(AB)}{\lambda^2} \tag{10.138}$$

Using (10.136), (10.137), and (10.138), (10.135) can be written as

$$\left(\sqrt{(2L_E)} - \frac{b}{\lambda}\right)^2 (2L_E - 1) = \left(\frac{D_\mathrm{p}}{2\sqrt{(2\pi L_E)}} - \frac{a}{\lambda}\right)^2 \left(\frac{D_\mathrm{p}^2}{18\pi^2 L_E} - 1\right) \tag{10.139}$$

where $L_E = l_E/\lambda$.

The only unknown in (10.139) is L_E, which can be solved by trial and error. Then B, A, and l_H follow from (10.137), (10.138), and (10.136), respectively. Finally, $R_E = R_H = R_\mathrm{p}$ is obtained from (10.132) or (10.133).

Experience has shown that, as a first trial for the solution of (10.139), L_E can be set to a value of $0.065\,D_\mathrm{p}$.

Example 3

To obtain some idea on the quantitative effects of aperture blockage, let us consider a circular aperture illuminated by a plane wave of the form

$$\mathbf{E}_\mathrm{s} = \begin{cases} 0 & 0 < r \leqslant a' \\ E_0\hat{\mathbf{y}} & a' < r \leqslant a \end{cases}$$

$$\mathbf{H}_\mathrm{s} = -|\mathbf{E}_\mathrm{s}|\hat{\mathbf{x}}$$

$2a'$ can be considered as the diameter of the object causing the blockage. What are the effects of such blockage?

Solution

By the principle of superposition, the far-zone electric field is equal to that due to a uniform aperture of radius a minus that due to a uniform aperture of radius

a'. From (10.34), we readily obtain

$$\mathbf{E}(r) = \frac{jk}{2r}\exp(-jkr)E_0[\hat{\boldsymbol{\theta}}\sin\phi(1+\cos\theta) + \hat{\boldsymbol{\phi}}(1+\cos\theta)]$$

$$\times\left(\frac{a^2 J_1(ka\sin\theta)}{ka\sin\theta} - \frac{a'^2 J_1(ka'\sin\theta)}{ka'\sin\theta}\right) \qquad (10.140)$$

In the y–z and x–z planes, the above equation reduces to

$$\mathbf{E}_{yz} = \frac{jk\exp(-jkr)}{2r}E_0(1+\cos\theta)\left(\frac{a^2 J_1(ka\sin\theta)}{ka\sin\theta} - \frac{a'^2 J_1(ka'\sin\theta)}{ka'\sin\theta}\right)\hat{\boldsymbol{\theta}} \qquad (10.141)$$

$$\mathbf{E}_{xz} = \frac{jk\exp(-jkr)}{2r}E_0(1+\cos\theta)\left(\frac{a^2 J_1(ka\sin\theta)}{ka\sin\theta} - \frac{a'^2 J_1(ka'\sin\theta)}{ka'\sin\theta}\right)\hat{\boldsymbol{\phi}} \qquad (10.142)$$

In the broadside direction,

$$|\mathbf{E}|_{\theta=0} = \frac{kE_0}{r}a^2\left(1 - \frac{a'^2}{a^2}\right) \qquad (10.143)$$

The ratio of the axial electric field with blockage to that without blockage is therefore

$$1 - a'^2/a^2$$

The maximum of the first side lobe occurs at approximately midway between the first and second zeros of the function $J_1(ka\sin\theta)$. Denoting this value of θ by θ_1, we have

$$ka\sin\theta_1 \simeq 5.43$$

or

$$\theta_1 = \sin^{-1}(5.43/ka) \qquad (10.144)$$

Substituting this value of θ_1 in (10.141) or (10.142), we obtain the first side-lobe maximum. The ratio of this value compared to that without blockage is

$$1 - \frac{a' J_1(ka'\sin\theta_1)}{a J_1(ka\sin\theta_1)}$$

As an example, let $a = 7\lambda$, $a' = \lambda$. Then

$$\frac{\text{axial electric field with blockage}}{\text{axial electric field without blockage}} = 1 - (1/7)^2 = 0.98$$

$$\frac{\text{maximum of first side lobe with blockage}}{\text{maximum of first side lobe without blockage}} = 1 - \frac{J_1(0.776)}{7J_1(5.43)}$$

$$= 1 - \frac{0.35}{7(-0.34)} = 1.147$$

Thus the axial electric field is decreased by 2% while the first side-lobe maximum is increased by almost 15%.

10.7 PROBLEMS

1. Show that the far field produced by a distribution of tangential magnetic field source $\hat{\mathbf{n}} \times \mathbf{H}_s$ over a closed surface S is the same as that produced by an equivalent surface conducting current density $\mathbf{J}_s = \hat{\mathbf{n}} \times \mathbf{H}_s$ over the same surface.

2. Maxwell's equations are often written in the following manner to allow for the effects of a fictitious magnetic charge density ρ_m and a fictitious magnetic current density J_m:

$$\nabla \times \mathbf{E} = -\left(\mathbf{J}_m + \frac{\partial B}{\partial t}\right) \qquad \nabla \cdot \mathbf{D} = \rho$$

$$\nabla \times \mathbf{H} = \mathbf{J} + \frac{\partial \mathbf{D}}{\partial t} \qquad \nabla \cdot \mathbf{B} = \rho_m$$

$$\frac{\partial \rho}{\partial t} + \nabla \cdot \mathbf{J} = 0 \qquad \frac{\partial \rho_m}{\partial t} + \nabla \cdot \mathbf{J}_m = 0$$

   (a) If $\rho = 0$ and $J = 0$, obtain expressions for the far-zone electromagnetic fields in terms of $\mathbf{J}_m$.
   (b) Show that the far field produced by a distribution of tangential electric field source $\hat{\mathbf{n}} \times \mathbf{E}_s$ over a closed surface S is the same as that produced by an equivalent surface magnetic current density $\mathbf{J}_{ms} = \mathbf{E}_s \times \hat{\mathbf{n}}$ over the same surface.

3. Use equation (10.2) to show that, for a tangential electric field source parallel (perpendicular) to the conducting plane, the image is in the same (opposite) direction.

4. Use equation (10.2) to show that, for a tangential magnetic field source parallel (perpendicular) to the conducting plane, the image is in the opposite (same) direction.

5. Verify equations (10.28) and (10.29).

6. The rectangular aperture of Figure 10.5 is illuminated by a plane wave of the form

$$\mathbf{E}_s = E_0[1 - (1 - \Delta)(2x/a)^2]\hat{\mathbf{y}}$$

$$\mathbf{H}_s = -\frac{|\mathbf{E}_s|}{\eta}\hat{\mathbf{x}}$$

   (a) Sketch the magnitude of the aperture electric field as a function of x.
   (b) Obtain expressions for the far-zone electric field in the y–z and x–z planes.
   (c) Assuming $ka \gg 1$ and $kb \gg 1$, obtain the widths of the main beams for

$\Delta = 0.8$ and 0.5 and compare them with the case of uniform illumination ($\Delta = 1$).

(d) Determine the aperture efficiency for $\Delta = 0.8$ and $\Delta = 0.5$.

7. The rectangular aperture of Figure 10.5 is illuminated by a plane wave of the from

$$\mathbf{E}_s = E_0 \exp(-j\beta 2x/a)\hat{\mathbf{y}}$$

$$\mathbf{H}_s = -\frac{|\mathbf{E}_s|}{\eta}\hat{\mathbf{x}}$$

Determine the aperture efficiency.

8. Verify equation (10.85)

9. Verify equations (10.115), (10.117), (10.128), and (10.129).

10. Derive an expression for the far-zone electric field on the axis of an H-plane sectoral horn and show that the directivity is given by

$$D_H = \frac{4\pi b R_2}{\lambda A}\{[C(p_1) - C(p_2)]^2 + [S(p_1) - S(p_2)]^2\}$$

where

$$p_1 = \frac{1}{\sqrt{2}}\left(\frac{\sqrt{(R_2/\lambda)}}{A/\lambda} + \frac{A/\lambda}{\sqrt{(R_2/\lambda)}}\right) \qquad p_2 = \frac{1}{\sqrt{2}}\left(\frac{\sqrt{(R_2/\lambda)}}{A/\lambda} - \frac{A/\lambda}{\sqrt{(R_2/\lambda)}}\right)$$

11. Derive equation (10.105) for the directivity of a pyramidal horn.

12. Design an optimum pyramidal horn for use with a 0.9 in × 0.4 inch rectangular waveguide (WR 90). It should have a gain of 20 dB at 10 GHz.

13. If the length of the slot antenna shown in Figure 10.18(a) is much less than a wavelength, obtain expressions for the far field and compare them with those of a Hertzian dipole.

14. The feed of the parabolic reflector shown in Figure 10.24 is a small circular loop antenna of radius a at the focus. Its magnetic moment $\mathbf{m} = (I\pi a^2)\hat{\mathbf{y}}$. Obtain an expression for the projected aperture electric field. Show that the y-component is opposite to that produced by an x-directed Hertzian dipole.

15. The feed of the parabolic reflector shown in Figure 10.24 consists of a Hertzian dipole of electric moment $\mathbf{p} = I_1\, dl\, \hat{\mathbf{x}}$ and a small circular loop of magnetic moment $\mathbf{m} = (I_2 \pi a^2)\hat{\mathbf{y}}$ at the focus. Obtain the condition for the ratio $|\mathbf{p}|/|\mathbf{m}|$ such that the projected aperture electric field has no cross-polarized component.

16. The feed of the parabolic reflector shown in Figure 10.24 is an open-ended rectangular waveguide operating in the dominant mode with its broad wall along the x-axis. Obtain an expression for the projected aperture electric field.

BIBLIOGRAPHY

Blake, L. V. (1966). *Antennas*, Willey, New York, Chapter 6.

Clarke, R. H. and Brown, J. (1980). *Diffraction Theory and Antennas*, Ellis Horwood, Chichester, Chapter 7.

Elliott, R. S. (1981). *Antenna Theory and Design*, Prentice-Hall, Englewood Cliffs, NJ, Chapter 3 and 10.

Love, A. W. (ed.) (1976). *Electromagnetic Horn Antennas*, IEEE Press, New York.

Love, A. W. (ed.) (1978). *Reflector Antennas*, IEEE Press, New York.

Silver, S. (1949). *Microwave Antenna Theory and Design*, McGraw-Hill, New York, Chapters 6, 9, 10, and 12.

Stutzman, W. L. and Thiele, G. A. (1981). *Antenna Theory and Design*, Wiley, New York, Chapter 8.

Appendix Some Useful Vector Relations

1. VECTOR INDENTITIES

$$\nabla \cdot (\nabla \times \mathbf{A}) = 0 \tag{A.1}$$

$$\nabla \times \nabla \Phi = 0 \tag{A.2}$$

$$\nabla \times (\nabla \times \mathbf{A}) = \nabla(\nabla \cdot \mathbf{A}) - \nabla^2 \mathbf{A} \tag{A.3}$$

$$\nabla \cdot (\mathbf{A} \times \mathbf{B}) = \mathbf{B} \cdot (\nabla \times \mathbf{A}) - \mathbf{A} \cdot (\nabla \times \mathbf{B}) \tag{A.4}$$

$$\iiint_{\tau} \nabla \cdot \mathbf{A} \, \mathrm{d}\tau = \oiint_{\sigma} \mathbf{A} \cdot \hat{\mathbf{n}} \, \mathrm{d}\sigma \tag{A.5}$$

$$\iint_{\sigma} (\nabla \times \mathbf{A}) \cdot \hat{\mathbf{n}} \, \mathrm{d}\sigma = \oint_{l} \mathbf{A} \cdot \mathrm{d}\mathbf{l} \tag{A.6}$$

2. VECTOR DIFFERENTIAL OPERATORS

Rectangular Coordinates

$$\nabla \Phi = \hat{\mathbf{x}} \frac{\partial \Phi}{\partial x} + \hat{\mathbf{y}} \frac{\partial \Phi}{\partial y} + \hat{\mathbf{z}} \frac{\partial \Phi}{\partial z} \tag{A.7}$$

$$\nabla \cdot \mathbf{A} = \frac{\partial A_x}{\partial x} + \frac{\partial A_y}{\partial y} + \frac{\partial A_z}{\partial z} \tag{A.8}$$

$$\nabla \times \mathbf{A} = \hat{\mathbf{x}} \left(\frac{\partial A_z}{\partial y} - \frac{\partial A_y}{\partial z} \right) + \hat{\mathbf{y}} \left(\frac{\partial A_x}{\partial z} - \frac{\partial A_z}{\partial x} \right) + \hat{\mathbf{z}} \left(\frac{\partial A_y}{\partial x} - \frac{\partial A_x}{\partial y} \right) \tag{A.9}$$

$$\nabla^2 \Phi = \frac{\partial^2 \Phi}{\partial x^2} + \frac{\partial^2 \Phi}{\partial y^2} + \frac{\partial^2 \Phi}{\partial z^2} \tag{A.10}$$

Cylindrical Coordinates

$$\nabla\Phi = \hat{\mathbf{r}}\frac{\partial\Phi}{\partial r} + \hat{\boldsymbol{\phi}}\frac{1}{r}\frac{\partial\Phi}{\partial\phi} + \hat{\mathbf{z}}\frac{\partial\Phi}{\partial z} \tag{A.11}$$

$$\nabla\cdot\mathbf{A} = \frac{1}{r}\frac{\partial}{\partial r}(rA_r) + \frac{1}{r}\frac{\partial A_\phi}{\partial\phi} + \frac{\partial A_z}{\partial z} \tag{A.12}$$

$$\nabla\times\mathbf{A} = \hat{\mathbf{r}}\left(\frac{1}{r}\frac{\partial A_z}{\partial\phi} - \frac{\partial A_\phi}{\partial z}\right) + \hat{\boldsymbol{\phi}}\left(\frac{\partial A_r}{\partial z} - \frac{\partial A_z}{\partial r}\right) + \hat{\mathbf{z}}\frac{1}{r}\left(\frac{\partial}{\partial r}(rA_\phi) - \frac{\partial A_r}{\partial\phi}\right) \tag{A.13}$$

$$\nabla^2\Phi = \frac{1}{r}\frac{\partial}{\partial r}\left(r\frac{\partial\Phi}{\partial r}\right) + \frac{1}{r^2}\frac{\partial^2\Phi}{\partial\phi^2} + \frac{\partial^2\Phi}{\partial z^2} \tag{A.14}$$

Spherical Coordinates

$$\nabla\Phi = \hat{\mathbf{r}}\frac{\partial\Phi}{\partial r} + \hat{\boldsymbol{\theta}}\frac{1}{r}\frac{\partial\Phi}{\partial\theta} + \hat{\boldsymbol{\phi}}\frac{1}{r\sin\theta}\frac{\partial\Phi}{\partial\phi} \tag{A.15}$$

$$\nabla\cdot\mathbf{A} = \frac{1}{r^2}\frac{\partial}{\partial r}(r^2A_r) + \frac{1}{r\sin\theta}\frac{\partial}{\partial\theta}(A_\theta\sin\theta) + \frac{1}{r\sin\theta}\frac{\partial A_\phi}{\partial\phi} \tag{A.16}$$

$$\nabla\times\mathbf{A} = \hat{\mathbf{r}}\frac{1}{r\sin\theta}\left(\frac{\partial}{\partial\theta}(A_\phi\sin\theta) - \frac{\partial A_\theta}{\partial\phi}\right) + \frac{\hat{\boldsymbol{\theta}}}{r}\left(\frac{1}{\sin\theta}\frac{\partial A_r}{\partial\phi.} - \frac{\partial}{\partial r}(rA_\phi)\right) \tag{A.17}$$

$$+ \hat{\boldsymbol{\phi}}\frac{1}{r}\left(\frac{\partial}{\partial r}(rA_\theta) - \frac{\partial A_r}{\partial\theta}\right)$$

$$\nabla^2\Phi = \frac{1}{r^2}\frac{\partial}{\partial r}\left(r^2\frac{\partial\Phi}{\partial r}\right) + \frac{1}{r^2\sin\theta}\frac{\partial}{\partial\theta}\left(\sin\theta\frac{\partial\Phi}{\partial\theta}\right) + \frac{1}{r^2\sin^2\theta}\frac{\partial^2\Phi}{\partial\phi^2} \tag{A.18}$$

3. RELATIONS BETWEEN UNIT VECTORS IN RECTANGULAR AND SPHERICAL COORDINATES

$$\hat{\mathbf{x}} = \hat{\mathbf{r}}\sin\theta\cos\phi + \hat{\boldsymbol{\theta}}\cos\theta\cos\phi - \hat{\boldsymbol{\phi}}\sin\phi \tag{A.19}$$

$$\hat{\mathbf{y}} = \hat{\mathbf{r}}\sin\theta\sin\phi + \hat{\boldsymbol{\theta}}\cos\theta\sin\phi + \hat{\boldsymbol{\phi}}\cos\phi \tag{A.20}$$

$$\hat{\mathbf{z}} = \hat{\mathbf{r}}\cos\theta - \hat{\boldsymbol{\theta}}\sin\theta \tag{A.21}$$

$$\hat{\mathbf{r}} = \hat{\mathbf{x}}\sin\theta\cos\phi + \hat{\mathbf{y}}\sin\theta\sin\phi + \hat{\mathbf{z}}\cos\theta \tag{A.22}$$

$$\hat{\boldsymbol{\theta}} = \hat{\mathbf{x}}\cos\theta\cos\phi + \hat{\mathbf{y}}\cos\theta\sin\phi - \hat{\mathbf{z}}\sin\theta \tag{A.23}$$

$$\hat{\boldsymbol{\phi}} = -\hat{\mathbf{x}}\sin\phi + \hat{\mathbf{y}}\cos\phi \tag{A.24}$$

References

Abramowitz, M. and Stegun, I. A. (eds) (1964). *Handbook of Mathematical Functions*, NBS Applied Math. Series No. 55, US Government Printing Office, Washington, DC, Chapter 5.

ARRL Antenna Book (1974). The American Radio Relay League, Newington, CT.

Boas, M. L. (1966). *Mathematical Methods in the Physical Sciences*, John Wiley and Sons, New York, p. 160.

Booker, H. G. (1946). 'Slot aerials and their relation to complementary wire aerials (Babiinet's principle)', *J. IEE* (London), Pt IIIA, **93**, 620–626.

Carrel, R. (1961). 'The design of log-periodic dipole antennas', *IRE Int. Convention Record*, Pt 1, 61–75.

Carter, P. S. (1932). 'Circuit relations in radiating systems and application to antenna problems', *Proc. IRE*, **20**, 1004–1041.

Chen, C. A. and Cheng, D. K. (1975). 'Optimum element lengths for Yagi–Uda arrays', *IEEE Trans. Antennas Propagation*, **AP-23**, 8–15.

Cheng, D. K. and Chen, C. A. (1973). 'Optimum spacings for Yagi–Uda arrays', *IEEE Trans. Antennas Propagation*, **AP-21**, 615–623.

Cheong, W. M. and King, R. W. P. (1967). 'Log-periodic dipole antennas', *Radio Sci.*, **2**, 1315–1325.

Collin, R. E. and Zucker, F. J.(eds) (1969). *Antenna Theory*, Pt 1, McGraw-Hill, New York, Chapter 11.

Cox, C. R. (1947). 'Mutual impedance between vertical antennas of unequal heights', *Proc. IRE*, **35**, 1367–1370.

De Vito, G. and Stracca, G. B. (1973). 'Comments on the design of log-periodic dipole antennas', *IEEE Trans. Antennas Propagation*, **AP-21**, 303–308.

De Vito, G. and Stracca, G. B. (1974). 'Further comments on the design of log-periodic dipole antennas', *IEEE Trans. Antennas Propagation*, **AP-22**, 714–718.

Dolph, C. L. (1946). 'A current distribution for broadside arrays which optimizes the relationship between beamwidth and side-lobe level', *Proc. IRE*, **34**, 335–348.

Dyson, J. D. (1962). 'A survey of the very wide band and frequency independent antennas—1945 to the present', *J. Res. NBS—D, Radio Propagation*, **66D**, 1–6.

Elliott, R. S. (1962). 'A view of frequency independent antennas', *Microwave J.*, November, 61–68.

Elliott, R. S. (1966). 'The theory of antenna arrays', in *Microwave Scanning Antennas*, Vol. II, *Array Theory and Practice* (ed. R. C. Hansen), Academic Press, New York, Chapter 1.

Elliott, R. S. (1981). *Antenna Theory and Design*, Prentice-Hall, Englewood Cliffs, NJ, Chapter 7, pp. 372, 375.

Green, H. E. (1966). 'Design data for short and medium length Yagi–Uda arrays', *Inst. Eng. (Aust.), Electr. Eng. Trans.*, 1–8.

Hallén, E. (1983). 'Theoretical investigations into the transmitting and receiving qualities of antennae', *Nova Acta Regiae Soc. Sci. Upsaliensis*, Serv. IV, **11** (No. 4), 1–44.

Hallén, E. (1948). 'Admittance diagrams for antennas and the relation between antenna theories', *Tech. Rep. No. 46*, Cruft Laboratory, Harvard University.

Hansen, W. W. and Woodyard, J. R. (1938). 'A new principle in directional antenna design', *Proc. IRE*, **26**, 333–345.

Harrington, R. F. (1968). *Field Computation by Moment Method*, Macmillan, New York, Chapters 4–6.

Harrison, C. W. (Jr) and King, R. W. P. (1961). 'Folded dipoles and loops', IRE Transactions on Antennas and Propagation, **AP-9**, 171–187.

Hollis, J. S., Lyon, T. J. and Clayton, L. (1970). *Microwave Antenna Measurements*, Scientific–Atlanta, Atlanta.

Isbell, D. E. (1960). 'Log-periodic dipole arrays', *IRE Trans. Antennas Propagation*, **AP-8**, 260–267.

Jasik, H. (ed.) (1961). *Antenna Engineering Handbook*, McGraw-Hill, New York, Chapters 5 and 24.

Jordan, E. C. and Balmain, K. G. (1968). *Electromagnetic Waves and Radiating* Systems, 2nd edn, Prentice-Hall, Englewood cliffs, NJ, p. 232, Chapters 8 and 16.

Jordan, E. C., Deschamps, D. A., Dyson, J. D., and Mayes, P. E. (1964). 'Developments in broadband antennas', *IEEE Spectrum*, **1**, 58–71.

Kammer, W. K. and Gillespie, E. S. (1978). 'Antenna measurements—1978', *Proc. IEEE*, **66**, 483–507.

Kerr, D. E. (ed.) (1965). *Propagation of Short Radio Waves*, Dover, New York, Chapter 6.

King, H. E. (1957). 'Mutual impedance of unequal length antennas in echelon', *Proc. IRE*, **5**, 306–313.

King, H. E. and Wong, J. L. (1980). 'Characteristics of 1 to 8 wavelength uniform helical Antennas', *IEEE Trans. Antennas Propagation*, **AP-28**, 291–296.

King, R. W. P. and Harrison, C. W. (Jr) (1943). 'The distribution of current along a symmetrical center-driven antenna', *Proc. IRE*, **31**, 548–567.

Kraus, J. (1950). *Antennas*, McGraw-Hill, New York, Chapters 7 and 12.

Lee, K. F., Wong, P. F., and Larm, K. F. (1982). 'Theory of the frequency responses of uniform and quasi-taper helical antennas', *IEEE Trans. Antennas Propagation*, **AP-30**, 1017–1020.

Mayes, P. E. (1963). 'Broadband backward-wave antennas', *Microwave J.*, January, 61–71.

Nakano, H. and Yamauchi, Y. (1979). 'The balanced helices radiating in the axial mode', *IEEE AP-S Int. Symp. Digest*, 404–407.

Ng, T. S. and Lee, K. F. (1982). 'Theory of corner-reflector antenna with tilted dipole', *IEE Proc.* (London), **129**, Pt H (No. 1), 11–17.

Radio Amateur's Bandbook (1974). The American Radio Relay League, Newington, CT.

Rumsey, V. (1966). *Frequency Independent Antennas*, Academic Press, New York.

Schelkunoff, S. A. and Friis, H. T. (1952). *Antennas, Theory and Practice*, Wiley, New York, pp. 528–529.

Sensiper, S. (1951). 'Electromagnetic wave propagation on helical conductors', *Tech. Rech. Rep. No. 194*, MIT Res. Lab. of Electronics.

Sinnott, D. H. (1974). 'Multiple-frequency computer analysis of the log-periodic dipole antenna', *IEEE Trans. Antennas Propagation*, **AP-22**, 592–594.

Stutzman, W. L. and Thiele, G. A. (1981). *Antenna Theory and Design*, Wiley, New York, p. 203.

Tai, C. T. (1948). 'Coupled antennas', *Proc. IRE*, **36**, 487–500.

Thiele, G. A. (1969). 'Analysis of Yagi-Uda type antennas', *IEEE Trans. Antennas Propagation*, **AP-17**, 24–31.

Thiele, G. A. (1973). 'Wire antennas', *Computer Techniques for Electromagnetics* (ed. R. Mittra), Pergamon, Oxford, Chapter 2.

Viezbicke, P. (1976). 'Yagi antenna design', *NBS Tech. Note 688*, US Government Printing Office, Washington, DC.

Weeks, W. L. (1968). *Antenna Engineering*, McGraw-Hill, New York, pp. 189, 190.

Whittaker, E. T. and Watson, G. N. (1962). *A Course of Modern Analysis*, Cambridge University Press, London, p. 170.

Wolff, E. A. (1966). *Antenna Analysis*, Wiley, New York, pp. 65, 88, 215 and 224.

Wolter, J. (1970). 'Solution of Maxwell's equations for log-periodic dipole antennas', *IEEE Trans. Antennas Propagation*, **AP-18**, 734–741.

Index